FAIRBROTHER'S TEXTBOOK OF BACTERIOLOGY

FAIRBROTHER'S TEXTBOOK OF BACTERIOLOGY

EDITED BY

R. L. VOLLUM,

M.A., D.Phil., Oxon.

Sometime Lecturer in Bacteriology, Oxford University

D. G. JAMISON,

M.A., B.M., Oxon.

University Demonstrator in Anatomy, Cambridge

C. S. CUMMINS,

Sc.D.

Sometime Reader in Bacteriology, London Hospital Medical College

TENTH EDITION

WILLIAM HEINEMANN · MEDICAL BOOKS · LTD

First Edition, February	1937
Reprinted July,	1937
Second Edition, May	1938
Reprinted, November	1938
Reprinted, April	1939
Third Edition, July	1940
Reprinted, April	1941
Fourth Edition, September	1942
Reprinted, June	1943
Reprinted, January	1945
Fifth Edition, September	1946
Reprinted, March	1948
Reprinted, September,	1948
Sixth Edition, September	1949
Reprinted, August	1950
Reprinted, February	1951
Reprinted, August	1952
Seventh Edition, August	1953
Reprinted, May	1955
Reprinted, October	1956
Eighth Edition, March	1959
Reprinted, April	1960
Reprinted, July	1961
Ninth Edition, July	1962
Reprinted, July	1964
Reprinted, July	1965
Reprinted, July	1966
Reprinted, April	1969
Tenth Edition Reset January	1970

SBN 433 10100 8

Printed in Great Britain by
The Whitefriars Press Ltd., London and Tonbridge

PREFACE TO THE FIRST EDITION

This book is an outline of the medical aspects of bacteriology. The field of bacteriology has been so greatly extended during the past 10 or 15 years that no account of the whole of it, even in outline, is possible in the short time available in the medical curriculum. The medical student must therefore restrict his attention to those branches of the subject that are of medical importance, viz., bacteria as agents of disease in man, and the application of bacteriological methods in the prevention, diagnosis and treatment of disease. Few of the available text-books stress these points adequately, and the presentation of principles is often interrupted by masses of technical detail needed only in the laboratory.

In the present book an attempt is made to meet these criticisms. In the absence of a universally accepted bacterial nomenclature, that recommended by the Society of American Bacteriologists is followed, and the biological relationship of allied organisms is emphasized.

The bacteriological works consulted during the writing of this book are too numerous to list in full, but my great indebtedness to the "Principles of Bacteriology," by W. W. C. Topley and G. S. Wilson, and to the "System of Bacteriology" published by the Medical Research Council, demands specific acknowledgment.

I wish to thank Professor H. B. Maitland, Dr. D. T. Robinson and Dr. R. Whitehead for advice and criticism, Dr. W. Susman and Mr. V. H. Moore for the various photographs, and my wife for the drawings, the coloured plates and much general assistance.

R. F. W.

DEPT. OF BACTERIOLOGY,
UNIVERSITY OF MANCHESTER. *January, 1937.*

PREFACE TO THE TENTH EDITION

Professor R. W. Fairbrother who was responsible for all previous editions of this Textbook of Bacteriology died in April 1969,, after a long illness. Professor Fairbrother did valuable work on viruses, particularly poliomyelitis, during the early days of his career and later worked on the use of antibiotics.

The present edition has been extensively revised to include new developments in several fields. The sections on antibiotics, immunity and immunoglobulins have been expanded. The chapters on mycobacteria, vibrios, viruses and the bacteriology of milk and water have been largely rewritten, and additions have been made to most other chapters. Obsolete material has been deleted to avoid any increase in the size of the volume.

In Part I of this edition, Chapter I was revised by Dr. Sylvia Reed, Chapters IX, X, XI by Dr. P. C. Wilkinson, the other Chapters by Dr. C. S. Cummins.

Parts II and III have been revised by Dr. R. L. Vollum and Dr. D. G. Jamison.

R. L. V.
D. G. J.

January, 1970.

TABLE OF CONTENTS

PART I

GENERAL BACTERIOLOGY

PART II

SYSTEMATIC BACTERIOLOGY

PART III

GENERAL TECHNIQUE

Part I

GENERAL BACTERIOLOGY

Chapter I

Historical Survey

ORIGINS

Although there is no doubt that bacteria were observed almost three centuries ago, their significance in the production of disease was not established until the brilliant work of Louis Pasteur, a French chemist, almost 100 years ago. Previous to this period much had been written about the nature of disease and the spontaneous generation of living things, but these views were largely the speculations of philosophers and in only a few cases were they subsequently proved to be correct. Nevertheless, these speculations were mainly responsible for the development of the science of bacteriology.

It has been long recognized that disease might be communicated from one individual to another; there are many references in the works of the earlier writers to indicate this, particularly in connection with plague and smallpox, both of which frequently appeared in epidemic form during the Middle Ages. The nature of the infecting material was not, however, determined; Fracastor of Verona (1546) put forward the view that disease was spread by particles too small to be seen and he named the responsible agent a *contagium vivum*. He divided the contagious diseases into two groups: those transmitted *per contactum*, i.e. from person to person, and those transmitted *per fomitem*, i.e. through the agency of some outside object. The first person to suggest, on experimental grounds, that disease might be caused by visible living bodies was a monk, Kircher. Kircher, in 1671, using simple lenses of low magnification, stated that he had observed peculiar "worms" in the blood of persons suffering from plague. Although he considered that these were the agents responsible for the disease, there seems to be little doubt that the objects he saw were simply arrangements of the red blood corpuscles.

FIRST DESCRIPTION OF BACTERIA

A short time afterwards the existence of minute living bodies was definitely established. Leeuwenhoek, a Dutch draper whose hobby was the preparation of lenses, introduced the simple microscope during the last quarter of the seventeenth century. Using these instruments, with which magnifications of from 40 to 300 diameters were obtained, he observed minute bodies which he named "animalcules". These were present in considerable numbers in rain water and scrapings collected from the spaces between the teeth. Some of the bodies were shaped like bent sticks and others like spirals, while many were actively motile. Leeuwenhoek described his results in numerous letters to the Royal Society of London, but he did not appreciate that the animalcules might have some connection with the production of disease. There seems to be no doubt that Leeuwenhoek actually saw the larger bacteria, which are normally present in water and in the mouth. An Austrian, Plenciz, in 1762 put forward the germ theory of disease when, as a result of experimental work of doubtful value, he concluded that each disease had its own specific agent which multiplied in the body and so produced disease.

THE CONTROVERSY OVER SPONTANEOUS GENERATION

At this time, the middle of the eighteenth century, there was considerable controversy on the question of spontaneous generation. After Leeuwenhoek had described the peculiar minute bodies, or animalcules, their presence in various situations was confirmed without difficulty. The adherents of the theory of spontaneous generation believed that by an investigation of the animalcules proof of this theory could be definitely established. Needham, about 1750, communicated to the Royal Society the results of his experiments on the generation of these minute bodies. He heated meat-broth at temperatures, as he thought, sufficient to destroy living things and left it tightly sealed for some days. On subsequent examination Needham found that the broth was swarming with the animalcules. By that and similar experiments he considered that the spontaneous origin of the animalcules was proved.

These views did not satisfy an Italian priest, Spallanzani, who prepared similar flasks of meat-broth and boiled them for varying periods of time. One set was sealed and the other corked; they were then stored for several days. On examination it was found that the animalcules were present in all the flasks except those which had been sealed after being boiled for some long time. Spallanzani considered that the minute bodies could resist boiling for a short time and that spontaneous generation had not taken place; the living bodies had simply developed from living bodies already present in the medium. Needham in reply to these criticisms stated that the experiments of Spallanzani were valueless, as by the prolonged heating of the meat-broth some hypothetical substance, "vegetative force" necessary for production of animalcules, had been destroyed. However, Spallanzani showed that animalcules could grow in broth which had been heated for prolonged periods. As a result of these and other experiments, the theory of spontaneous generation gradually became discredited.

The development of the microscope was also advancing rapidly, and higher magnifications were soon available for the study of minute bodies. It must be realized that, at this stage, the only means of examination was by the observation through these early microscopes of unstained material. In spite of these difficult technical conditions definite progress was made; Ehrenberg (1838) described various forms and classified four groups: vibrio, bacterium, spirillum and spirochaeta. A short time previously, in 1837, Latour and Schwann had found yeasts in beer and wine and considered that they were responsible for the fermentation process.

Although some of these early observations were undoubtedly accurate, many conflicting statements were also made, based on most unsatisfactory experimental evidence. This haphazard manner of investigation was criticized appropriately by Henle, a famous anatomist, in 1840. Henle emphasized that before aetiological relationship of a micro-organism to a disease process could be established it must be found constantly in cases of the disease and it must be isolated and tested by inoculation into healthy animals. These arguments were amplified later by Koch in his now famous postulates (see page 7).

At this time the question of spontaneous generation again came to the fore. The observations of Schwann and Latour, that yeasts were responsible for the fermentation process in beer and wine, were much criticized. Liebig, a celebrated German chemist, stated that fermentation was purely a chemical process and that the ferment was obtained from the dead protein already present in the reacting system. In view of his reputation, Liebig's views were almost generally accepted; this did much to retard progress.

PASTEUR'S WORK ON FERMENTATION

It was at this stage, 1856, that Pasteur entered the realms of bacteriology. Originally a chemist, Pasteur undertook an investigation into the cause of some faulty fermentation which was seriously affecting the wine industry of Lille. As a result of his work, Liebig's theory of fermentation was finally and completely disproved; Pasteur found that yeasts were responsible for the fermentation process and that sugar could be converted into alcohol in the absence of any protein in the medium. He also showed that faulty fermentation was the result of the presence of some contaminating rod-shaped micro-organisms. The theory of spontaneous generation was also finally crushed by further researches of Pasteur, in which innumerable flasks with wide, constricted, twisted and elongated necks were made in order to show that the contaminating micro-organisms had their origin in the air. The effect of heat was also examined, and he found that certain organisms were able to withstand boiling for some time; these were doubtless the spore-bearing bacteria. Some organisms were found to grow in the absence of oxygen while others required its presence. The use of cotton-wool plugs in order to prevent organisms from entering flasks already sterilized was also demonstrated about this time by Schroeder and Dusch. The relationship of micro-organisms to the processes of fermentation and putrefaction thus became definitely established, but it is interesting to note that up to this time conclusive evidence that these minute bodies were agents of disease had not been obtained. After his brilliant work on fermentation, Pasteur (1865) was next persuaded to investigate a disease of silkworms, pébrine, which was crippling the silk industry of France. Although mishaps and setbacks were frequent, Pasteur, after five years' work, had settled the problem and re-established the silk industry on a prosperous basis. He found that the disease of the worms was due to some living micro-organism and that the disease was transmitted from one animal to another. By obtaining healthy silkworms and subsequently removing any showing signs of infection Pasteur was able to maintain a healthy stock, although surrounded by infected farms. The relationship of living microscopic objects to the disease was thus proved; it is a point of interest that this disease was of protozoal and not bacterial origin.

LISTER—THE RELATIONSHIP OF BACTERIA TO WOUND SEPSIS

These fundamental observations of Pasteur did not pass unnoticed by the outside world. Although there were many who decried these new and revolutionary ideas, there were others who realized the true significance of this work. Lister, then Professor of Surgery at Glasgow University, considered that the micro-organisms, so prevalent in the air, might be responsible for the frequent occurrence of the inflammation and suppuration after surgical operations. In spite of much criticism and opposition, in 1867 he revolutionized surgery by introducing an antiseptic technique. In this method sprays or washings of carbolic acid were directed during the operation onto the wound, which was subsequently protected from the air by suitable dressings. The results which followed this procedure were remarkably successful and the practice became increasingly popular; there followed a striking reduction in the incidence of post-operative sepsis. This was an important demonstration of the role played by micro-organisms in the causation of disease. A further important development was the replacement of antiseptic surgery by aseptic surgical techniques.

The study of the micro-organisms, although a matter of some difficulty, was now becoming increasingly popular. Many varied forms were seen and described. A further and more systematic method of classification was elaborated by Cohn (1872); in this were included the following generic names — micrococcus, bacterium, bacillus, vibrio and spirochaeta. The term bacterium was also used for the first time to designate the micro-organisms multiplying by fission. This classification was based on morphological observations and it is interesting that many of the terms are used today in the sense originally employed by Cohn, who also described the appearance of spores in a non-pathogenic bacillus and noted the increased resistance to heat and disinfectants possessed by the spore-bearers.

ANTHRAX: PROTECTIVE INOCULATION

About this time the relationship of a micro-organism to an important animal disease, splenic fever or anthrax, was being investigated. Pollender, in 1849, and Davaine in 1850, had described the presence of a rod-shaped organism in the blood of animals suffering from splenic fever. Some time later Davaine reproduced the disease in healthy animals by injecting them with blood collected from infected animals. However, the connection between the organism and the disease was not definitely established until the thorough and brilliant work of a medical practitioner in Germany, Robert Koch. Koch examined the organisms present in the blood of infected animals, reproduced the disease in healthy mice by inoculating them with infected blood and also prepared cultures by inoculating, on a hollow-ground slide, the aqueous humour of a bullock's eye with the infected blood. The growth of the organisms in culture was observed, and after several sub-cultures the organism was still found to be virulent for the mouse. The production of spores by the rods, and the vegetation of the spores into rods were also observed and described. In this way, in 1876, the aetiology of anthrax was established beyond any shadow of doubt. This observation was confirmed and elaborated shortly afterwards by Pasteur, who introduced a method of active immunization whereby the resistance of animals to the disease was increased. Pasteur had previously found by accident that the inoculation of chickens with relatively avirulent cultures of the organisms of chicken cholera did not produce the disease, and that a later inoculation of these birds with a virulent strain was also unsuccessful. This was a discovery of great fundamental importance since it indicated to Pasteur that the original avirulent culture had protected the chickens against subsequent infection with virulent organisms.

Pasteur then endeavoured to apply his discovery to splenic fever. Here the difficulty was to obtain a relatively avirulent culture of the anthrax bacillus. This was overcome by growing the organisms at a temperature higher than that optimal for growth and at which spores did not develop. This attenuated culture was given, and was then followed after a short interval by a second inoculation of a less attenuated culture. The value of these injections in protecting sheep was demonstrated publicly by Pasteur at Pouilly-le-Fort in 1881; the demonstration was a striking success; the control animals died of anthrax or splenic fever, whereas the protected sheep were quite unaffected by the inoculation of the virulent test-culture. Although subsequent trials were less satisfactory, these observations began the study of protective inoculation against infectious diseases.

THE WORK OF ROBERT KOCH

Koch meanwhile had become a full-time bacteriologist and was making further important advances in bacteriological technique. The methods employed in the study of bacteria at this time were very primitive; microscopical observations were made on unstained material, while cultures had been prepared only in fluid media. Many would consider that, under these conditions, the study of bacteria was impossible; the valuable results obtained by Koch and Pasteur consequently give some indication of the brilliance of these workers. Koch soon made several important improvements; in 1877 he prepared dried films which he stained with certain aniline dyes; he also improved the microscope by using an Abbé condenser and oil-immersion lenses, which had been employed in spheres other than bacteriology. Shortly afterwards he introduced another very important new technique, by developing the use of solid media for growing bacteria. The preparation of pure cultures was a matter of considerable difficulty when only fluid media were used. The introduction by Koch of gelatin as a solidifying agent was an important advance in bacteriological technique, as growth was now obtainable in the form of colonies which could be readily selected and picked off for further study. A simple means of separating individual organisms from contaminated material was thus available. Subsequently, serum and later agar-agar were used in the preparation of a solid medium. Potato and bread had been employed previously, but the results were uncertain and unreliable. A further improvement was the introduction of satisfactory containers for solid media by Petri; "Petri" dishes are still used extensively in bacteriology.

In 1882 Koch published his classical investigations on tuberculosis, by which he proved that the tubercule bacillus was the causal agent of the disease. In his reports of this work, Koch formulated his famous postulates on the relationship of an organism to the disease it causes. He suggested that in order to prove that an organism causes a particular disease, (a) the organism should always be associated with the disease, (b) the organism should be isolated in pure culture, and (c) the cultured organism should reproduce the disease in susceptible animals. Only if all three postulates are fulfilled is the conclusive proof of a causal relationship between the organism and the disease obtained. However, there are some instances of infections in which it has not been possible to fulfil all three postulates, even though the available evidence strongly suggests that a causal relationship does in fact exist.

The years 1880–1900 saw an enormously rapid advance in bacteriology, as the methods of Pasteur and Koch were applied to the study of human disease. In this relatively short period of 20 years the bacteria causing typhoid, tuberculosis, cholera, dysentery, diphtheria, gonorrhoea and many other infectious diseases were discovered and described. Naturally enough, the early work on bacteriology was directed to solving the immediate practical problem posed by the common infectious diseases of man and animals. More recently, the scope of bacteriology has widened. To medical and veterinary bacteriology are now added agricultural and industrial microbiology; we have become aware of the existence of innumerable species of non-pathogenic micro-organisms and we now know something of their significance.

ORIGINS OF IMMUNOLOGY

Following immediately on early studies of bacteria came investigation of mechanisms of immunity to disease, and from these beginnings the science of immunology has developed. A hundred years earlier, in 1794, Jenner had used material from cow-pox lesions to vaccinate people against smallpox, though he had no clear idea of the mechanism by which the protection was produced. Pasteur successfully immunized sheep against anthrax and human beings against rabies, in each case using injections of the appropriate organism treated to reduce its virulence. Like Jenner, he had no precise ideas about how the protection was produced. During the last decade of the nineteenth century, other investigators took up the study of reactions between bacteria or their products and the sera of immunized animals. Jules Bordet (1870–1961) a Belgian, and Paul Ehrlich (1854–1915) were two outstanding early workers in immunology. Both Ehrlich and Bordet studied the antigen–antibody reaction, and both evolved theories about its mechanisms.

During the twentieth century the scope of immunology has steadily broadened. It was soon recognized that an understanding of antigen–antibody reactions could be applied not only to prophylaxis of infections, but also to diagnosis and to treatment. Serum therapy was outstandingly successful in diphtheria, though it proved less satisfactory in some other infections. On the other hand the success of prophylactic immunization has been such that one of the present practical problems in this field is to devise ways in which immunization against some half dozen important infections can be achieved within the short space of a child's first year of life.

The discovery in 1902 by Portier and Richet of the phenomenon of anaphylaxis was one of the first observations of allergic and hypersensitive reactions, which are still the field of much active research. In 1900, Landsteiner's discovery of the ABO blood group system formed the foundation of the present day science of blood group serology.

ORIGINS OF VIROLOGY

The existence of infectious agents smaller than bacteria was first demonstrated in 1898 when Loeffler and Frosch found that foot and mouth disease could be transmitted in fluid which had passed through a bacteria-retaining filter. In 1915 Twort demonstrated that a filterable agent, now known to be bacteriophage, was capable of lysing staphylococcal cultures.

It soon became clear that some infectious diseases were caused by agents too small to be seen with the light microscope. These "filterable viruses" could at first only be handled in the laboratory by using cumbersome techniques involving inoculation of animals.

Nevertheless, the viral aetiology of such important conditions as poliomyelitis (Landsteiner and Popper, 1909) and influenza (Smith, Andrewes and Laidlaw, 1933), were established. A significant advance was the use, in 1931, of the developing chick embryo and the membranes surrounding it for growing viruses. These methods were developed by Burnet and his associates, and provided a great stimulus to experimental virology. Since about 1950 tissue cultures have been increasingly used for growing viruses. One of the chief factors making this possible was the development of antibiotics. Addition of antibiotics to tissue culture medium made it very much easier to suppress bacterial contamination of the cultures. Largely as a result of tissue culture methods, virology has recently expanded in a way reminiscent of the expansion of bacteriology during the years 1880–1900. In the last few years many viruses previously entirely unrecognized have been cultured in the laboratory for the first time.

CHEMOTHERAPY AND ANTIBIOTICS

During the first years of the twentieth century Ehrlich searched systematically for drugs which could be used for specific treatment of infectious diseases. He hoped to find a "magic bullet" which, while highly toxic for the infecting organism, would be relatively harmless to the host. His efforts were rewarded by his discovery in 1910 of the arsenical drug Salvarsan 606, which was highly toxic to *Treponema pallidum*, and which, if carefully used, proved effective in treating syphilis. Despite this success, most bacteriologists at this time were concerned with the possibility of treating infectious diseases by serological means. It seemed logical to attempt to reinforce the natural defences of the body. Eventually, however, it became clearer that very few infections could be treated really effectively by the administration of antiserum and the search for "magic bullets" was intensified.

In 1935 the German chemist Domagk found that a red dye, "prontosil", would protect mice against streptococcal infections. This discovery led to the production first, of the active component of the prontosil molecule, sulphonamide, and later a whole series of highly active sulphonamide compounds. An even more important result of the discovery was the stimulation of interest in chemotherapy; and this interest was even more sharply intensified by the outbreak of the Second World War, and the urgent need for treating infected wounds. Some years earlier, in 1929, Fleming discovered that a substance produced by the common mould *Penicillium notatum* would inhibit growth of staphylococci and other bacteria. He called the substance penicillin. Fleming attempted to use penicillin for treating patients, but his efforts were hampered by the difficulty of obtaining sufficiently pure and concentrated preparations. In 1940 Florey and Chain, investigating known antibacterial compounds, chose to study penicillin. The more concentrated, purified preparations which they succeeded in producing were soon recognized to be much more effective and much less toxic than any previously known chemotherapeutic drug. This was the beginning of the "antibiotic era" which has so changed the practice of medicine.

Examination of fungi and other micro-organisms for antibacterial properties resulted in the discovery of streptomycin (1944), chloramphenicol (1947), chlortetracycline (1948) and since then many other antibiotics. Chemists and microbiologists working together are continuing to develop new drugs. The semi-synthetic penicillins such as methicillin and ampicillin are examples of this approach. The success of research in the fields of vaccine prophylaxis and chemotherapy has meant that some of the practical problems of control of bacterial disease have now been largely solved. If tuberculosis, syphilis and diphtheria are still public health problems in this and other countries this is not due to the ignorance of how to control these conditions, but rather to an inability for various reasons to apply the knowledge.

In recent years the development of new techniques and the wider introduction of chemical and physical methods into microbiology have produced valuable results. Studies of microbial biochemistry now link with the newer science of microbial genetics, and with our growing knowledge of the detailed anatomy of bacterial cells. What was once simply bacteriology, has enlarged and extended, producing the newer sciences of virology and immunology. These in turn are now linking with other fields of research, such as the study of carcinogenesis and of tissue transplantation. We are now accumulating a detailed knowledge of the structure and function of a very wide range of micro-organisms, and of the ways in which the body reacts not only to the presence of micro-organisms but to unfamiliar substances in general. The gathering of this kind of information is often regarded as "pure" research, having no obvious or immediate relevance to practical problems. Currently, the practical problems we are facing include drug resistance of micro-organisms, the specific treatment of virus infections, and the control of hypersensitive and auto-immune reactions. An intimately detailed knowledge of micro-organisms and of host reactions will undoubtedly be necessary if such problems are to be solved.

Chapter II

Bacterial Anatomy

Bacteria are the smallest of free-living organisms, the average diameter of a bacterial cell being about 1/1000 mm. In discussing objects of this size it is inconvenient to use fractions of a millimetre, and the unit usually used is 1 μ (mu) = 1/1000 mm. For viruses a still smaller unit is needed, 1 mμ (millimu) = 1/1000 μ. Frequently also the dimensions of various parts of the bacterial cell which can be seen in the electron microscope are given in Ångström units (Å) which are normally used to describe the wavelength of light, and are smaller still, so that 1 Å = 1/10 mμ.

The units of measurement commonly used to describe bacteria and viruses are therefore related as follows:

$$1\ \mu = 0{\cdot}001\ \text{mm} \quad (= 10^{-3}\ \text{mm}) \quad (= 1/25{,}000\ \text{in})$$
$$1\ \text{m}\mu = 0{\cdot}001\ \mu \quad (= 10^{-6}\ \text{mm})$$
$$1\ \text{Å} = 0{\cdot}1\ \text{m}\mu \quad (= 10^{-7}\ \text{mm}).$$

METHODS OF EXAMINATION

Because of their very small size, a microscope is essential for examining bacteria. There are two broad methods of microscopy in use, *light microscopy*, using visible or ultraviolet light, and *electron microscopy*, using a beam of electrons. The resolution of a microscope, i.e. the smallest distance apart at which two objects can still be seen as separate entities, depends on the wavelength of the light used and with the light microscope using visible light the resolution is in theory 0·2 μ, and in practice about 0·25 μ. Since this is about one quarter of the diameter of the average bacterial cell, it is not surprising that there has been much disagreement in the past about the details of the internal structure of bacteria. It is possible to increase the resolution (i.e. to distinguish objects closer together than 0·25 μ) by using light of shorter wavelength, e.g. ultra-violet light, but the image can then only be photographed or made visible on a fluorescent screen. In the electron microscope, because of the extremely short effective wavelength of the electron beam as opposed to visible light, the resolution is much greater than in the light microscope, and under good conditions elements as close together as 5 Å can be distinguished. The main disadvantage of the electron microscope in biological work is that the material must be examined dry and *in vacuo.*

LIGHT MICROSCOPY

The main techniques using visible or ultra-violet light are:

1. Microscopy by transmitted light
2. Dark ground microscopy
3. Phase contrast microscopy
4. Fluorescent microscopy

1. Microscopy by transmitted light. This is the normal way of using the light microscope. Light from a lamp is concentrated on the object by the condenser lens system, a real image is formed by the objective lens and this image is further magnified by the eyepiece to give a large virtual image.

In the case of ultra-violet light, a special light source (mercury vapour lamp) is used in conjunction with filters to eliminate light of other wavelengths and quartz lenses are used to allow the ultra-violet light to pass. As already mentioned, the image must be recorded photographically.

2. Dark ground microscopy. In this method a special condenser is used which directs all light at the object from a very oblique angle and allows no rays to pass directly through it. Thus the only light which passes up the microscope is that reflected from the object, which stands out brilliantly illuminated against a velvety black background if the light has been correctly adjusted. The method can also be used with ultra-violet light.

3. Phase contrast microscopy. When light passes through an unstained object, diffraction patterns are formed due to differences in refractive index between different parts of the object, but the light and dark areas of the pattern are normally almost invisible in the image because of the very large amount of undiffracted light. Hence the contrast in the unstained specimen is usually very slight. In phase contrast microscopy, the diffracted rays are retarded 1/4 wavelength (1/4) with respect to the direct beam, by being made to pass through the thicker part of a glass plate while the direct beam passes through a thinner part. The result is to cause maximum interference between the two beams and to accentuate the differences in intensity between different parts of the object. This gives *positive phase contrast*, in which most parts of the object appear in varying degrees darker than the background: negative phase contrast, where the object appears bright against a dark background, is obtained if the direct beam is retarded with respect to the diffracted rays.

FLUORESCENT MICROSCOPY

A number of dyes will fluoresce in the presence of ultra-violet light, that is, the dye molecules will give out visible light of a characteristic colour when ultra-violet light falls on them, and this property can be used in microscopy in two ways.

1. Staining by fluorescent dyes. Certain dyes such as auramine or acridine orange can be used to stain bacteria, and the bacteria will then fluoresce if examined in ultra-violet light. (The method is mainly applicable to acid fast organisms (see Chap. XXIII).) Special precautions must be taken to see that ultra-violet light does not enter the observer's eye.

2. Immunofluorescence. Fluorescent dyes such as fluorescein can be attached chemically to antibody globulins without interfering with their ability to react specifically with antigen. If such fluorescein-treated antibody combines with its antigen at the surface of a bacterium, the fluorescein will be bound to the bacterial surface, and the organisms will show a fluorescent outline when examined by ultra-violet light. Since the method has the specificity associated with all antigen–antibody reactions (see Chap. IX) it can be used to identify different immunological varieties of micro-organisms by direct examination of material from the body, e.g. in specimens of faeces. It has not, however, so far supplanted traditional methods of identifying bacteria in faeces (Chap. XVII).

ELECTRON MICROSCOPY

In this instrument a beam of electrons from a tungsten filament is used to form an image on a fluorescent screen or photographic plate. The electron beam, being composed of negatively charged particles, can be bent in an electric field, and the lenses in the electron microscope are therefore composed of electromagnets which are focused by altering the strength of current passing through the winding. The arrangement of lenses is comparable to that in the light microscope, so that there is usually a condenser lens to focus the beam on the object, and intermediate (= objective) and projector lenses to form a greatly magnified image which can be focused on the screen or photographic plate. The electron beam cannot be maintained except in a high vacuum (10^{-3} mmHg or better), so that the apparatus must be continuously pumped out while in use.

The specimen for examination is mounted on a very thin membrane of cellodion or formvar, the membrane being held on a copper grid for support. In the case of bacteria and viruses the samples can either be whole organisms which are placed on the membrane in the form of dilute suspensions which are allowed to dry, or else more concentrated suspensions can be fixed and embedded in methacrylate or some other resin. The resin is then hardened and ultra thin sections (*c.* 5 mμ) are cut by a process analogous to that used for the paraffin sections used in light microscopy. Sections for electron microscopy cannot be stained in the ordinary way, but if they are treated with solutions of salts of heavy metals such as lanthanum, uranium or osmium, some parts, e.g. cell membranes, will take up the metal atoms preferentially. These parts will show up quite strikingly when the specimen is examined, because the metal is "electron dense" (i.e. impedes the passage of electrons more than the rest of the section). Alternatively, with a suspension of whole organisms, a form of negative staining can be used in which the preparation is treated with uranium salts or phosphotungstic acid (both very electron dense) to fill in the background and thus outline the bacteria.

THE SHAPE AND ARRANGEMENT OF BACTERIAL CELLS

Variations in the appearance of different bacteria were recognized by the earliest observers. For example van Leeuwenhoek (1683) described clearly bacilli and spiral forms. The main forms now recognized are the coccus, bacillus, vibrio, actinomycete, spirillum and spirochaete.

A *coccus* is a spherical form, although in some cases such as the pneumococcus or the enterococcus organisms which are classified as cocci are in fact distinctly oval. Depending upon the mode of division cocci may appear to be arranged in clusters (staphylococci), in chains (streptococci) or less frequently in pairs (diplococci), in groups of four (tetrads) or packets of eight (sarcina).

A *bacillus* is a rod-shaped organism: great variation is found among different bacilli in the proportion of length to transverse diameter. The form is variable and may be straight rod with square ends, as in the aerobic sporing bacilli, or show an irregular often curved outline and variable width, as in corynebacteria. The size of bacilli varies considerably: some are very large, while others approximate to coccal forms and may be referred to as "coccobacilli". As well as being used as a general term for any rod-shaped bacterium, *Bacillus* is also the generic name for the group of Gram positive aerobic sporing organisms. However, when used in this way the word is given a capital letter and printed in italics.

Vibrios are slightly curved rods, usually with terminal flagella. A single cell of this form may frequently present the shape of a comma.

An *actinomycis* is an organism which grows predominantly in the form of long branching filaments or mycelia. The term actinomycete is also sometimes used.

(*a*) (*b*) (*c*)

Cocci arranged (*a*) in clusters, (*b*) in chains and (*c*) in groups of four.

(*d*) (*e*) (*f*) (*g*)

(*d*) and (*e*)—Bacilli, (*f*)—Vibrios, (*g*)—Spirilla.

(*h*) (*i*) (*j*)

(*h*) and (*i*)—Spirochaetes, (*j*)—Actinomyces.

Fig. 1.II.

A *spirillum* is an actively motile organism with a number of rigid spirals about the long axis and a tuft of flagella at each end.

A *spirochaete* is an elongated organism twisted spirally about the long axis. Spirochaetes differ from spirilla in being flexible and in being motile without possessing flagella.

Cell division in bacilli occurs by the formation of cross-walls at right angles to the long axis, and where the organisms do not separate after division long chains of bacilli may be formed. In other cases, especially in some corynebacteria, the organisms as seen in smears lie at angles to each other forming V or L shapes, and groups of bacilli may resemble Chinese letters.

VIBRIO, SPIRILLUM, ACTINOMYCETES, SPIROCHAETES

The basic forms, e.g. coccus, bacillus and so on remain broadly constant for any particular strain of bacteria, but under certain conditions the morphology of some organisms can show considerable variation or *pleomorphism*. Thus in old cultures of some Gram-negative bacilli such as *Haemophilus influenzae* both cocco-bacilli and long rods may be present at the same time. With other organisms such as the diphtheria bacillus (*Corynebacterium diphtheriae*) this morphological variation is a feature of cultures of all ages.

METHODS OF EXAMINING BACTERIA

Bacteria may be examined by light microscopy either as (1) unstained, usually living organisms suspended in a water medium, or (2) fixed and stained preparations.

In diagnostic work it is more usual to examine stained preparations, because of the general importance of differential staining methods such as the Gram stain.

Unstained preparations. The various parts of the bacterial cell have refractive indices almost the same as that of water and little detail is visible in preparations of living organisms examined by *transmitted light*. Apart from some special cases, therefore, such as the determination of motility or the presence of capsules, this method is little used in routine bacteriology. *Phase contrast* microscopy, by increasing contrast, does enable details of morphology in living organisms to be seen and photographed, but has been used more as a research tool than a routine method of examination. *Dark-ground* illumination makes organisms readily visible and is used as a diagnostic method in the case of certain groups such as spirochaetes whose morphology and type of movement are very distinctive.

Fixed and stained material. Only the general features of fixation and staining are considered here, the detailed technique of the various processes being given later (Chap. XXXV).

The usual way of fixing smears of bacteria which have been made on glass slides is by heat. The smear is allowed to dry in air, and then fixed on the slide by passing it once or twice rapidly through the Bunsen flame. Fixation may also be carried out by treating the smear with such substances as alcohol, mercuric chloride or formalin.

Simple stains. Bacteria are usually stained readily by the aniline dyes. These are derivatives of the coal-tar product aniline $C_6H_5.NH_2$, and are divided into basic and acid dyes. The basic dyes, salts of monovalent acids and dye bases, are more satisfactory for staining bacteria, since they combine with the ribonucleic acid in the bacterial cytoplasm. Those most frequently employed are:

Basic fuchsin	
Methylene blue	
Safranin	
Crystal violet	Gentian violet
	Methyl violet
Neutral red	
Nigrosin as background stain	

These dyes are dissolved in either water or alcohol and may be kept as concentrated solutions, as these are more stable than weaker ones. For staining, dilute aqueous solutions are applied to fixed films for varying times depending on the dyes being used—a few seconds in the case of crystal violet to several minutes for methylene blue. The stain is then washed off with water, the film gently dried between layers of blotting paper and examined under the oil immersion lens.

The colloidal dye nigrosin is sometimes used to demonstrate bacteria by negative staining. Bacteria are suspended in a drop of nigrosin (1 per cent) on a slide and the drop spread out and allowed to dry. If such a preparation is examined in the microscope the bacteria are seen unstained against a greyish-purple background.

The staining of bacteria by these basic dyes is largely due to combination between the dye molecules and the acidic ribonucleic acid of the bacterial cytoplasm. The bacterial cell wall does not stain with these simple stains and special methods must be used to make it visible. Furthermore the staining of the bacterial cytoplasm with these dyes is usually so intense that internal details such as chromatinic bodies cannot be distinguished.

Differential staining. The simple stains will colour almost all bacteria so that they can be seen more easily. However, they do not distinguish one type of organism from another except to make their size and shape more easily recognizable. Differential stains which allow some types of bacteria to be distinguished from others by their colour are much more useful, and the one most generally used in bacteriology is that described by Gram in 1884. The original technique has been modified by different bacteriologists, but the principles involved are essentially the same. Smears are stained for a short time, with a pararosanilin dye, such as gentian violet or methyl violet, treated with an iodine solution (Gram's or Lugol's iodine), decolourized with alcohol, methylated spirit or acetone, washed well with water and counterstained with weak safranin or carbol-fuchsin.

Stained by this method the bacteria fall into two groups: some retain the violet stain and are termed Gram-positive, others, being decolourized, take the counterstain and are called Gram-negative. This simple method of subdivision of the bacteria forms an important step in the examination of material containing bacteria. It is important to note that, when correctly stained, a Gram-negative micro-organism is always Gram-negative; a Gram-positive organism, however, may, at certain stages of growth, appear to be Gram-negative. This change in reaction is not infrequently observed in old cultures when dead and degenerating bacteria are present; under such conditions both Gram-positive and -negative forms may be observed in the same field. Faulty technique is frequently responsible for irregular results.

The reason for the differential behaviour of various bacteria when stained by the Gram technique is not yet fully understood. However, it seems certain that alcohol or acetone alters the cell wall of Gram-positive bacteria in such a way that the crystal violet–iodine complex which is responsible for the purple-black colour cannot be dissolved out by these solvents. The cell walls of Gram-negative organisms are not affected in this way, possibly because they contain a much higher proportion of lipid, and the purple dye complex can be extracted by the solvents. Subsequent staining with safranin will then colour the Gram-negative cells red.

For organisms to stain Gram-positive the cell wall must be intact. Crushed or broken bacteria or dead or partly digested cells in old cultures of known Gram-positive strains always stain Gram-negative. This means that to determine the Gram reaction of a particular strain it is important to examine young, vigorously growing cultures.

Ziehl–Neelsen stain for acid-fast organisms. Another important differential stain is that described independently by Ziehl and Neelsen, and generally known as the Ziehl–Neelsen method. Certain organisms, such as the tubercle bacillus do not stain readily by the methods already mentioned; they require prolonged exposure to the dye, frequently with the application of heat, but when stained they are resistant to the decolourizing action of acids. They have consequently been termed "acid-fast bacteria". Some, including the tubercle bacillus itself, are also resistant to decolourization by alcohol, and are therefore "acid- and alcohol-fast". This increased resistance of tubercle bacilli and similar organisms to dyes is due to the high proportion of lipid in their cell walls.

In carrying out the Ziehl–Neelsen stain the fixed film is covered with carbol-fuchsin and heated until steam rises, for 5 min, then washed with water and decolourized in 15–25 per cent H_2SO_4. It is then washed again with water and counterstained with methylene blue. The acid-fast bacilli, which retain the carbol-fuchsin, are bright red, while other bacteria and cells stain blue.

In some kinds of actinomycetes, e.g. many strains of *Nocardia*, the strong acid used in the usual Ziehl–Neelsen technique will decolourize the organisms, but the carbol-fuchsin is not removed by weaker acid, e.g. 1 per cent H_2SO_4. Such organisms are described as "weakly acid fast". A similar method using weak acid can be used for heat-resistant spores (see Chap. XXXV).

Staining of spirochaetes. These organisms do not generally stain at all well with simple stains or by Gram's method, and it is necessary to use either Giemsa's stain (as used for blood film) or one of the special silver stains such as Fontana or Levaditi (see Chap. XXXV).

THE BACTERIAL CELL

The light microscope, even under the most perfect conditions, can resolve few details of bacterial structure. However, the much higher resolution of the electron microscope has made it possible to give a reasonably full account of the anatomy of bacteria and to compare it with that of other living cells.

The bacterial cell consists of *cytoplasm* which is bounded externally by a thin *cytoplasmic membrane*. Outside this is a *cell wall* which is considerably thicker and more rigid. Within the cytoplasm there are the *chromatin bodies* which constitute the bacterial nucleus, and there may also be *granules* or *inclusion bodies*. Around the cell wall in some cases may be found a *capsule* composed of thick viscid material, usually acidic polysaccharide. Many bacteria also exhibit motility due to the action of one or more *flagella*, long protein threads which beat with a wave-like motion. Some organisms also possess curious hair-like structures called *fimbriae*, external to the cell wall. These are not connected with motility. Finally, in some groups of bacteria we find the very characteristic *endospores* which enable these organisms to survive adverse conditions fatal to the vegetative cells.

These various elements of the bacterial cell are described in more detail below, starting with external features such as flagella and capsules, and working inwards to the cytoplasm.

FLAGELLA

Flagella are long filamentous structures which can be seen to arise in the cytoplasm and pass out through the cell wall. The filaments show regular sinusoidal curves and are composed of a fibrin protein similar in structure to keratin and myosin. Individual flagella are very fine, about 20 mμ in diameter, but they can be seen in the light microscope when suitably stained because they tend to aggregate into bundles, and also because the stains used precipitate on their surface and increase their size.

The arrangement of flagella on the surface of the cell is constant for any particular organism and may be either *polar* (at one or both ends) or *peritrichate* (all over the surface). Organisms with the polar arrangement may have a single flagellum or a tuft composed of several. The flagellar protein is strongly antigenic and a large number of different serological types occur which are important in the classification of some groups of bacteria, especially Gram-negative bacilli.

Although it has been argued that flagella are the result and not the cause of bacterial movement, the experimental evidence is conclusively against this. For example it is possible to remove flagella from living organisms by very vigorous shaking: bacteria so treated are made non-motile but if the culture is left for a time they will grow new flagella and become motile again. However, the exact way in which flagella movement drives a bacterium through a fluid medium is not certain.

It is possible for strains of normally motile species to lose the ability to produce flagella, and thus give rise to non-motile variants. However, the reverse does not happen under natural conditions, i.e. species which are normally non-motile do not give rise to variants which possess flagella. Organisms showing motility occur in most bacterial groups, but are uncommon among the cocci.

FIMBRIAE

Like flagella, these are also long thread-like appendages on the outside of bacteria, but they are straight and not curved and are not connected with motility. They appear to arise from the cell wall itself and can only be seen in the electron microscope as they are very fine, only 5–10 mμ in diameter. The significance of fimbriae for bacteria is not known, but fimbriate organisms will adhere to red cells, causing haemagglutination.

CAPSULES, MICROCAPSULES AND SLIME LAYERS

Many bacteria, when actively growing produce viscous substances of high molecular weight which tend to collect on the surface of the bacterial cells and in the surrounding culture fluid. In some cases this viscous material forms a well-defined persistent layer, often several times the diameter of the cell it surrounds, and is then referred to as a *capsule*. Such typical capsules may readily be demonstrated by suspending the organism in Indian ink; the particles of carbon in the ink cannot penetrate the gelatinous material of the capsule which appears as a clear halo around the bacteria when these are sharply in focus. Capsules may also be demonstrated by special staining methods (see page 403).

In other organisms the viscous material produced by the cells does not form a well-defined visible layer on the surface of individual bacteria, but is present more diffusely in the medium. Such ill-defined material is usually referred to as a *slime-layer* rather than a capsule, and is generally very easily removed by washing.

The term microcapsule is most generally used to describe the presence on the outer surface of the cell wall of a layer of material too thin to be seen by the light microscope, but whose presence can be inferred from other evidence, usually serological. The *Vi* antigen of the typhoid bacillus is a good example of a layer of this kind (see Chap. XVII).

The chemical nature of the material forming capsules, microcapsules or slime layers varies very much from one organism to another, but it is most often polysaccharide. However, in a number of cases such as *Bacillus anthracis* (the anthrax bacillus) and *Pasteurella pestis* (the plague bacillus) the capsular material is polypeptide.

In pathogenic organisms which are capsulated it is usually found that the presence of the capsule enables the bacteria to resist phagocytosis and thus establish themselves in the tissues. However, the production of capsules is very dependent on the conditions of growth and organisms which were capsulated when isolated from a patient may easily lose the ability to form capsules when subcultured on artificial media.

THE CELL WALL

Bacteria, like plant cells, possess a well-defined cell wall 10–20 mμ thick, which can be made visible in the living cells of some strains by plasmolysis; when these organisms are placed in hypertonic solutions the cytoplasm retracts and the outline of the wall is exposed. The cell wall in bacteria is not stained by methods such as the Gram stain, but if smears are first treated with a mordant (e.g. tannic acid) and then stained with crystal violet the wall is made visible. It is also easily seen when ultra-thin sections of bacteria are examined in the electron microscope.

Chemically the cell wall is a complex mucopolysaccharide and contains amino sugars and amino acids not normally found in the animal body. The walls of Gram-negative organisms generally differ from those of Gram-positives in being thinner and having a much higher proportion of lipid, the most striking exception to this being the mycobacteria which, although Gram-positive, have very large amounts of fatty material in their walls, up to 60 per cent in some cases.

Physically the bacterial cell wall is a tough inelastic membrane which controls the shape of the cell. The presence of an intact cell wall is essential for the proper functioning of the cell, since without the restraint of the wall the cytoplasm would swell up and burst due to osmotic pressure. Some antibiotics, notably penicillin, act by preventing the synthesis of new cell wall material, thus seriously interfering with bacterial multiplication.

CYTOPLASMIC MEMBRANE

The cytoplasmic membrane is a specialized lipoprotein layer about 1–2 mμ thick which forms the outer boundary of the cytoplasm. Whereas the cell wall, whose purpose is to hold the cell together, normally allows free passage of small molecules of all kinds, the cytoplasmic membrane is an osmotically active membrane which will preferentially allow through substances required for cell metabolism. Damage to this membrane allows soluble constituents of the cytoplasm to leak away, and this soon causes death of the cell.

The cytoplasm of bacteria, especially in young actively growing cultures, contains a large amount of ribonucleic acid, which the electron microscope shows to be collected into granules called ribosomes. These are only about 10–20 mμ in diameter.

There may also be present in the cytoplasm *inclusion bodies* of different kinds which are large enough to be seen by ordinary microscopy, and may have characteristic staining reactions. The best known of these are the *Metachromatic granules* (= Volutin granules), so called because they stain reddish with blue dyes like methylene blue. They are composed of polymetaphosphate and are very characteristic of some species such as *C. diphtheriae*. In other cases inclusion bodies may be composed of lipid droplets, or polysaccharides such as glycogen.

BACTERIAL SPORES

Certain kinds of bacteria, principally Gram-positive bacteria belonging to the two genera *Bacillus* and *Clostridium*, produce at the end of their period of active growth intracellular bodies called *spores* or *endospores*. These are specialized cells which are much more resistant to heat and drying than normal bacteria.

In the light microscope the developing spore appears near one end of the bacillus as a round or oval area which does not stain by the usual methods. In unstained preparations the spores are more refractile than the rest of the organism and appear brighter. Electron microscopic observations have shown that the first stage of spore formation is the development of a cross-septum from the cytoplasmic membrane. This cuts off part of the nuclear material of the cell, and around this the spore develops, forming a complex outer coat composed of several layers.

Eventually the bacterial cell which produced the spore dies and disintegrates, and the spore is set free. These free spores will normally remain dormant unless placed in fresh medium, when they germinate, each giving rise to one bacterium. It should be noted that the organisms so obtained have all the properties of the original bacteria which gave rise to the spores.

Although bacterial spores do not stain with Gram stain or other simple stains, they can be stained by a modification of the method used for acid-fast bacilli. If smears are stained with hot carbol-fuchsin and then treated with alcohol, it is possible to remove all the stains from the bacteria, leaving the spores stained red (see Chap. XXXV).

The position of the spore in the cell in which it develops is usually near one end, i.e. *subterminal*, but it may sometimes be *terminal* and occasionally it appears to be *central* or *equatorial*. The developing spore may be oval or spherical and may or may not be wider than the parent cell. These variations in size, shape and position of the spore give rise in some cases to characteristic appearances by which organisms can be recognized, for example the "drum-stick" shape produced by the large spherical terminal spores of *Clostridium tetani* (the tetanus bacillus).

The mechanisms of spore formation are not understood, but the advantages to the organisms are obvious since spores can remain viable in the dry state for years, and in some cases will stand 100°C for several hours: they are also frequently much more resistant than the vegetative cell to the various substances used as antiseptics or disinfectants. The increased resistance of spores to adverse conditions is due to the impermeability of the spore membrane, and the low water content of their cytoplasm.

NUCLEUS

It seems evident that some form of nuclear apparatus must be responsible for the transfer of hereditary characters in bacteria but it has not been possible to demonstrate its presence satisfactorily until recently. However, by special staining methods such as the Feulgen technique, which stains nuclear chromatin in plant and animal cells, intra-cellular bodies called *chromatin bodies* can be detected which go through a regular cycle of division during the growth of the bacterial cell. They can also be seen in living bacteria by phase contrast microscopy.

These chromatin bodies are now generally accepted as representing the bacterial nucleus, but there are certain characteristic differences between them and the nuclei of higher cells. For example, the electron microscope shows that there is no nuclear membrane in bacteria, and no nucleolus. Moreover the nuclear material does not show any of the classic stages of mitotic division (prophase, metaphase, etc.) which are so distinctive in animal and plant cells, and it is not possible to distinguish separate chromosomes in the chromatin bodies.

Chapter III

Bacterial Physiology

Bacteria are very widely distributed in nature and their metabolic abilities vary equally widely. In some cases, as with the true *autotrophic* bacteria, they are capable of synthesizing all their requirements from simple inorganic materials such as CO_2 and ammonium salts, and these highly independent organisms can exist in almost any situation, provided it is moist and the conditions of temperature are not too extreme. However, most bacteria are *heterotrophic* organisms which need to be provided with their carbon and nitrogen in a more highly organized form, and often require in addition a number of accessory growth factors such as members of the vitamin B complex. The heterotrophes therefore usually need organic material from other cells for their growth.

In the case of bacteria of medical interest it is usual to make a distinction between *saprophytes*, which live on dead organic matter, and *parasites*, which can exist only in or on the living body of man or animals. These terms are, however, only relative. It is impossible absolutely to separate one group from the other, since under some conditions a parasitic organism can live a saprophytic existence and vice versa. In a few cases parasitic bacteria have become so dependent on their host that they can only live inside host cells, e.g. *Mycobacterium leprae*.

MATERIALS NEEDED FOR BACTERIAL GROWTH

The materials needed or bacterial growth are not basically different from those required by other cells. They are:

(1) **A source of energy.** This is provided by carbohydrate or protein material. During the process of breaking down these large molecules the bacteria obtain energy which is used for growth and multiplication, or for other things such as flagellar activity. The process of breakdown also provides low molecular weight compounds of carbon and nitrogen which the cell can use to build up its own proteins.

(2) **Growth factors.** These are organic substances of relatively low molecular weight which some organisms require in their metabolism but cannot synthesize for themselves. Examples are

p-aminobenzoic acid, nicotinic acid, folic acid and thiamine. Many if not all of these substances are necessary for the growth of *all* types of bacteria, i.e. they are *essential metabolites*, but they are only considered as growth factors in the case of strains which cannot synthesize them, and for which they must be provided in the growth medium. They are normally available to parasitic organisms in the body of the host.

(3) **Essential trace elements.** Certain elements such as magnesium, manganese and iron are needed by all bacteria in very small amounts. These will of course also be present in the body of the host, but may need to be added to artificial media. These trace elements are necessary for the functioning of certain enzyme systems, for example cytochromes.

In addition to the main food supply, there are some general factors which exert a great influence on bacterial growth and multiplication. The most important of these are the *composition of the atmosphere* in which growth occurs, the *pH of the medium* and the *temperature of incubation*.

COMPOSITION OF THE ATMOSPHERE

Although all organisms ultimately obtain their energy by oxidation, only the final stage of the process involves molecular oxygen, and the intermediate steps are relative oxidations performed by the transfer of hydrogen from one substance to another so that the donor substance is oxidized and the acceptor substance reduced. Some organisms are in fact unable to grow in the presence of free oxygen, and are termed *anaerobes*. If their growth is inhibited by even minute traces of oxygen they are referred to as *obligate* or *strict anaerobes*. Other strains will tolerate oxygen but in lower concentrations than that present in air, and are termed *microaerophiles* or *microaerophilic organisms*. At the other end of the scale, a few organisms are *strict aerobes*, for whom the presence of free oxygen is obligatory. Most bacteria, however, can grow under a wide range of oxygen tensions, and are described as *aerobic and facultatively anaerobic*, although their metabolism will not necessarily be the same under anaerobic conditions as under aerobic. It is not absolutely clear why free oxygen prevents the growth of strict anaerobes, but it has been suggested that in the presence of oxygen these organisms produce hydrogen peroxide which is toxic to them as they do not possess the enzyme catalase which breaks it down again to water and oxygen. Another possibility is that in strict anaerobes some enzymes combine readily with oxygen to form inactive compounds.

The only other gas whose concentration markedly affects growth is carbon dioxide. This is necessary in small amounts for all bacteria, but the amount normally present in air (0·04 per cent) is sufficient. However, in a few cases, e.g. *Brucella abortus* or *Neisseria gonorrhoeae* there is little or no growth on first isolation from the host unless 5–10 per cent of CO_2 is present in the atmosphere of incubation. After repeated subculture on artificial media such strains can usually dispense with the added CO_2.

TEMPERATURE

Multiplication of bacteria will only take place within a given range of temperature, outside which no growth occurs. For all species there are therefore not only maximum and minimum temperatures, but also one at which growth is most rapid, the *optimum temperature.*

The optimum for various bacterial species shows marked variation and depends largely on the environment to which they are accustomed. In the case of human or animal parasites, it is generally about 37°C, i.e. about body temperature: in the case of birds, however, which have a higher body temperature (40°–42°C), parasitic bacteria have a rather higher optimum temperature of growth.

On the other hand many saprophytic organisms normally present in water and soil grow best at 20°–25°C, while some *thermophilic bacteria*, which have been isolated from warm springs, decomposing compost, or other similar situations may have an optimum as high as 60°C.

In the case of most bacteria of medical importance, multiplication will take place over quite a wide range of temperatures, i.e. about 18°–42°C. However, some pathogens have much narrower range about the optimum: the gonococcus (*Neisseria gonorrhoeae*), for example, will grow only between 34°C and 39°C.

If the temperature is raised above the maximum at which growth can occur, bacteria are fairly rapidly killed, and except for those that form heat-resistant spores no strain can withstand 60°C for more than about 20 min. Higher temperatures, e.g. 100°C, will kill all non-sporing organisms very quickly (1–2 min).

Low temperatures, on the other hand, are much less destructive than high ones. Growth seldom occurs below 10°C, although a few strains will grow at 4°C. At 0°C many bacteria remain viable for a considerable time, although they will eventually die. At very low temperatures, i.e. −70°C and below, cultures may be preserved indefinitely, and will grow again when warmed up and placed in a fresh supply of suitable medium. Cultures frozen at −70°C, and then dried by evaporation from the frozen state, can be preserved indefinitely at room temperature in the dry state in sealed ampoules. Such *freeze-dried* or *lyophilized cultures* are used to maintain reference strains of bacteria.

pH OF THE MEDIUM

The hydrogen ion concentration has a very marked effect on the growth of bacteria, and most organisms will grow poorly, if at all, outside the range pH 6·5–8·0. A few types of bacteria will grow at more extreme pH values. For example lactobacilli can grow well in conditions as acid as pH 4·0–4·5, and the *Vibrio cholerae* will grow in alkaline peptone water at pH 9·0. However, it is safe to say that the optimum pH value for almost all pathogenic bacteria is around that of the body, i.e. pH 7·3–7·4, slightly on the alkaline side of neutrality.

METABOLIC ACTIVITIES OF BACTERIA

Most bacteria, and certainly all those of medical interest, obtain energy by the breakdown of either carbohydrates or nitrogenous materials. In most cases, only a small part of the available energy can be used and many intermediate products of metabolism accumulate in the growth medium.

CARBOHYDRATE METABOLISM

The complete oxidation of carbohydrate would lead to the production of nothing but carbon dioxide and water, but in the case of bacteria, as in other living cells, most carbohydrates which are attacked undergo a partial breakdown known as *fermentation* which does not require free oxygen, and which was described by Louis Pasteur, who first investigated it thoroughly, as "La vie sans air". The most frequent end-products of bacterial fermentations are various organic acids such as pyruvic, lactic, propionic, formic, acetic and butyric, together with small amounts of ethyl alcohol, glycerol, 2,3-butanediol, acetoin and many other substances. Gas is frequently formed during fermentation and consists usually of a mixture of CO_2 and H_2.

Not all of the substances mentioned are produced by every organism, and one of them may be so predominant that it gives its name to the type of fermentation exhibited. For example, lactic acid is produced in large amounts by lactobacilli, hence *lactic* (or homolactic) *fermentation*, while in the case of brewers yeast ethanol (ethyl alcohol) is the principal product, hence *ethanolic fermentation*.

The types of carbohydrate substrate which bacteria are able to attack vary widely, ranging from single hexoses or pentoses to complex polysaccharides like starch and glycogen. However, individual strains or species usually show a constant *fermentation pattern* or range of substances which they can ferment. In the case of disaccharides and more complex substances these must first be hydrolysed to simple sugars before they can be broken down further. Thus lactose, a disaccharide containing one molecule of glucose and one of galactose, must first be split into these two hexoses, and any organism unable to do this will not ferment lactose, although it may be able to attack both glucose and galactose.

Although the range of substrates which may be attacked by bacteria is very wide, the properties of individual strains or groups of strains in this respect are usually constant, and can be used as a means of identification. This is done either by a series of *fermentation tests*, to establish which substrates are attacked, or by *biochemical tests* of different kinds to detect specific fermentation products such as lactic acid or acetyl methyl carbinol (acetoin). In practice, fermentation tests are more commonly used in diagnostic bacteriology. The substances to be tested are added as a sterile solution to a basal medium (usually peptone water) to give a final concentration of 0·5–1·0 per cent; an indicator is also added to show acid production. Any organism which produces acid when grown in such a medium is said to *ferment* the sugar which was added. A small inverted tube (a Durham tube) is usually also present in the tube of medium to show whether or not gas is produced.

Although the carbohydrate substances used to test the fermentative abilities of bacteria are usually referred to as "*sugars*", this is not always strictly true in the chemical sense. The following carbohydrates are in frequent use:

(a) *Monosaccharides.* Glucose, fructose, galactose, mannose, arabinose, xylose, rhamnose.

(b) *Disaccharides.* Lactose, maltose, saccharose.

(c) *Polysaccharides.* Starch, glycogen, dextrin, inulin.

(d) *Polyols and sugar alcohols.* Glycerol, dulcitol, mannitol, sorbitol.

(e) *Glycosides.* Salicin, amigdalin.

Many other related substances have been used.

METABOLISM OF PROTEINS AND OTHER NITROGENOUS COMPOUNDS

Bacteria can break down nitrogenous compounds such as proteins and nucleic acids to amino acids, purines, pyrimidines and other small molecules which can serve as sources of energy for growth.

PROTEINS

Most bacteria can hydrolyse proteins and peptides to amino acids, the enzymes concerned being called *proteases* and *peptidases*. These are very similar in action to the proteolytic enzymes trypsin and pepsin found in the alimentary tract of man. In some cases where particular proteins are specifically attacked, the enzymes are named accordingly, e.g. *gelatinase*, *collagenase*, *fibrinolysin*, and so on. In the laboratory the ability to attack protein is usually tested by noting the power of various bacteria to digest gelatine, meat particles, or the clot formed in milk. Many strains of anaerobic spore-forming bacilli (Clostridia) are especially active in this respect and are referred to as *actively proteolytic*, e.g. *Clostridium sporogenes* and *Cl. histolyticum*.

The amino acids produced by the hydrolysis of proteins are broken down by bacteria in various ways, with the production of NH_3, CO_2, H_2S, fatty acids and a variety of volatile compounds such as mercaptans which have a very unpleasant odour. The latter are only found when amino acids are broken down under anaerobic conditions, a process generally referred to as *putrefaction*. If amino acids are metabolized under fully aerobic conditions, mercaptans and other foul-smelling products are not formed.

In some cases the breakdown products of individual amino acids are of value in the identification of bacteria. For example some Gram-negative bacilli, e.g. *Escherichia coli*, produce the substance indole from the amino acid tryptophane, while others, e.g. *Salmonella* species, do not.

NUCLEIC ACIDS

Many organisms produce both *ribonuclease* (RNase) which hydrolyses ribonucleic acid (RNA) and *deoxyribonuclease* (DNase) which hydrolyses deoxyribonucleic acid (DNA). Although it is not known whether these reactions are of value to the organisms concerned from the point of view of supplying energy, such enzymes are important medically because they are at least partly responsible for the liquefaction of purulent exudates. In the case of streptococcal deoxyribonuclease the enzyme has been purified and made available for clinical use.

SYNTHETIC ABILITIES OF BACTERIA

The processes of breakdown are only the necessary preliminary to growth and multiplication, which implies the synthesis on a large scale of all the components of the bacterial cell. This includes the capsules, flagella, cell walls, toxins, pigments and all the other elements of the cell which make it recognizable by morphological or biochemical investigation.

Chapter IV

The Cultivation of Bacteria

Although most bacteria are able to multiply in nature under a large variety of conditions, a detailed study of their activity is only possible by the examination of the growth characteristics in artificial media. In the majority of instances the nutritional needs of the bacteria are not very specialized, but some parasitic members are very exacting in their requirements. Even these, however, usually display considerable adaptability; on first isolation from the tissues some grow very poorly, but after several sub-cultures they gradually adapt themselves to the new conditions and growth occurs in a wider range of media.

The basis for almost all bacteriological media is some kind of extract of meat, which provides most of the substances required for bacterial growth. The media may be in the form of either (1) *fluid media* or (2) *solid media*. In the case of the latter agar, or less frequently gelatine, is added to give a solid gel. Innumerable media have been described; details of those in general use are given later (Chap. XXXVI).

(1) FLUID MEDIA

The early studies of the role played by bacteria in spontaneous generation and fermentation were carried out by observations of their activity in fluid media. This method was both tedious and inaccurate as, by the use of a fluid medium alone, pure cultures, i.e. cultures of one species only, are extremely difficult to obtain unless a complicated dilution technique is used. The introduction of a solid medium by Koch was a step of the greatest importance in the isolation of pure cultures, and it is difficult to see how any great advance in bacteriology could have been made without it. Nevertheless fluid media serve a useful purpose, and they are used extensively in the study of pure cultures of bacteria. The meat extracts which form the basis of fluid media may be made either as an *infusion broth* by simply extracting meat with water, or as a *digest broth* in which the meat is first partly digested with proteolytic enzymes.

In the case of *infusion broth* the extract is prepared by standing minced meat with water overnight, straining, and boiling the resulting straw-coloured liquid to coagulate the protein, which is then filtered off (see p. 408). Commercial meat extracts such as Lab Lemco may be used instead of extracts of fresh meat. The meat extract alone is not very suitable for the growth of bacteria since most of the proteins have been removed by the filtration process. Digested and uncoagulable protein is therefore added to infusion broth in the form of commercial peptone (acid hydrolysed protein). In the case of digest broth this is not necessary since adequate amounts of protein breakdown products are available as a result of the preliminary enzyme digestion. It is usual to add 0·5 per cent sodium chloride to both infusion and digest broth.

Another common liquid medium, used alone or as the basis for more elaborate media, is *peptone water*. This is simply an aqueous solution of peptone and sodium chloride. It is used in testing bacteria for their capacity to produce indole and also as a basis for the sugar solutions used in fermentation tests.

ADJUSTMENT OF THE pH

The hydrogen ion concentration of the medium is a matter of great importance if good growth is to be obtained, and it must be adjusted before the medium is used. The majority of bacteria grow best in the range pH 7·2–7·6. The pH of a bacteriological medium is usually estimated by adding an indicator such as phenol red to a sample of the medium of known volume (e.g. 5 ml), and comparing the colour produced with the colours of a set of standards of known pH. The amount of N/10 HCl or N/10 NaOH to correct the pH of the 5 ml sample is then determined and the total quantity needed to adjust the reaction of the whole bulk of medium can then be calculated. Phenol red exhibits a definite range of colours from yellow to purplish-pink through the pH range 6·8–8·4 and is consequently a common indicator for routine bacteriological purposes. The indicator technique is satisfactory for most routine procedures, but where very precise readings are essential, electrometric methods of estimating pH should be used.

The effect of bacterial growth is usually to alter the pH of the medium away from the optimum range, and unless precautions are taken growth may be prematurely arrested. This is especially the case with media containing much fermentable carbohydrate, where large amounts of acid may be produced. These pH alterations during growth may be avoided to some extent by the addition of phosphate, bicarbonate or other buffer salts to the medium.

(2) SOLID MEDIA

Solid media are usually prepared by the addition to broth of some substance such as gelatin or agar, which gives consistency to the medium.

Gelatin, an albuminoid derived from tendons and cartilage, was the substance originally employed by Koch in the preparation of solid medium. It has the big disadvantage in that it melts at about 26°C and, as the optimum temperature for the multiplication of parasitic bacteria is usually 37°C, the use of gelatin for solid media is somewhat limited. The proportion of gelatin employed is generally 10–15 per cent by weight; this is added to broth. In hot climates the amount should be increased to 20–25 per cent. High temperatures interfere with its power of solidifying and care is therefore required in sterilization; steaming on successive days is the usual procedure. Gelatin, if too concentrated, tends to split when inoculated

Agar is prepared from dried seaweed collected originally around the shores of China and Japan; a useful product has now been prepared from other seaweeds, including some from the shores of this country. It is added to broth to give a final concentration of 1–2 per cent, and this constitutes an excellent solid medium, which is known as *nutrient agar*. After the addition of the agar the broth is placed in the steamer for 1 hr to effect solution. The medium may be cleared by the addition of egg-white at a temperature of 55°C; the egg-albumen coagulates and removes gross particles, which are easily held back on filtering through paper.

Nutrient agar, or agar, liquefies only on heating to 98°C and sets at 45°C; it is not damaged by temperatures of the autoclave and is thus an extremely satisfactory and serviceable medium.

ENRICHED MEDIA

Simple media, such as nutrient broth or nutrient agar, are sometimes unsuitable for the isolation and subsequent growth of certain parasitic bacteria, such as the gonococcus, the pneumococcus and the influenza bacillus. It is consequently necessary to enrich them by the addition of appropriate substances; those usually employed are serum and ascitic fluid (10 per cent), blood (5–10 per cent) and glucose (1–2 per cent). The resulting media, e.g. serum-, blood-, ascitic fluid-, glucose-broth or agar, are termed "enriched media". The final dilution of the enriching substance usually varies from 1–10 per cent for ordinary purposes.

Body fluids used for enrichment, such as blood, serum, plasma or ascitic fluid must of course be sterile. They cannot be added to media which are subsequently to be sterilized by heat, since this would coagulate the protein they contain, and they are therefore added after the medium has been sterilized. In the case of agar this is done by melting the agar and cooling it to 50–55°C at which temperature agar is still molten, but proteins in the added blood or serum are not denatured.

Blood agar, that is nutrient agar with 5–10 per cent whole blood added, is made in this way, and is one of the most commonly used general-purpose media in clinical bacteriology.

Carbohydrates such as glucose may be added before sterilization, but since there is a danger of their being partly destroyed if the medium is heated much above 100°C, they are more often added later in the form of a sterile solution.

Selective and differential media. In order to have the best possible chance of isolating a particular type of organism it is important to use a medium which favours its growth and if possible also discourages the growth of unwanted types. Such a medium is described as *selective*. For example in MacConkey's medium for intestinal bacteria the presence of bile salts is inhibitory to non-intestinal organisms. Again, the addition of small amounts of penicillin to blood agar or other media make them inhibitory to Gram-positive organisms such as staphylococci and streptococci, while allowing Gram-negative organisms such as *Haemophilus influenzae* or *Bordetella pertussis* to grow. This allows the latter organisms to be isolated more readily from sputum or other material which may contain large numbers of other bacteria which interfere with the isolation of the pathogens.

In addition to being selective, most media in common use are also designed to be to some extent *differential* in that the colonies of the different organisms which grow on the medium can be distinguished from each other in some way easily visible to the eye. To take MacConkey's medium again as an example, the presence in the medium of lactose, together with an indicator to show acid production, enables the distinction to be made between the pink colonies of *Escherichia coli* which ferments lactose, and the colourless colonies of non-lactose-fermenting intestinal pathogens such as *Shigella sonnei* or *Salmonella typhi*.

Sterilization of media. All media used for the cultivation of bacteria must obviously be free from contaminants and therefore must be sterilized before use. For nutrient or digest broth or agar it is usual to sterilize in the autoclave at 10 lb sq in (= 115°C) for 15–20 min. However, glucose or other sugars may undergo partial breakdown at this temperature, and media containing them must be sterilized for a longer time at a lower temperature, e.g. by steaming at 100°C for 30 min. As already explained, body fluids containing protein cannot be sterilized at these temperatures, and so are added later to the sterile media with suitable precautions to prevent contamination.

Liquid media, or solutions of sugars can also be sterilized by filtration.

Methods of sterilization are discussed in more detail later (Chap. XXXVI).

For storage, solid and liquid media are usually sterilized in bulk in screw-capped bottles. The medium is dispensed into test tubes, smaller bottles or Petri dishes for use.

THE USE OF CULTURE MEDIA

In order to derive the maximum information from the cultivation of bacteria it is essential that the cultures should be prepared in a satisfactory manner. Under ordinary circumstances the procedure varies with the type of medium.

Fluid media are prepared either in ordinary test-tubes ($6 \times \frac{1}{2}$ in) containing 5–10 ml, small screw-cap bottles or flasks of varying sizes (50–250 ml). The flasks and test-tubes are both stoppered with plugs of cotton-wool.

Solid media are prepared either in test-tubes, small screw-cap bottles or Petri dishes. The medium in the test-tube is either upright or sloped. An ordinary upright tube contains about 10 ml. The form most commonly employed is a sloped tube or *slope*. About 5 ml of melted agar are placed in a test-tube and allowed to cool in a slanted position. In this way a large surface area is obtained from a relatively small quantity of medium.

When a large and readily accessible surface is required, about 15 ml of the melted medium are allowed to cool to 5°–10°C above the setting temperature and are then poured into a Petri dish. Condensation of steam on to the medium or the lid is avoided by allowing the medium to set at 37°C with the lid not completely covering the medium but resting on one side of the dish. Plates are particularly useful in the isolation of pure cultures from grossly contaminated material.

The inoculation of media. The instruments commonly used to inoculate or seed a medium are the platinum loop and needle. The platinum loop consists of a piece of platinum wire, 3 in long, one end of which is fixed into an aluminium holder while the free end is bent in the form of a loop. The loop must be closed and quite flat. The needle is similar, but without the loop. Owing to the high cost of platinum, resistance wire is usually used for class purposes. This cools slower than platinum and it is important after flaming to allow the wire adequate time for cooling before use.

Fluid material such as serum may be added to a medium by means of Pasteur pipettes. These are made from a piece of glass tubing about 10 in long, which has been plugged with cotton wool at both ends and then sterilized. The glass tubing is heated in the middle until soft and then drawn out to form two pipettes.

The inoculation or seeding of a medium requires practice before a satisfactory technique is acquired. The method varies with the medium used. When the medium is contained in a test-tube, this should be held at its lower end in the left hand between the thumb and the first two fingers. The platinum loop is held in the right hand and is sterilized in a bunsen flame, while the plug of the tube is removed by the little finger and the palm of the same hand; the mouth of the tube is passed through a bunsen flame and the inoculation carried out, after which the plug is replaced. By the use of such technique it is possible to observe all the stages of the procedure. A common fault is to grip the middle of the tube with one hand and then to plunge the loop blindly inside. In inoculating from one tube to another, both tubes should be held between the thumb and the first two fingers of the left hand. One plug is removed by pressure applied by the little finger and the palm, and the second between the little and ring fingers of the right hand.

The plugs must not be placed on the bench during the seeding of the tube. Such a practice is certain to result in contamination.

In seeding fluid media the charged loop should be passed immediately below the surface and rubbed gently against the side of the tube. Vigorous agitation of the loop should be avoided.

In seeding solid media the arrangement of the medium must be considered. For a slope culture the loop should be pressed gently but firmly on the surface of the lowest part of the medium and drawn up the surface by passing from side to side. In this way the inoculum is thinned as a result of the streaking. Upright media may be used either as stab or shake cultures. In a *stab* culture the charged needle is passed vertically down the centre of the medium. In a *shake* culture the medium is melted and allowed to cool to a few degrees above the setting temperature, at which stage the inoculation is made either by a loop or a capillary pipette. The tube is then rotated between the hands, to distribute the inoculum, and allowed to cool.

The platinum loop or needle must always be flamed before and after use; after the initial heating it must naturally be cooled to a temperature not injurious to the bacteria. This is a detail often forgotten by the student and may be responsible for the failure of organisms to grow in culture.

A medium which has been successfully inoculated is termed a *culture*. When only one species of bacterium is present it is termed a *pure culture*. The inoculation of a second medium from a previous culture is termed a *subculture*. It is important that all cultures should be carefully labelled and the source of the material and date clearly indicated.

The incubation of cultures. Inoculated media are placed in an incubator or hot room whose temperature is controlled by a thermostat. For parasitic bacteria the usual temperature is 37°C. Agar plates are incubated in the inverted position, so that the drops of condensation water which may form on the inside of the lid will not fall on the surface of the medium. Gelatine plates, however, cannot be incubated at 37°C since the medium is liquid at that temperature. They are usually incubated at 22°C, but are not inverted, since some organisms produce enzymes which liquefy gelatine.

Anaerobic methods. The atmospheric conditions of cultivation also require attention. The anaerobic bacteria will not grow in the presence of free oxygen, while microaerophiles require a greatly reduced O_2 tension. It is therefore necessary to remove or exclude the atmospheric O_2 in order to obtain multiplication of these bacteria. The following methods may be used for this purpose:

(1) Exclusion of air from the medium. In the case of a tall column of medium in a test tube, the deeper layers contain relatively little oxygen if the medium is undisturbed. With liquid media any dissolved oxygen can be driven off by heating the tube and then allowing the medium to cool undisturbed. After inoculating the medium to the bottom of the tube, the surface may be sealed with sterile vaseline or liquid paraffin. With solid media the material containing bacteria can be mixed with the molten agar at about 50°C and then allowed to solidify (shake culture) or inoculated by stabbing into the depths of the solid medium with a straight wire (agar stab). In either case, anaerobic organisms will grow in the deeper parts of the agar.

(2) Addition of reducing substances. The addition of reducing substances to ensure the absence of free O_2 is a satisfactory way of growing anaerobic organisms in liquid media. It is ineffective in the case of surface growths on solid media (e.g. a nutrient agar plate), since the organisms are directly in contact with the air.

The substances used may be glucose or some other reducing sugar, compounds containing sulphydryl groups (–SH) such as thioglycollic acid, or other types of reducing compound such as ascorbic acid. A piece of soft iron (e.g. a nail) may also be used. If the medium is made semi-solid by the incorporation of 0·1 per cent agar, re-aeration by convection currents is prevented, and it is easier to maintain anaerobic conditions. A frequently-used medium employing this principle is Brewer's thioglycollate broth, which is essentially a semi-solid nutrient medium containing 0·1 per cent thioglycollic acid and 0·5 per cent glucose. A small amount of methylene blue is also added: this dye is colourless in the reduced state and thus acts as an indicator of anaerobiosis. Another very common medium for the growth of anaerobes is Robertson's cooked meat medium, which consists of coarsely ground meat particles in a nutrient broth. The reducing substances in the meat produce anaerobic conditions in the medium.

(3) Removal of oxygen from the atmosphere of incubation. The most satisfactory method of cultivating the anaerobes on solid media is by incubating the cultures in a closed jar from which all oxygen has been removed. This can be done by simple displacement, i.e. by passing an inert gas such as nitrogen through the jar until all oxygen has been washed out, but the more usual method is to replace most of the air by hydrogen, and remove the remaining oxygen by making it combine with hydrogen under the influence of a catalyst. The best apparatus for this purpose is the McIntosh and Fildes' anaerobic jar, or some modification of it. The jar is made of glass or metal, with a well-fitting lid which can be tightly clamped down. On the under surface of the lid is fixed a piece of asbestos wool impregnated with finely divided palladium, which is surrounded by wire gauze. The lid is also pierced with two taps. To operate the jar, it is partly evacuated by a pump, and hydrogen is allowed to flow in; under the influence of the catalyst the residual oxygen combines with the hydrogen. In the older type of anaerobic jar the reaction would not start unless the catalyst was first heated. This was done by having an electrical resistanc wound round it, through which a current was passed. However, in modern jars the catalyst will initiate the reaction without being heated. Hydrogen may be obtained either from a cylinder or from a Kipp's apparatus. If the latter source is used, it is important to wash the gas to remove impurities such as sulphuretted hydrogen which might be inhibitory to growth.

To make sure that anaerobic conditions have been achieved in the jar, a small tube may be included containing methylene blue in alkaline glucose solution. This is boiled and cooled before being placed in the jar, and should remain colourless if all oxygen has been removed.

ISOLATION OF PURE CULTURES

In order to study the cultural characteristics of a bacterium it is essential to obtain it in *pure culture*, uncontaminated by other organisms. Before the introduction of solid media by Koch in 1881, the isolation of pure cultures was a matter of considerable difficulty. Material containing a mixture of organisms had to be diluted in saline or broth so that only one type of organism was likely to be present if a small volume of the diluted material was

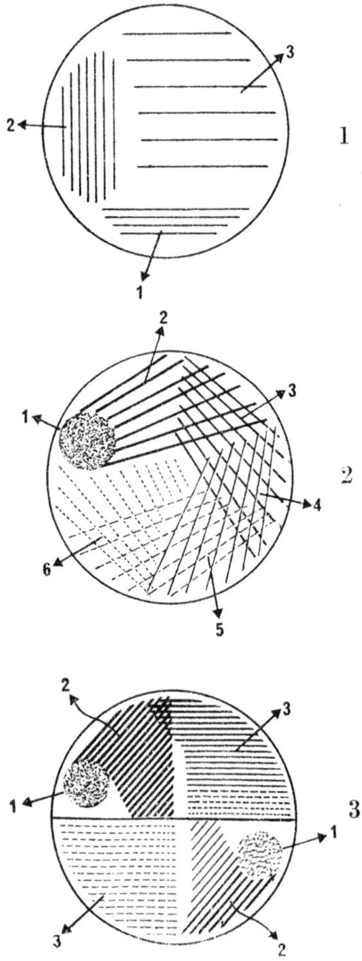

Fig. 1.IV. Plating on solid media.

inoculated into fresh medium. The method is unreliable and in any case will only give a pure culture of the organism present in largest numbers.

However, if such a mixture is diluted by spreading it over the surface of a solid nutrient medium, each organism grows uncontaminated by its neighbours, producing a visible *colony* consisting of bacteria of only one kind, which can be *subcultured* into other media to give a pure culture.

Plating on solid media. Cultures are usually prepared on solid media in Petri dishes by streaking with a platinum loop. The loop is sterilized in the flame, and when cool is charged with material and a series of strokes made on the surface of the medium. Several methods of doing this are illustrated in Fig. 1.IV; the primary inoculation is at 1, and successive streaks are made at 2, 3, etc. It is important that the surface of the medium should be dry, otherwise separate colonies may not be obtained since the growth will be confluent.

In plating out to obtain pure cultures it is obviously important to use a medium which will give good growth of the organism one wishes to isolate, and a suitable selective or differential medium should be chosen. It will also be necessary to take account of other properties of the organism such as oxygen requirements and optimum temperature for growth. In some cases the most satisfactory method of isolating bacteria from infected material is by *animal inoculation.* For example, the mouse is extremely susceptible to infection by the pneumococcus, and after inoculation of a mouse with material containing virulent pneumococci, a pure culture of this organism can generally be obtained from the heart's blood after the death of the mouse.

In the case of organisms which form heat-resistant spores, non-sporing contaminants can be eliminated by heating the material at 80°C for 10 min. This method is largely used for the isolation of anaerobic spore-bearing organisms from contaminated material.

Chapter V

Bacterial Growth and Multiplication

TOTAL AND VIABLE COUNTS

Under optimum conditions, bacteria are capable of multiplying indefinitely at a very rapid rate, so that their numbers may double every 20 min or so. However, under normal conditions in a confined volume of medium, this maximum rate cannot be kept up for long because of the exhaustion of food supply and the accumulation of metabolic products such as organic acids which alter the pH. If a small inoculum of an actively growing culture is inoculated into a bottle or flask of fresh medium, the subsequent events can be rather arbitrarily divided into four phases of growth: (a) the lag phase, (b) the logarithmic phase, (c) the stationary phase and (d) the phase of decline.

(a) Lag phase. During this phase there is no increase in the number of living organisms, and there may in fact be a slight decrease due to death of some of the inoculated bacteria. The length of the lag phase varies very widely. If young actively growing cells are transferred to fresh medium which has been warmed to optimum temperature, then there may be virtually no lag in growth. On the other hand if cells from an old culture are inoculated into cold medium which warms up only slowly, then the lag phase may last for 8–10 hr.

(b) Logarithmic phase. This is the phase of rapid and steady multiplication. The name derives from the fact that during this phase a straight line graph is obtained if the logarithm of the number of organisms is plotted against time (see Fig. 1.V). After a time, the rate of growth declines due to the accumulation of metabolic products and other factors, and the culture passes into the stationary phase.

(c) Stationary phase. During this phase any bacterial multiplication which takes place is balanced by an increased death rate and the number of living organisms remains more or less constant, although the *total* bacterial count (dead and living) will rise slowly.

(d) Phase of Decline. The number of viable organisms decreases until all are dead. The length of this phase is extremely variable, and if precautions are taken against drying up of the medium, many types of organisms will remain alive in old cultures for years.

It should be realized that cultures go through these phases because of the limited conditions under which they are forced to grow, and not because of any inherent properties of the bacteria themselves. If precautions are taken to neutralize or remove waste products and to provide a constant supply of fresh nutrient medium, it is possible to keep bacteria in the logarithmic phase for months on end in a *continuous culture apparatus*.

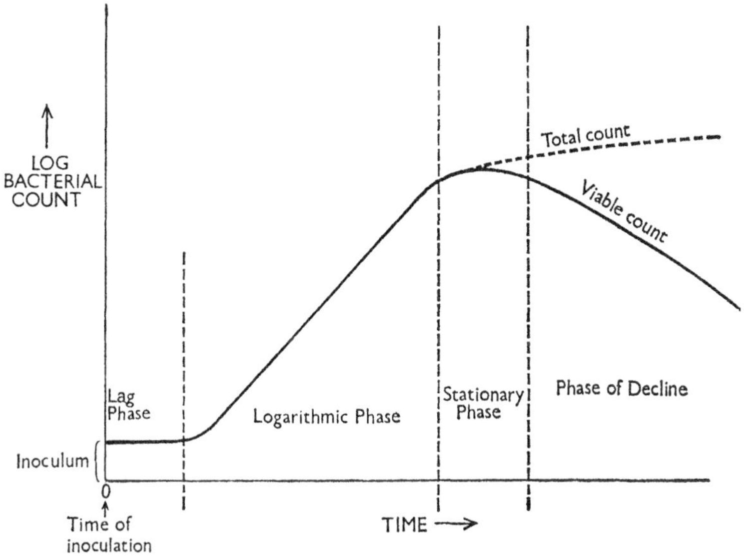

Fig. 1.V. Bacterial growth curve of a hypothetical organism showing the four phases of growth.

In order to follow the progress of bacterial growth and illustrate it by graphs like that shown in Fig. 1.V it is necessary to have some method of counting the number of bacteria present in the culture at different times. Two kinds of estimate can be made.

(1) The total count, i.e. the total number of bacteria, both living and dead.

(2) The viable count, i.e. the number of living organisms present.

(1) TOTAL COUNT

This is usually estimated in one of two ways:

(a) by making a direct count of the numbers of bacteria in a known volume of the culture, using a special slide and coverslip. The method is similar to that used for red or white blood cells;

(b) by estimating the turbidity of the suspension and comparing this with the turbidity due to bacterial suspensions of known strength or with a set of artificial barium sulphate standards.

The latter are usually known as "Brown's tubes" and are available from Messrs. Burroughs Wellcome & Co. The turbidity of bacterial suspensions may also be estimated photoelectrically, and the galvanometer readings can if necessary be calibrated directly in terms of bacterial numbers for any particular organism.

(2) VIABLE COUNT

This is usually done either by (a) the pour plate method or (b) the Miles and Misra dropping technique.

(a) Pour plate method. Serial decimal dilutions are made of the bacterial suspension in sterile saline or peptone water, from 10^{-1} to 10^{-7} or higher if thought necessary and a fixed amount of each dilution (e.g. 0·5 ml) is placed in a sterile Petri dish. Melted agar cooled to 50–55°C is added and the plate is gently rocked to distribute the organisms in the sample evenly throughout the agar before it sets. After incubation the plates are examined, and one containing a countable number of colonies is selected and counted. A simple calculation gives the number of viable organisms in the original sample.

(b) Miles and Misra dropping technique. Dilutions of the sample are made as before and a standard drop of each dilution is placed on the surface of a well-dried plate of medium. After incubation, the number of organisms in the original is calculated from the number of colonies formed where the various drops were deposited.

Note (1) Errors inherent in the methods.
(2) Uses of estimating numbers.
(a) Vaccines.
(b) Blood cultures.
(c) Milk and water counts.

ERRORS INHERENT IN THE METHODS

In estimations of the total count of bacteria errors may arise due to the tendency of some organisms to aggregate in clumps which may be difficult to disperse. This affects all methods. Direct microscopic counting is inaccurate due to the difficulty of determining the number of organisms in a clump. Turbidity estimations are inaccurate unless the organisms are evenly dispersed. Variation in the size of organisms at different stages of growth also affects turbidity. In the viable count by either the pour plate method or the Miles and Misra dropping technique, a visible colony may arise from a single organism or from a group of organisms such as a chain of streptococci. Unless the microscopic examination of the specimen shows that the organisms are evenly dispersed, the results must be regarded as an approximation.

USES OF ESTIMATING NUMBERS

(a) Vaccines. Bacterial vaccines are generally prepared to contain a certain number of organisms per ml. Fortunately, most Gram-negative organisms from which vaccines are usually prepared are easily dispersed and reasonably accurate estimations of numbers can be made. In the case of mycobacteria, the growth can be weighed and the dosage of vaccine is expressed in weight rather than in numbers of organisms.

(b) Blood cultures. The numbers of organisms in blood may be estimated by the pour plate method. In chronic diseases such as sub-acute bacterial endocarditis successive cultures may give some indication of the effect of therapy.

(c) Milk and water. See Chapter XXXIV.

Chapter VI

Sterilization and Disinfection

The term *sterilization* means the complete destruction or removal of all micro-organisms, and this may be brought about in a number of ways. Broadly speaking, the agents which destroy bacteria may be subdivided into *physical* and *chemical*. Of these the physical agents such as heat are more frequently used, since the rather drastic chemical treatments necessary to ensure complete sterilization may have undesirable side-effects on the material being sterilized. Also, the substances used for chemical sterilization must generally be applied in solution, and there is the further technical problem of removing the sterilizing agent after destruction of bacteria is complete. In the case of fluids (liquids or gases) complete removal of bacteria can be brought about by *filtration*, and this is often the method of choice for fluids which would be damaged by heating.

The terms *antiseptic* and *disinfectant* were in use before the bacteriological era, and it is therefore not surprising that their definition and differentiation in relation to bacterial disease is not always clear. Both are applied to the general process of using chemical substances to eliminate infective agents. However, by convention it is usual to confine the use of the words *disinfection* and *disinfectant* to substances such as lysol and other corrosive phenolic compounds, which are used to sterilize inanimate objects (e.g. bedpans), while *antisepsis* and *antiseptic* are used for substances such as antiseptic dyes which can be applied to living tissues. This convention is rather arbitrary and the same substance may be used in a relatively high concentration as a disinfectant and at a lower concentration as an antiseptic (e.g. halogenated xylenes such as Dettol).

Antiseptics and disinfectants are normally considered separately from antibiotics or chemotherapeutic substances which are given by mouth or injection.

PHYSICAL AGENTS USED IN STERILIZATION

The physical agents usually used to kill bacteria are (1) radiation and (2) heat.

STERILIZATION BY RADIATION

The radiations of practical value in sterilization are of two kinds. There are firstly the *high energy ionizing radiations* such as X-*rays*, *gamma rays*, *cathode rays* and *β-rays*. Of these X-rays and gamma rays are electromagnetic waves of very short wavelengths (down to 10^{-4} Ångstroms), while cathode rays and β-rays are in fact beams of high energy electrons. Cathode rays are emitted from the cathode (−ve) when a high voltage potential is established between two electrodes in an evacuated tube. The name β-radiation is given to similar electron beams emitted from naturally decaying radioactive elements. The other kind of radiation which may be used in sterilization is *ultra-violet light*, especially that with a wavelength between 2400 and 2800 Å. *Infra-red rays* (wavelength 2·5–16 μÅ) have also been used successfully to sterilize glassware (and other objects), but their sterilizing power is due to the fact that they heat up the glass, rather than to any direct primary effect of the radiation on individual bacteria.

MECHANISM OF KILLING BY HIGH ENERGY RADIATION

The primary effect of high energy radiation is to activate the molecules they hit, causing them to lose electrons and become positively-charged ions. The lost electrons become attached to other molecules turning them into negative ions. It is for this reason that these radiations are called *ionizing radiations*. The effect of the ionization is to make the molecules abnormally reactive, and to cause chemical reactions which disrupt the normal working of the cell. If this process affects the cell nucleic acids to any great extent (especially the DNA) it is likely to be lethal.

These high energy radiations are measured either in terms of the amount of ionization they produce (*roentgens*), or of the amount of energy absorbed by the substance being irradiated (*rads*). In the latter case, *1 rad* represents an energy absorption of 100 ergs/gram of material.

The killing power of ionizing radiation varies with a number of factors, of which the most important is the *oxygen tension*. Most organisms, except anaerobes, are more sensitive to irradiation in the presence of oxygen, probably because of the ionization of water and the formation of peroxide under these conditions. On the other hand, reducing substances and sulphydryl compounds may give protection against radiation effects. The physiological state of the organisms at the time of irradiation also affects their resistance; the rapidly-dividing cells of the logarithmic phase are much more susceptible than those in the lag or stationary phases. Also organisms in the dry state, e.g. freeze-dried cultures, are less susceptible than those suspended in a watery medium.

LETHAL ACTION OF ULTRA-VIOLET LIGHT

The penetrating power and lethal effect of ultra-violet light is much less than that of the ionizing radiations, because the energy carried by the waves is much lower, and no ionization is produced in the molecules which are hit. However, certain kinds of molecules, especially proteins and nucleic acids, absorb ultra-violet very strongly, and the energy absorbed makes them unduly reactive and interferes with the normal metabolism of the cell.

PRACTICAL USES OF RADIATION STERILIZATION

The high energy radiations are increasingly being used in the sterilization of such things as disposable syringes, needles and transfusion equipment. They have the advantage of combining great penetrating power with little or no heating of the material and the objects to be sterilized can therefore be sealed in plastic bags before irradiation, and will remain sterile until opened.

The method is also used to sterilize biological material such as arterial grafts, which are stored in the dry state.

Radioactive elements such as *Cobalt 60* or *Caesium 137* are frequently used as sources of gamma rays for this type of sterilization. These concentrated sources of radiation are of course highly dangerous to personnel, and must be suitably shielded.

Ultra-violet light is a relatively inefficient sterilizing agent of poor penetrating power, and very long treatment with it is necessary for complete sterilization, although it may produce a significant reduction in bacterial numbers in quite a short time. Over 90 per cent of *Myco. tuberculosis* are killed in 30 min by ultra-violet light, but 15 hr exposure is required to kill all. The main use of ultra-violet light is therefore in reducing the number of bacteria, or preventing their increase, rather than in complete elimination. It has been used in attempts to prevent aerial cross-infection in schools, hospitals, operating theatres, and so on. One of the difficulties is that an ultra-violet light source of sufficient intensity to kill useful numbers of bacteria in air is strong enough to irritate the eyes, hands and skin of people using the room in which it has been placed. This can be avoided if the air is continuously recirculated past the ultra-violet source. Ultra-violet light has also been used in industrial processes such as filling ampoules with sterile powders, as an additional precaution to keep the risk of contamination to a minimum.

REMOVAL OF BACTERIA BY FILTRATION

Filtration is widely employed in the sterilization of solutions of substances likely to be damaged by heat: it is thus used in the preparation of toxins, antisera, solutions of sugars used in fermentation tests, solutions of heat-labile drugs, fluids for intravenous use, and so on.

The main types of bacterial filter in use are:

(1) Diatomaceous earth or Kieselguhr filters (Berkefeld type).
(2) Unglazed porcelain filters (Chamberland type).
(3) Asbestos pad filters (Seitz type).

Other types which are used for some purposes are:

(4) Sintered glass filters.
(5) Membrane filters (Gradocol or Millipore filters).

Diatomaceous earth filters. These are usually made in the form of a hollow candle by forming a thick paste of the filter material (sifted Kieselguhr and asbestos) into a suitable shape and then heating in a kiln. The two best-known makes are the Berkefeld and the Mandler. The filters are prepared in three grades:

(i) "V", the coarsest grade, will not hold back the smaller bacteria, such as *Chromobacter prodigiosum.*
(ii) "N", the intermediate grade, is the most commonly used type. It does not allow bacteria to pass through.
(iii) "W", the finest grade.

These filters are efficient but are fragile and need careful handling. The Mandler filter is an American modification of the Berkefeld.

Unglazed porcelain filters. Filters of this type were first produced by Pasteur and Chamberland in 1884 by heating a mixture of quartz sand and kaolin. As with the Berkefeld filters, they are traditionally made in the form of a hollow candle, but are also available in the form of discs. Chamberland filters are made in a number of different porosity grades, L1, L2, L3, L5 and so on up to L13, which is the finest. These filters are mechanically stronger than the Berkefeld type.

Both Berkefeld and Chamberland filters may be cleaned after use. The candles are scrubbed and washed in hot water. For a more thorough cleaning, hypochlorite solution may be used, or the filters may be treated with trypsin to remove protein, and then washed thoroughly.

Seitz filters. In these the filter proper is a pad made of asbestos fibres with other material to bind them together. This pad is fragile when wet and for use in filtration is supported by a wire gauze and clamped in a metal holder. The metal holder with the pad in position is sterilized before use. The asbestos pads are available in varying diameters (up to 14 cm) and usually in two grades of porosity, "K" for clarifying and "EK" for removing bacteria. The pads are discarded after use, so that the problem of cleaning does not arise.

Sintered glass filters. These are made in the form of discs by heating pulverized glass in a mould to a temperature just below its melting point. This causes the glass particles to adhere, leaving pores between them. They are available in a variety of porosities from 00 (coarsest) to 5/3 which will remove most bacteria, although it may not be effective with slender organisms such as cholera vibrios or leptospira. These glass filters may be cleaned with strong acid to which a little oxidizing agent has been added.

Membrane filters. These are thin membranes of collodion (nitrocellulose) with pore sizes from 0·5 μ downwards. The finest of them will hold back viruses and large protein molecules; the coarser ones may be used for bacteria. They have been used in the bacteriological investigation of water since organisms retained on the filter will subsequently grow if the filter disc is placed on nutrient medium.

STERILIZATION AND USE OF FILTERS

Fluid will not normally pass through bacterial filters unless pressure is applied. This is most usually done by fitting the filter to a flask with a sidearm which is attached to a vacuum pump. A negative pressure of 100–200 mmHg is usually enough to draw fluid through the filter. A higher filtration pressure may clog the filter rather than increase the filtration rate. If a water pump is used, it is advisable to have a "trap" between pump and vacuum flask in case water is sucked back from the pump. Filtration under positive pressure is equally effective, although less frequently employed for small filters.

The filters and the receiver for the filtered fluid must of course be sterilized before use. This is normally done by heat (autoclave) but in the case of some membrane filters the membranes are kept sterile in alcohol.

PHYSICAL BASIS OF FILTRATION

Filtration is by no means a simple mechanical process governed only by the relative sizes of the pores of the filter and the particles to be filtered. Quite apart from the fact that the filter pores are probably very tortuous and irregular in diameter, electrostatic attraction between the particles being filtered and the material of which the filter is made plays an important part in filtration. Other factors are the pH of the material being filtered, the amount of protein in the solution, the filtration pressure applied and so on. Seitz asbestos filter pads in particular are strongly adsorptive and protein or virus particles may be completely adsorbed and so lost if only small amounts of material are handled. Sintered glass filters or membrane filters on the other hand, adsorb very little of the material passed through them.

FILTRATION OF AIR

Filtration is the only satisfactory method of supplying air which is virtually free from all bacteria to operating theatres, dressing rooms, ampoule-filling units and so on. Although cotton wool is the material traditionally used for laboratory air filters, e.g. cotton wool plugs on tubes of media, the filter material in large air filters is generally slag wool, glass wool, or paper, asbestos and glass-fibre mixtures. Such filters will remove 99 per cent or more of the infective particles from the air, which is sufficient for practical purposes. The air may first be passed through a coarse filter, to avoid the fine filter becoming clogged rapidly with large particles.

CHEMICAL AGENTS USED IN STERILIZATION

A very large number of chemical agents have bactericidal properties in some degree, but unfortunately the majority of them are also extremely toxic to body cells. This means that bactericidal action *in vitro* is no guarantee of useful application *in vivo*. The substances used in chemical sterilization fall into a number of widely different chemical classes, the most useful of which are given below.

(1) OXIDIZING AGENTS

Substances which give up oxygen readily are generally active disinfectants. The most widely used are probably *hydrogen peroxide* (H_2O_2) and *potassium permanganate* ($K_2Mn_2O_7$). Their exact mode of action is not certain, but it is probably by producing irreversible changes in essential proteins such as enzymes.

(2) HALOGENS

Of the four halogen elements, only chlorine and iodine are useful disinfecting agents, since bromine and fluorine are too irritating or difficult to handle.

Chlorine is active against both spores and vegetative organisms and is very extensively used as a disinfectant, especially for water supplies and swimming baths. It is also used for dairy equipment, babies' bottles, washing-up crockery or glasses in restaurants, and so on. The compounds usually used are: (1) hypochlorites and (2) chloramines, but in the large-scale treatment of water supplies chlorine gas from cylinders is employed.

Chlorine rapidly attacks most organic compounds, and the presence of much organic material in water supplies or swimming baths reduces its bactericidal efficiency. The organic chloramines such as Chloramine T are less effective.

Iodine is most widely used in the form of tincture of iodine, and 2 per cent iodine in 70 per cent alcohol remains one of the most effective agents for skin sterilization.

A number of iodine compounds are available as antiseptics or disinfectants, but none are very widely used. A recent development has been the use of iodine in solution in non-ionic detergents. These mixtures are called "iodophors" and have been used quite extensively in the United States, especially for dairy hygiene.

ALCOHOL AND RELATED COMPOUNDS

Ethyl alcohol is quite actively bactericidal against vegetative bacteria, although completely inactive against spores, which have been shown to survive unharmed in it for periods of several years. Alcohol is generally used in the form of methylated spirits, and there is a good deal of evidence that 70 per cent spirit in water is the most effective bactericidal concentration. Other alcohols, e.g. isopropyl alcohol also have bactericidal activity but have been little used.

Formaldehyde (HCHO), an irritating gas, is a derivative of methyl alcohol (CH_3OH). It is a very effective disinfecting agent, and is generally used as *formalin* which is a 40 per cent solution of formaldehyde gas in water. However, the gas itself may be used for disinfecting rooms, or for objects such as rubber catheters, which may be damaged by heat. Twenty-four hours in an atmosphere saturated with formaldehyde vapour will usually sterilize.

Formalin, as 5 per cent formol saline, is of course also used extensively as a histological fixative and lower concentrations (1–2 per cent) are often used to kill bacterial cultures which are to be used for immunization or for chemical investigations.

PHENOLIC COMPOUNDS

Derivatives of *phenol* ($C_6H_5.OH$) are often powerful disinfectants, although phenol itself is only rather weakly germicidal.

The general effect of phenolic disinfectants is to denature protein.

The principal compounds in this group are *cresols* and *xylenols*, in which one or more alkyl groups (CH_3, C_2H_5, etc.) have been introduced into the phenol molecule. The addition of a halogen atom also (usually chlorine), increases the antibacterial activity. The best known of these halogenated phenolic compounds is probably *Dettol*, which is a chlorinated xylenol.

Phenolic compounds are on the whole not very soluble in water, but can be made more soluble by being mixed with soaps (usually sodium and potassium salts of long chain fatty acids). The resulting preparations contain a large proportion of the phenolic substance and only enough soap to act as a solubilizing agent.

Many commercial phenolic disinfectants ("coal-tar disinfectants") are based on coal-tar distillates, composed largely of cresols. These are generally prepared either as "white" fluids where the phenolic compounds are emulsified with gelatine or casein, or "black" fluids in which they are dissolved in soap. The commonest of the latter type is *lysol*, which is a very useful general disinfectant, but irritating to the skin. In bacteriological laboratories a jar of 2 per cent lysol is generally used to receive contaminated pipettes, tubes, cotton wool plugs, etc.

HEAVY METAL COMPOUNDS

A considerable number of compounds of heavy metals have a marked disinfectant action, especially salts and organic complexes of mercury, silver and copper; copper compounds, however, are of more value as antifungal than as antibacterial agents.

MERCURY COMPOUNDS

Salts of mercury have been used for several centuries in the treatment of skin diseases and in syphilitic infections. As antibacterial substances they are bacteriostatic in high dilution but have poor bactericidal power, and hence are not satisfactory as sterilizing agents. They are quite inactive against spores. *Mercuric chloride* or *oxycyanide*, or the complex salt *potassium mercuric iodide* are the most common compounds used.

Organic mercury compounds such as *merthiolate* (Na-ethylmercurithiosalicylate) or *phenylmercuric nitrate* are less toxic and more highly bacteriostatic than the inorganic salts, but they too are inactive against spores.

All mercury compounds, especially the inorganic salts, are very much less effective in the presence of protein. Their effect is to combine with and inactivate sulphydryl groups (–SH) and this action can be specifically neutralized by substances such as glutathione and thioglycollic acid which are rich in groups of this kind.

COMPOUNDS OF SILVER

These are usually used as colloidal preparations of silver or silver oxide such as argyrol (e.g. in eye infections) although salts such as *silver nitrate* or *silver picrate* have also been used. As with mercury salts, they are bacteriostatic in high dilution, but only poorly bactericidal, and inactive against spores.

DYES

A number of organic dyes are very inhibitory to bacteria, especially *triphenylmethane dyes* (aniline dyes) and various *acridine compounds*.

TRIPHENYLMETHANE DYES

These include *malachite green, brilliant green* and *crystal violet* (methyl violet or gentian violet), all of which are actively bacteriostatic against Gram-positive organisms in concentrations of the order of 1 : 500,000 or more, although less active against Gram-negatives. They are also fungistatic. The presence of blood, serum or other body fluids considerably reduces their activity. The strong staining they produce reduces their clinical usefulness,but they have been much used in bacteriological media for allowing selective growth of Gram-negative bacteria (e.g. brilliant green broth) or to select streptococci at the expense of staphylococci (1 : 500,000 crystal violet in blood agar).

ACRIDINE COMPOUNDS

These are more useful clinically than the triphenylmethane dyes, since they stain less and their activity is little affected by the presence of tissue proteins. The most commonly used are *acriflavine* (2:8-diamino-10-methyl acridinium chloride) and *proflavine* (2:8-diamino-acridine sulphate), which is rather less toxic to tissues than acriflavine. As with the other dyes, the acridines are more active against Gram-positive than Gram-negative bacteria. The acridines are strongly bacteriostatic but only slowly bactericidal, and have no activity against bacterial spores, or against fungi. They have been largely used in the local treatment of infected wounds and burns.

DETERGENTS

These are substances which lower surface tension and have cleansing properties. There are three classes of detergents, non-ionic, anionic and cationic, depending on whether or not the substances ionize in solution in water, and on the type of ions produced. The non-ionic compounds (e.g. Tween 80) and the anionic detergents (e.g. sodium lauryl sulphate) are good cleansing agents, but have little antibacterial activity. However, the cationic detergents—the *quaternary ammonium compounds* such as *CTAB* (cetavlon, cetrimide), *Zephiran* (Zephirol Roccal) or *Bradosol*—are relatively poor cleansers but are actively bactericidal against a wide range of Gram-positive and Gram-negative organisms. These substances are called quaternary ammonium compounds because they are derived from the ammonium radicle (NH_4–) by substituting other groups for the four hydrogen atoms, e.g. CTAB is cetyl-trimethyl-ammonium bromide.

The quaternary ammonium compounds are inactivated by tissue proteins, by phospholipids (e.g. lecithin), by anionic detergents and by soaps. Because they are non-irritant and practically non-toxic they have been very widely used. However, as in the case of the dyes, they are less active against Gram-negative bacteria than against Gram-positives. *Pseudomonas* species may be particularly resistant, and have been reported as growing in a 1 per cent solution of CTAB.

Ethylene oxide is a colourless liquid with a boiling point of 10·7°C; it is therefore gaseous at ordinary room temperature. It has been used for many years as a pesticide and disinfecting agent for foodstuffs, but only recently as a sterilizing agent in medical practice. The vapour is mildly toxic, and is inflammable so that it is frequently used as a 10 per cent mixture in CO_2. Ethylene oxide is bactericidal, and will kill both bacteria and spores. It has been used successfully to sterilize catheters, cystoscopes, components of a heart-lung machine and disposable plastic syringes packed in polythene containers. However, its effectiveness depends on careful control of the temperature, humidity and time of application.

STERILIZATION BY HEAT

Heat is the most widely applied bactericidal agent available, and is to be preferred unless there is some good reason against its use, such as the need to sterilize material which will be damaged by heating. The destructive action of heat can be utilized in two forms, *moist heat* and *dry heat*, of which moist heat is by far the more effective.

It is important to remember the great difference in heat resistance between *vegetative organisms* and *spores*. Non-sporing vegetative bacteria are killed with increasing rapidity under moist conditions as the temperature is raised above 55–60°C. For example, at 60°C it may take 1 hr to sterilize a bacterial suspension, while at 70°C it may take a few minutes, and at 80°C probably only a few seconds, once these temperatures have been attained in all parts of the fluid. The spores of some organisms on the other hand will withstand boiling for many hours.

The actions of moist heat and dry heat in killing bacteria are probably different: moist heat acts primarily by its effect in denaturing proteins, while death by dry heat is probably largely due to an oxidative process.

STERILIZATION BY MOIST HEAT

This may be done by *pasteurization*, by *boiling* or *steaming at 100°C*, or by steam under pressure in the *autoclave*. The method of *intermittent steaming* or *tyndallization* may also be used.

PASTEURIZATION

This process is named after Louis Pasteur who discovered that mild heating at 50–60°C would prevent the spoilage of wines and beers. The method was subsequently applied to milk. The milk is either held at 62·8°C (145°F) for 30 min ("holding" method) or else heated at 71·7°C (161°F) for 15 sec ("flash" or "high temperature-short time" method). The latter is applied by a continuous flow technique. Both are equally satisfactory provided the plant is properly maintained, and either method will kill pathogenic organisms such as *Mycobacterium tuberculosis* and *Brucella abortus*. Pasteurization does not completely sterilize milk as spores and certain heat-resistant (thermoduric) vegetative forms are not killed.

BOILING

Boiling in a water bath is a traditional method of sterilizing instruments such as scissors, knives, syringes, etc., and is perfectly satisfactory provided its limitations are realized. The most serious limitation is of course that it fails to kill spores. However, apart from this there are a number of technical reasons why boiling may be ineffective. Although at 100°C the time taken to kill vegetative bacteria is probably only a fraction of a second, it must be remembered that all parts of the object to be sterilized must reach a sufficiently high temperature, and this may take some minutes if a cold syringe wrapped in gauze is placed in boiling water. The addition of 2 per cent sodium carbonate to boiling water increases the bactericidal effect.

STEAMING AT 100°C

A steamer, in which bottles are stood on perforated shelves above boiling water, is frequently used in bacteriological laboratories for sterilizing media which might be damaged by heating above 100°C. The most certain method is that of intermittent sterilization or tyndallization, which involves steaming for 30 min on each of three successive days. The principle is that, after heating, the medium is incubated to allow the spores to germinate and the vegetative organisms so formed are killed by the next steaming. The third heating is a safety precaution. The method is useless if the spores cannot germinate in the fluid.

STERILIZATION IN AN AUTOCLAVE

An autoclave is a closed vessel in which steam can be used at pressures above atmospheric. Steam under pressure is a very effective sterilizing agent for three reasons. (1) The temperature is raised above 100°C. (2) Steam penetrates materials such as surgical dressings or theatre gowns very effectively since the condensation of the first part of the steam causes a lowering of pressure which tends to draw more steam in. (3) A large amount of latent heat is given out when steam condenses to water. The times necessary for complete sterilization are shown in Table I (taken from "Hospital Infections", R. E. O. Williams *et al.*, Lloyd-Luke, 1960, p. 224, slightly modified). The table also shows the temperatures obtained at different pressures, provided that the autoclave contains only steam.

Time Required for Absolute Sterilization by Pure Steam at Different Temperatures

Pressure (lb/sq in above atmospheric)	*Temperature* °F	°C	*Time for sterilization (min)*
0	212	100	hours
10	240	116	30
12	245	118	18
15	250	121	12
19	257	125	8
27	270	132	2

If an autoclave is to function efficiently it is essential to replace all the air in it by steam because the temperatures of mixtures of air and steam are lower than those of pure steam at the same pressure. The air may be removed either by downward displacement, where steam admitted at the top of the chamber drives out air at the bottom, or by preliminary evacuation of the air by a pump. It is important also not to use superheated steam, which does not easily condense and give up its latent heat.

The autoclaves used in bacteriological laboratories for sterilizing media or killing discarded cultures are usually of the simple vertical type in which steam is produced by heating water in the bottom of the autoclave. Those used in operating theatres are larger and rely on a steam main. They also usually have an outer jacket through which steam is passed to prevent loss of heat from the sterilizing chamber, a vacuum pump to remove air before steam is admitted, and a filtration system to prevent contamination when air is readmitted after sterilization.

Autoclaves may fail to sterilize materials fully for a number of reasons such as faulty packing of the load or failure to make sure that all air has been replaced by steam, and it is important to test their efficiency at frequent intervals. This can be done by including a packet containing paper impregnated with live spores. These are afterwards tested for growth. Alternatively, tubes containing a chemical indicator may be used (Browne's tubes), which change colour from red to green if the correct temperature has been held for a sufficient time.

The operating conditions will vary to some extent depending on the type of material being sterilized and the necessity for penetration by the steam. For bottles of media 15 min at 15 lb (121°C) is usually sufficient, but for objects such as towels, gowns and drums of dressings it is usual to give 20–30 min at 121°C.

HEAT RESISTANCE OF PATHOGENIC SPORES

It is fortunate that the spores of the pathogens *B. anthracis*, *Cl. tetani* and *Cl. welchii* are not very resistant to temperatures much above 100°C and are quite rapidly killed at 121°C. However, *Cl. botulinum* spores are more heat resistant, and have been reported as withstanding 2 hr at 115°C.

The spores usually used in testing autoclaves are those from *Bacillus stearothermophilus*, a non-pathogenic thermophilic organism which grows best at about 50°C: these spores may stand up to 30 min at 121°C.

STERILIZATION BY DRY HEAT

Dry heat is not such a powerful bactericidal agent as moist heat owing to its lower power of penetration. Its practical applications are also rather more limited, because the higher temperatures which must be used are harmful to many materials.

The main applications of dry heat are:

(1) Incineration for the destruction of carcases or contaminated material. This is usually done in a gas or electric furnace.

(2) The bunsen or other naked flame for sterilizing platinum loops, needles and other non-inflammable substances.

(3) Hot air oven for sterilizing glassware such as Petri dishes, flasks, pipettes, all-glass syringes and so on. It may also be used for sterilizing powders or high boiling point liquids such as oils.

Because of the relative inefficiency of dry heat as opposed to moist heat, complete sterilization in a hot air oven requires considerably longer than in the autoclave, and it needs 3 hr at 140°C, or 1 hr at 160°C to kill all spores. The two principal causes of faulty sterilization in a hot air oven are failure to allow time for the load to heat up to sterilizing temperature, or failure of part of the load to reach this temperature because of bad distribution of heat. The temperature in different parts of a hot air oven may vary as much as 40°C unless air is circulated by a fan during the operation of the oven.

As in the case of the autoclave, the efficiency of a hot air oven may be checked either by a living spore suspension or by a chemical test. A special type of tube, Browne's tube No. 3, is available which requires a temperature of 160°C for 1 hr to make it change colour.

INFRA-RED IRRADIATION

As a sterilizing agent, this acts by causing heating of the material irradiated, and has been successfully applied to such things as syringes in sealed metal containers. It is not suitable for large objects.

Chapter VII

Chemotherapy

The earliest application of a specific chemical substance to the treatment of microbial disease is probably the use of mercury in syphilis. This was introduced soon after the disease first appeared in Europe in 1492 and remained the standard treatment for it until the introduction of salvarsan (606) by Ehrlich in 1910. The success of salvarsan raised hopes that other bacterial diseases would respond similarly, but except in the case of spirochaetal infections the results were disappointing, although chemotherapy had been quite widely successful in protozoal diseases, e.g. quinine in malaria, arsenical compounds in trypanosomiasis, and antimony compounds in schistosomiasis and leishmaniasis. Quinine affords another example of the early use of chemotherapy, since in the crude form of chincona bark it was used in the treatment of malaria as early as the seventeenth century.

In 1935 the position was completely revolutionized by Domagk's demonstration that streptococcal infections could be successfully treated by a diazo dye called Prontosil. This was the beginning of the modern chemotherapy of bacterial infections, and progress since then has been rapid. In the early 1940s a further striking advance was made with the introduction of *penicillin* and subsequently a large number of specific antibacterial substances have become available. Many of these, like penicillin, were originally isolated from culture filtrates of fungi or soil actinomycetes, and are now called *antibiotics* because they are substances produced by one organism which are inhibitory to others. However, there is no essential difference between antibiotics and other antibacterial agents, and at least one antibiotic (chloramphenicol) is now manufactured synthetically.

A most essential feature of a good chemotherapeutic agent is that it must show a high degree of *selective toxicity*, so that it can be given by mouth or injection in sufficient doses to inhibit or kill bacteria throughout the body without harming the body cells. Penicillin most nearly approaches the ideal in this respect, being virtually non-toxic except in those patients who become hypersensitive to it. Other antibiotic and chemotherapeutic substances do show toxic effects, but usually only if given in large doses. This relative lack of toxicity to the cells of the human and animal body is one of the main features which distinguish chemotherapeutic substances such as sulphonamides from disinfectants such as the phenolic compounds, which may be more toxic to leucocytes than they are to bacteria. However, some substances classed as antibiotics are in fact too toxic for systemic use and are only of value for local applications in creams or ointments.

BACTERIOSTATIC AND BACTERICIDAL DRUGS

Two broad kinds of antibacterial effect can be distinguished, depending on whether bacteria are actively killed by a drug, or merely prevented from multiplying. Drugs such as *streptomycin* which exert an active killing effect are termed bactericidal, while those like *tetracycline* which merely prevent multiplication but leave the organisms still capable of recovery are called *bacteriostatic*. As in the case of substances used as disinfectants and antiseptics, an antibiotic may be bacteriostatic at low concentrations, and bactericidal at higher concentrations, for example *erythromycin*. Since some combinations of bactericidal and bacteriostatic drugs are less effective than either drug given alone, it may be important to know whether a particular substance is bacteriostatic or bactericidal.

SUBSTANCES USED IN CHEMOTHERAPY

The different groups of substances most commonly used in the chemotherapy of bacterial infections are shown in Table I. Each group in reality comprises a number of compounds whose broad chemical structure and mode of action are very similar, but which differ often quite widely in degree of activity, solubility, toxicity, distribution within the body and so on.

THE SULPHONAMIDES

Sulphonamide preparations were used as dyes some 50 years ago, but they were not considered as therapeutic agents until Domagk in 1935 reported the excellent results produced in streptococcal infection by *Prontosil*, a red diazo dye. It was soon proved that the active agent was in fact the simpler substance *sulphanilamide* or *p*-aminobenzene sulphonamide which was produced from Prontosil in the body. The structural formulae of these compounds and the related substance *p*-aminobenzoic acid are shown in Fig. 1.VII.

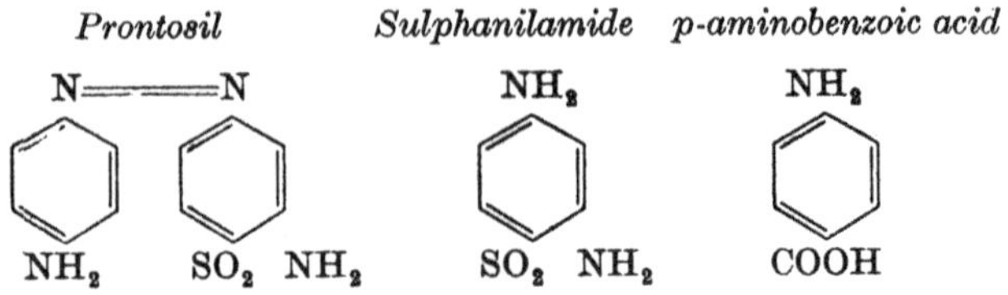

Fig. 1.VII

The original sulphanilamide was of value only in a rather limited range of bacterial diseases, being most active in infections due to *β-haemolytic streptococci*, *gonococci*, *meningococci* and *E. coli*. Numerous modifications of the sulphanilamide molecule have been made in attempts to increase anti-bacterial activity or extend its range. The only derivatives of therapeutic value are those in which one of the hydrogen atoms of the *sulphonamido* ($-SO_2NH_2$) *group* is substituted. Alteration of the *amino group* ($-NH_2$) at the other end of the benzene ring produces inactive compounds. Some examples of substituted sulphonamides are shown in Fig. 2.VII.

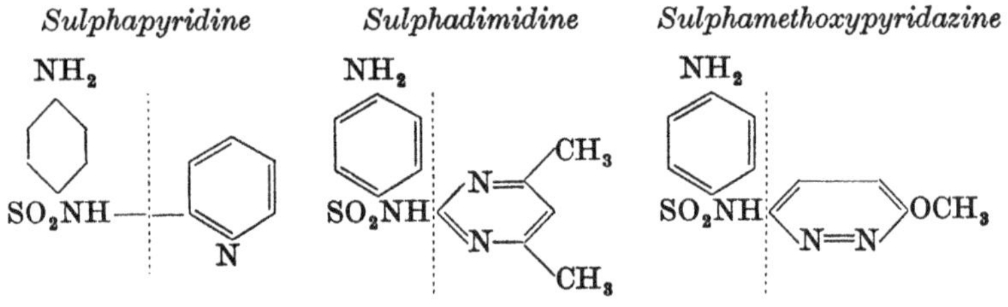

Fig. 2.VII

In each case the group to the right of the dotted line has been substituted for one of the hydrogen atoms of the sulphonamido group. Unlike Prontosil, these substances do not break down to sulphanilamide in the body, but act as the whole molecule.

Sulphapyridine (M & B 693) was the first of these compounds which represented a real advance since it was active against pneumococci as well as haemolytic streptococci. It is, however, very toxic and is now only of historic interest. Subsequently a very large number of other sulphonamide derivatives have appeared. These include *sulphathiazole*, *sulphadiazine*, *sulphadimidine*, *sulphafurazole*, *sulphasomidine* and many others. Some of them, such as *sulphaguanidine*, *succinyl-sulphathiazole* (sulphasuxidine) and *phthalyl-sulphathiazole* (sulphathalidine) are valuable in gastrointestinal infections, because they are very poorly absorbed from the intestinal tract and therefore high concentrations of these drugs are obtained in the intestinal contents.

Sulphadimidine is one of the best sulphonamides for general use since it is readily absorbed and has a relatively low toxicity. Sulphafurazole and sulphasomidine are valuable in urinary infections because they are rapidly excreted and highly soluble in urine. Sulphanilamide, sulphapyridine and sulphathiazole are not now used because of low activity or high toxicity.

LONG-ACTING SULPHONAMIDES

A recent important development has been the introduction of sulphonamide compounds of low toxicity which are readily absorbed, but very slowly excreted. *Sulphamethoxypyridazine* (Lederkin, Midicel), *sulphasomizole* and *sulphamethoxazole* are examples of this type of compound. The two latter are valuable for urinary infections because of their higher solubility.

PHARMACOLOGY AND TOXIC REACTIONS

With the exception of sulphaguanidine, succinyl-sulphathiazole and phthalylsulphathiazole, the sulphonamides are rapidly absorbed from the stomach and small intestine, and are therefore usually given orally. They readily diffuse, and are found in tissue and body fluids, including C.S.F., soon after administration by mouth. In plasma, a fairly high proportion (70 per cent or more) of most sulphonamides is in loose combination with plasma proteins, and not immediately available. However, with adequate dosage sufficient of the drug is present in diffusible form for antibacterial purposes. A dose of 2 g of sulphadimidine will produce in about 4 hr a total blood level of 8–10 mg/ml, of which 2·5–3 mg/ml will be available.

If parenteral administration is considered necessary, either for very severe infection or because oral dosage is not tolerated, the sodium salts of the various preparations may be given by intravenous or intramuscular injection. These solutions are very alkaline (pH more than 9·0) and may cause pain and possibly necrosis at the site of injection. They cannot therefore be given intrathecally.

Excretion of sulphonamides is mainly in the urine, and in most cases about 50 per cent of the drug is acetylated. These acetyl derivatives of sulphonamides are not only inactive against bacteria but are also much less soluble than the original drug, and especially in acid urine are apt to crystallize out in the renal tubules causing severe and perhaps fatal anuria. To avoid this it is usual to keep the urine alkaline by administering sodium citrate or bicarbonate and to make sure that the patient's urinary output of fluid is sufficiently high. These measures are especially important in hot climates. Large doses of sulphonamides should not be given to patients with impaired renal function.

Apart from renal complications due to crystalluria, a variety of other toxic symptoms may result from sulphonamide therapy Mild symptoms are usually nausea, vomiting, cyanosis, rashes and headache. Sulphonamide rashes are most likely to be due to the development of hypersensitivity. The most serious toxic effect of sulphonamides is on the bone-marrow and may result in thrombocytopenia, aplastic or haemolytic anaemia or fatal agranulocytosis. These effects are fortunately rare: the majority of cases have followed prolonged administration of large doses.

MODE OF ACTION

The sulphonamides are essentially bacteriostatic, preventing the growth of bacteria, and so rendering them a ready prey for the normal defence mechanisms of the body.

The most generally accepted explanation of their mode of action is that sulphonamides interfere with the utilization of *p*-aminobenzoic acid (PABA) which, as can be seen from Fig. 1.VII, is very similar in chemical structure to sulphanilamide. Sir Paul Fildes and D. D. Woods in 1940 showed that PABA would neutralize the antibacterial activity of sulphonamides. They suggested that PABA was in fact an essential metabolite for bacteria, and this has subsequently been shown to be true. In bacteria, PABA is used to synthesize folic acid, and it is thought that the sulphonamides, because of their close structural similarity to it, block one of the enzymes involved in the reaction. Sulphonamides are therefore good examples of *competitive inhibitors*, which prevent bacterial growth by competing with an essential metabolite for a site on a vital enzyme. It must be noted that although host tissues contain folic acid, bacteria are unable to avail themselves of this for some reason; if they could, sulphonamides would presumably not be active *in vivo*.

It is usual to add PABA (5 mg/100 ml) to media used for culturing material collected from patients receiving sulphonamide therapy. In this way any inhibitory action of the sulphonamides is eliminated.

The antibacterial activity of sulphonamides may also be antagonized by various unidentified substances (not PABA) in bacteriological media, or in pus, and this can give rise to errors in testing the sulphonamide sensitivity of bacteria (see p. 426).

Fortunately these antagonists are neutralized by adding lysed horse blood to the media used for such tests.

CLINICAL USES OF SULPHONAMIDES

Although in the test-tube the sulphonamides are inhibitory to a wide range of bacteria, clinically their usefulness is more restricted, especially in view of the wide range of other chemotherapeutic substances now available. Indeed at one time it appeared as if there might no longer be a place for sulphonamides in chemotherapy. However, the development of less toxic, more soluble and longer-acting sulphonamides has considerably increased their value. They are also much cheaper than many of the antibiotics in common use. They are probably most often used by themselves in urinary infections, and in gastroenteritis or bacillary dysentery. In many infections sulphonamide treatment may be of particular value in cases due to antibiotic-resistant strains of *Proteus* and *Pseudomonas*. The sulphonamides may also be used in combination with antibiotics.

RESISTANCE TO SULPHONAMIDES

Bacteria can develop increased resistance to sulphonamide drugs both *in vitro* and *in vivo*. In the latter case resistance usually develops as the result of long-continued treatment (three weeks or more) or inadequate dosage, or both. Strains which have become resistant to one sulphonamide are resistant to all drugs of this type. However, this problem is unlikely to arise with adequate courses of treatment which last no longer than 5–7 days.

PENICILLIN

The history of penicillin starts with the observation by Fleming in 1929 that, on an agar plate of staphylococcus contaminated with a mould, the colonies in the neighbourhood of the contaminant had faded away. He investigated this phenomenon in detail and found that culture-filtrates of the mould, *Penicillium notatum*, contained a soluble substance which he called *penicillin* and had a markedly inhibitory action on many other organisms as well as the staphylococcus. The substance proved very labile and attempts at that time to concentrate it for therapeutic use were unsuccessful. Fleming, however, used it in the preparation of a selective medium for the isolation of *H. influenzae* which was not inhibited by weak solutions of penicillin.

Further progress was not made until Florey, Chain and their co-workers (1940) prepared penicillin in a concentrated and stable form. They were consequently able to demonstrate the therapeutic properties of the substance in experimental infections with streptococci and staphylococci, and also, with great success, in staphylococcal infections in man.

Although originally discovered by Fleming in culture-filtrates of *Penicillium notatum*, the cultures now used to produce penicillin commercially are high-yielding variants of the related organisms *Penicillium chrysogenum.*

GENERAL PROPERTIES OF PENICILLINS

It became evident in the early days of commercial production that a number of different penicillins were being produced. These have different chemical side chains in position Ⓡ in Fig. 4.VII.

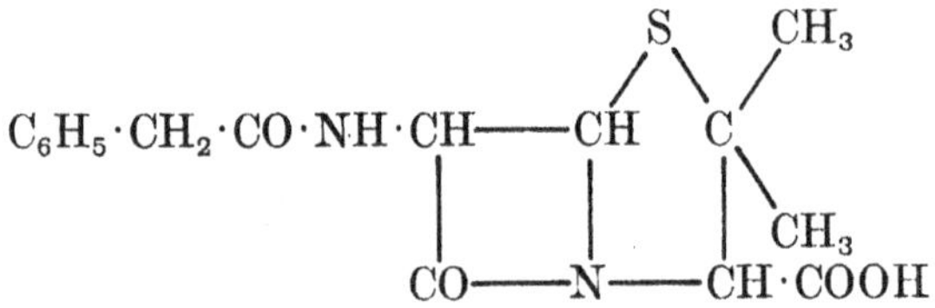

benzyl penicillin-Penicillin G

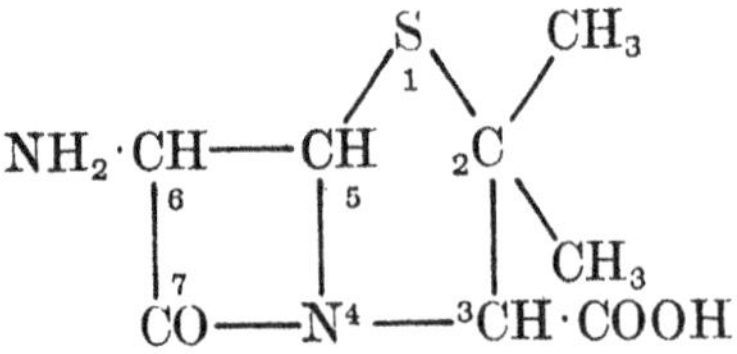

6-Aminopenicillanic acid.

Fig. 3.VII

For example *penicillin G, benzylpenicillin* has $C_6H_5CH_2$—in place of R, while *penicillin V, phenoxymethylpenicillin,* has $C_6H_5O.CH_2$.

The more recent discovery of methods by which large amounts

of 6-aminopenicillanic acid can be obtained has opened the way for the preparation of the so-called *synthetic penicillins* by the addition of different side chains to the NH_2 group in the 6- position (see Fig. 3.VII).

The penicillins are organic acids and are usually prepared commercially as the sodium or potassium salts. They are in general relatively unstable, being readily destroyed by mineral acids, alkalies, heavy metals, oxidizing agents and heat. They are also normally destroyed by the enzyme *penicillinase* which is produced by a variety of bacteria.

DIFFERENT TYPES OF PENICILLINS AVAILABLE

BENZYLPENICILLIN. "SOLUBLE" PENICILLIN

The potassium salt of benzylpenicillin is a highly soluble crystalline powder. It is rapidly distributed in the body after injection, but is also very rapidly excreted and frequent injections are necessary to maintain adequate blood levels. Absorption by mouth is unsatisfactory as the substance is largely destroyed by stomach acid.

"LONG ACTING" OR "DEPOT" PENICILLINS

Compounds such as *procaine penicillin* or *benzathine penicillin* release benzylpenicillin slowly over periods of 24 hr or longer. These may be combined with a soluble penicillin preparation to give a high initial level together with sustained action over 24–48 hr.

ACID-RESISTANT PENICILLINS

These substances were devised to overcome the difficulties of giving penicillin by mouth. The first was *phenoxymethyl penicillin*, penicillin V. Others are *phenethecillin* and *propicillin*. These acid-resistant penicillins will produce adequate blood levels if given orally.

PENICILLINASE—RESISTANT PENICILLINS

One of the more important developments of recent years has been the production of penicillinase—resistant penicillins. These have been synthesized from 6-amino penicillanic acid (Fig. 3.VII) by the introduction of side-chains which protect the molecule from penicillinase action. For example, *methicillin* is 2-6-dimethyl-oxybenzyl penicillin and *cloxacillin* is a more complex compound (5-methyl-3-orthochlorophenyl-4-isoxazolyl-penicillin) which is both penicillinase and acid resistant, and can therefore be given by mouth.

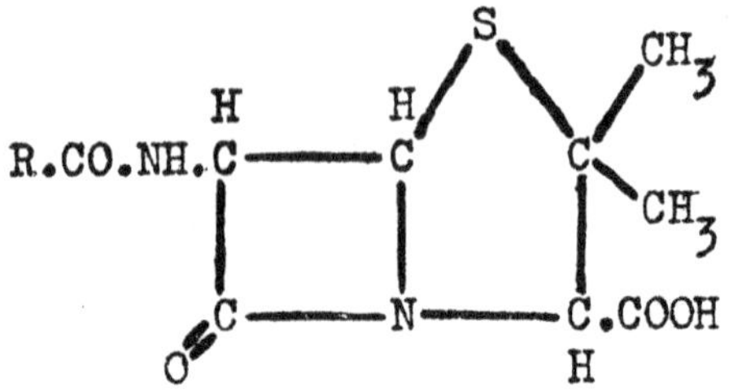

Fig. 4.VII

BROADSPECTRUM PENICILLINS

Ampicillin, which has an amino-benzoyl side chain, is highly active against some Gram-negative organisms, such as *Haemophilus influenzae*, and also against salmonella and shigella strains. It is acid resistant, but not resistant to penicillinase, and this makes it ineffective against Gram-negative organisms which are penicillinase producers, e.g. many strains of *Proteus*.

Toxicity and hypersensitivity. Penicillin compounds seem to be virtually without any primary toxic effect and very large doses (several hundred megaunits; 10 g or more of penicillin) have been given without ill effect.

Penicillin units. The original Oxford unit was entirely arbitrary since at that time penicillin had not been purified. When pure crystalline penicillin G became available, it was found that 1 mg of it contained 1666 units (i.e. 1 unit = 0·6 μg). Doses are often stated in megaunits (1 megaunit = 1 million units), but with the more recently introduced synthetic penicillins dosage is usually given by weight, e.g. a 500 mg dose. This is more logical and in line with the methods of dosage used for other antibiotics.

Sensitivity of bacteria to penicillin. The common pathogens vary in their sensitivity to the penicillins over a very wide range, but it is possible to give a rough grading into *very sensitive*, *relatively sensitive* and *insensitive*.

Organisms normally highly sensitive to penicillin (inhibited by 0·1 unit/ml or less). *Staph. pyogenes*, *Strep. pyogenes*, *Strep. pneumoniae*, *N. meningitidis*, *N. gonorrhoeae*, *C. diphtheriae*, *Cl. welchii*, *Cl. oedematiens*, *Cl. septicum*, *Cl. tetani*, *A. israelii*, *Trep. pallidum*, *Lepto. icterohaemorrhagiae*, *Lepto. canicola*.

Organisms relatively sensitive to penicillin (usually inhibited by 2–4 units/ml). *Strep. viridans*, *Strep. faecalis*, *H. influenzae*.

Organisms insensitive to penicillin (inhibited by 8–250+ units/ml). Strains of *E. coli*, *Ps. pyocyanea*, *Proteus*, *Brucella*, *Klebsiella*, *Salmonella*, *Shigella* and other Gram-negative organisms.

It can be assumed that infections due to organisms in the first group will respond to penicillin treatment unless tests have shown the particular strain in question to be resistant.

On the other hand, it must be remembered that a broad spectrum penicillin such as ampicillin will be active against strains of *E. coli*, *Salmonella* and *Shigella*.

MODE OF ACTION OF PENICILLIN

Penicillin is primarily a bactericidal substance but only exerts its killing effect on growing organisms. As a result largely of the work of Strominger and Park, it is now clear that penicillin acts by preventing the synthesis of new cell wall material. This causes organisms growing in the presence of penicillin to assume bizarre shapes and ultimately to swell up and burst because the weakened cell wall cannot withstand the internal osmotic pressure of the cells. Cell wall material already synthesized is not destroyed and resting organisms are therefore not killed. The substance whose synthesis is most affected by penicillin is an amino-sugar polymer called mucopeptide which forms about 50 per cent of the wall in Gram-positive bacteria, but only about 1–5 per cent in Gram-negatives. The fact that mucopeptide is less important in the walls of Gram-negatives probably explains why in general these organisms are less severely inhibited by penicillin than are Gram-positive bacteria.

PENICILLIN-RESISTANT ORGANISMS

Bacteria may exhibit two kinds of resistance to penicillin:

(1) RESISTANCE DUE TO PENICILLINASE PRODUCTION

This is especially important in staphylococci, but other organisms such as Gram-positive and Gram-negative bacilli also produce penicillinase. In such organisms, the presence of penicillin merely causes more penicillinase to be produced, and treatment of such infections with penicillin is impossible unless a penicillinase-resistant penicillin is used.

(2) RESISTANCE NOT DUE TO PENICILLINASE PRODUCTION

Some organisms are naturally more resistant to penicillin than others. For example, with benzylpenicillin it may take 2–4 units/ml to inhibit *Strep. faecalis*, while *Strep. pyogenes* is inhibited by only 0·01 units/ml. On the other hand some Gram-negative bacilli will grow in the presence of 100 units/ml or more, even though they do not produce penicillinase.

The reasons for these innate differences in resistance to penicillin are not clear. They may be due to the variations in the type and amount of the mucopeptide substance present in the cell walls of different bacteria.

If the organism infecting a patient is sensitive to penicillin at the start of treatment, it is very unusual for resistance to develop during treatment provided that adequate doses of the drug are given and the course is not prolonged (over a week). In conditions where treatment must be continued for several weeks, as in sub-acute bacterial endocarditis, resistance is likely to develop unless another drug is given at the same time (see page 81).

STREPTOMYCIN

Streptomycin was originally isolated by Waksman and his colleagues in 1944 from a strain of *Streptomyces griseus* from soil.

GENERAL PROPERTIES

Streptomycin is made up of three components, streptidive, streptose and N-methyl-L-glucosamine, linked by glycosidic bonds. It is strongly basic and forms salts with acids. *Streptomycin sulphate*, a white powder very soluble in water, is the salt most commonly available. *Dihydrostreptomycin*, prepared by the hydrogenation of streptomycin, is sometimes used, also in the form of the sulphate.

Neither streptomycin nor dihydrostreptomycin are absorbed to any extent from the intestinal tract, and they are therefore given by injection except for intestinal infections. The drug diffuses readily into most body tissues except the C.S.F. In meningitis, there is some diffusion from the blood into the C.S.F., but intrathecal injection may be necessary. The dosage is normally measured in grams or milligrams, although at one time units were used (1 unit = 1 μg; 1 g = 1,000,000 units).

Streptomycin is most active against mycobacteria, staphylococci and Gram-negative bacilli. Its activity against mycobacteria makes it an important drug in the treatment of tuberculosis. The inhibitory concentrations for normally sensitive organisms range from 0·5 μg/ml in the case of *Mycobacterium tuberculosis* to about 4 μg/ml in the case of most Gram-negative bacilli, although some may be more resistant. This is well below the levels (15–25 μg/ml) present in the blood 4 hr after intramuscular injection of 1 g of streptomycin sulphate.

DEVELOPMENT OF RESISTANCE TO STREPTOMYCIN

Unfortunately all bacteria rapidly develop resistance to streptomycin. This is due to the fact that a very few (about 1 in 10^8) cells in any bacterial culture are normally resistant to the antibiotic, and in the presence of streptomycin, the culture soon comes to consist entirely of the descendants of these resistant cells. It is even possible to obtain strains which require streptomycin for growth and will not grow in its absence. Strains which have become resistant to the antibiotics neomycin and kanamycin are often found to have become resistant to streptomycin as well.

Because of this rapid development of resistance, streptomycin is almost always used in combination with other antibiotics.

TOXIC EFFECTS OF STREPTOMYCIN

Pain at the site of injection, skin rashes and fever may occur during streptomycin treatment, but the really serious toxic effects are on the eighth cranial nerve, resulting in vertigo or deafness. Dihydrostreptomycin is particularly liable to cause permanent deafness, and the drug is little used for this reason. Patients on streptomycin suffer from vertigo rather than deafness, but this is not usually permanent unless high doses (over 1 g per day) are given. Patients over 40 years of age are more likely to suffer permanent eighth nerve damage from streptomycin than younger patients.

MODE OF ACTION OF STREPTOMYCIN

Streptomycin is primarily a bactericidal drug and, unlike penicillin, is capable of killing bacteria even when they are not multiplying. A combination of penicillin and streptomycin is often particularly effective in clinical conditions in which bactericidal action is important, for example, in subacute bacterial endocarditis. Streptomycin is considerably more active in alkaline than in acid media, and this point must be remembered if it is used to treat urinary infections.

ANTIBIOTIC SUBSTANCES RELATED TO STREPTOMYCIN

A number of basic substances chemically similar to streptomycin and with a similar antibiotic spectrum are produced by other strains of streptomyces. Two of the more commonly used are *neomycin* and *kanamycin*. These resemble streptomycin in being bactericidal and acting best in alkaline media, but are considerably more toxic and cause renal as well as eighth nerve damage. Their relationship to streptomycin is also shown by the fact that neomycin- or kanamycin-resistant strains are often resistant to streptomycin as well.

NEOMYCIN

Neomycin is too toxic for systemic use, but is used for topical application in infected wounds or burns, or in the treatment of staphylococcal carriers. To prevent the development of resistant strains, it is usually combined with other substances such as bacitracin, polymyxin or chlorhexidine. It can also be used in gastrointestinal infections since it is not absorbed by the gut.

KANAMYCIN

Although this antibiotic is more toxic than streptomycin, it is less so than neomycin, and can be given parenterally if the daily dosage is not more than 1 g, and the patient has no pre-existing renal disease. It may be particularly useful in generalized infections due to Gram-negative bacilli such as *Proteus*. Kanamycin, like neomycin, may be valuable in the treatment of gastrointestinal infections, since it is not absorbed from the intestine.

Other antibiotics in this group are *framycetin* and *paromomycin*. The latter is active against entamoeba as well as bacteria, and has been used to treat amoebic dysentery.

THE ERYTHROMYCIN GROUP

The four related substances *erythromycin*, *carbomycin*, *oleandomycin* and *spiramycin* are the most commonly used antibiotics in this group. All are derived from different species of streptomyces isolated from soil. Chemically these substances are complex, consisting of large lactone rings to which are attached dimethylamino sugars. The group is also referred to as *macrolides* (macrocyclic lactone ring compounds). Erythromycin and oleandomycin are the most active of the four.

GENERAL PROPERTIES

These substances are sensitive to acid and are usually given in enteric-coated capsules, from which they are absorbed in the intestine. Erythromycin can be given as *erythromycin stearate*, which is resistant to acid, but is broken down by the intestinal enzymes to liberate erythromycin. Preparations of *erythromycin glucoheptonate* are used for intravenous injection. *Erythromycin estolate* (the lauryl sulphate compound of propionyl erythromycin) is better absorbed than other compounds when given by mouth and gives higher blood levels.

These drugs are well distributed in the body, except in the C.S.F., where only very low levels are found.

MODE OF ACTION AND TOXIC EFFECTS

The antibiotics of the erythromycin group are primarily bacteriostatic, although bactericidal at high concentrations. Their range of antibacterial action is similar to penicillin in that they are mainly of value against Gram-positive organisms. Erythromycin has a low toxicity, although it may give rise to diarrhoea. However, erythromycin estolate and also triacetyloleandomycin, which is a similar ester of oleandomycin, may give rise to toxic hepatitis.

Resistance. Bacteria quite readily becomes resistant to these antibiotics, and erythromycin-resistant staphylococci have been found to spread rapidly in hospitals where this antibiotic was used extensively. Strains resistant to erythromycin are usually, but not always, resistant to the other members of the group, although the reverse is not always the case.

THE TETRACYCLINES

The tetracyclines are a group of closely related and widely used antibiotics, whose basic chemical structure is shown in Fig. 5.VII.

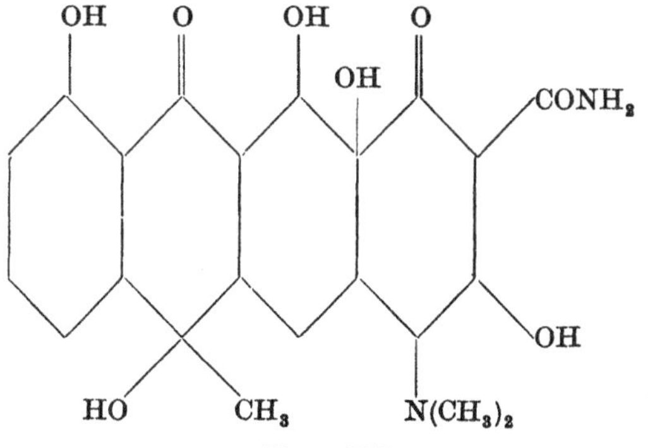

Fig. 5.VII

SUBSTITUENT GROUPS IN DIFFERENT TETRACYCLINES

	R_1	R_2	R_3
Tetracycline	H	CH_3	H
Aureomycin (7-chlortetracycline)	Cl	CH_3	H
Terramycin (5-oxytetracycline)	H	CH_3	OH
Demethyltetracycline (6-demethyltetracycline)	H	H	H

The first to be discovered, *Aureomycin*, was isolated from a strain of *Streptomyces aureofaciens*, in 1948, and *Terramycin* was isolated a few years later from a strain of *Streptomyces rimosus*. These substances are now normally called *chlortetracycline* and *oxytetracycline* because of their chemical formulae (see Fig. 5.VII). Subsequently, tetracycline was synthesized: it can be isolated from other strains of streptomyces. More recently, *demethylchlortetracycline* has been prepared by demethylation at position R_2.

GENERAL PROPERTIES

The tetracyclines are yellowish crystalline substances generally used in the form of the hydrochlorides since these are more soluble than the parent drug. They are given by mouth and are absorbed from the intestine, although an appreciable amount of the dose is excreted unchanged in the faeces. After absorption they are found in all body tissues, including the C.S.F.

The different tetracyclines all have a similar wide range of activity, and the term *broad spectrum antibiotic* originated with these substances because they are active not only against both Gram-positive and Gram-negative bacteria, but also against spirochaetes, rickettsia and organisms of the psittacosis group.

Demethylchlortetracycline is better absorbed and more slowly excreted than the other tetracyclines and therefore smaller doses of it are needed to produce adequate blood levels.

MODE OF ACTION

The tetracyclines act on bacterial cells by inhibiting protein synthesis, and are classic examples of bacteriostatic substances, since bacteria will remain viable for many hours in concentrations of the drugs which prevent growth and multiplication. Because they merely prevent growth without killing the cells, the tetracyclines interfere with the bactericidal action of penicillin, which can act only on growing organisms: they should, therefore, not be used clinically in combination with penicillin.

DEVELOPMENT OF RESISTANCE

Organisms can become resistant to the tetracyclines, and strains of tetracycline-resistant staphylococci and Gram-negative bacilli are not infrequently isolated in hospitals if these drugs are used extensively.

TOXIC EFFECTS

The tetracyclines do not produce serious toxic effects, although gastrointestinal disturbances such as nausea and vomiting are common. However, a more serious side effect of treatment with these broad-spectrum drugs is due to their suppression of the normal flora of the gastro-intestinal tract. In these circumstances, tetracycline-resistant strains of coliform organisms may cause diarrhoea and, occasionally, tetracycline-resistant *Staph. pyogenes* may give rise to a fatal *Staphylococcal enterocolitis*. The absence of the normal flora may also permit extensive growth of the yeast *Candida albicans* in the mouth and intestine, giving rise to thrush, enteritis or pruritis ani.

CHLORAMPHENICOL

This substance is yet another of the antibiotics produced by members of the genus *Streptomyces*, having first been isolated from a strain of *Streptomyces venezualeae*. Its chemical nature was soon established (Fig. 6.VII) and it is now prepared synthetically.

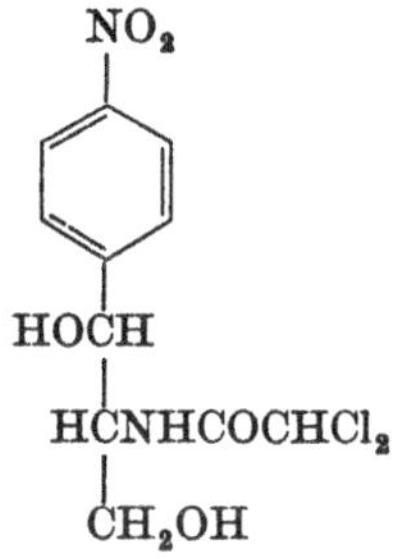

Fig. 6.VII. Structure of Chloramphenicol.

GENERAL PROPERTIES

Chloramphenicol is a yellowish compound with an extremely bitter taste. It is given by mouth either in capsules (250 mg), or as a suspension of *chloramphenicol palmitate*, which is tasteless: the latter preparation is useful for children. The drug is very stable, readily absorbed and diffuses widely in the tissues, including the C.S.F.

MODE OF ACTION

Chloramphenicol, like the tetracyclines, is an entirely bacteriostatic drug, and like them should, therefore, not be given in combination with penicillin. It is a broad-spectrum antibiotic, but has been especially useful in typhoid fever and in rickettsial infections. Because it merely inhibits but does not kill the infecting organisms, relapses are rather frequent after chloramphenicol therapy, e.g. in typhoid. However, these can usually be controlled by a second course of treatment. The bacteriostatic action of chloramphenicol is due to its inhibition of protein synthesis.

TOXIC EFFECTS

Chloramphenicol, like the tetracyclines, may cause stomatitis, nausea, vomiting and diarrhoea, due to its broad inhibitory effect on the flora of the alimentary tract; these effects are not usually serious. However, in common with other nitrobenzene compounds, it is toxic to bone marrow, and has been responsible for a significant number of deaths due to *marrow aplasia.* Because of this risk, its use should be strictly limited to serious infection, such as typhoid fever, in which it is particularly effective, and it should not be used prophylactically or for minor conditions. Even in typhoid it may well be replaced by ampicillin.

THE POLYPEPTIDE ANTIBIOTICS

These antibacterial substances are produced by various aerobic sporing Gram-positive bacilli belonging to the genus *Bacillus.* A large number of such antibiotics are known but most are too toxic for use clinically, since they cause severe damage to the epithelium of the renal tubules. However, in the case of the *bacitracins* and the *polymyxins,* the toxicity is sufficiently low for them to be used clinically.

BACITRACINS

This group consists of three closely related substances, bacitracins A, B and C, produced by a strain of *Bacillus licheniformis.* The bacitracins are strongly inhibitory to Gram-positive organisms but not to Gram-negative bacilli. They are not absorbed from the intestine, and therefore must be given by injection. However, on account of their nephrotoxicity, they are mainly of value in the local treatment of Gram-positive infections, e.g. as a constituent of antibacterial creams for treating staphylococcal carriers.

POLYMYXINS

These are basic polypeptides attached to the fatty acid actanoic acid. Five chemically distinct polymyxins have been distinguished called A, B, C, D and E. *Polymyxin A* is identical with the substance called *aerosporin,* and *polymyxin E* is probably identical with the antibiotic called *colomycin* or *colistin.* All the polymyxins are toxic to the renal epithelium, but the nephrotoxicity of *polymyxins B* and *E* is much less than that of A, C or D.

In contradistinction to bacitracins, the polymyxins have little action on Gram-positive organisms, but are very inhibitory to Gram-negative bacteria, with the important exception of *Proteus*, which are resistant. Polymyxins B or E are of greatest value in the treatment of infections due to *Ps. pyocyanea*. They are not absorbed from the gut, and must therefore be given by injection; if necessary, the drug can be given intrathecally. The polymyxins have also been used locally in the treatment of infected burns.

It is very rare for organisms to become resistant to polymyxin.

VANCOMYCIN AND RISTOCETIN

These are two more of the many antibiotics produced by strains of soil streptomycetes. They are toxic substances, which must be given intravenously, and are active against Gram-positive but not Gram-negative organisms. Because of their toxicity their use is seldom justified except in cases of severe staphylococcal infection, which have failed to respond to other antibiotics.

NOVOBIOCIN

This substance was isolated independently from two different strains of streptomyces. It has been marketed under such names as Cathomycin and Albamycin. Novobiocin is a bacteriostatic substance which is active against Gram-positive rather than Gram-negative organisms. It has been mainly used in the treatment of staphylococcal infections.

FUCIDIN

Fucidin, produced by the mould *fusidium coccineum*, is another antibacterial substance which is mainly of value in the treatment of infections by staphylococci which have become resistant to other antibiotics. Resistance to fucidin develops rather rapidly and to prevent this it has been suggested that fucidin and novobiocin be given together.

CHEMOTHERAPY OF TUBERCULOSIS

Although many of the antibiotics used against other types of bacteria will also inhibit the growth of the tubercle bacillus to some extent *in vitro*, none of them, except streptomycin (and some related drugs like kanamycin), is sufficiently active *in vivo* to be of value in the treatment of tuberculosis. There are, however, a number of more specifically antimycobacterial substances which are used in the treatment of tuberculous infections, usually in combination with streptomycin.

ISONIAZID (isonicotinic acid hydrazide)

The antituberculous action of this substance became known in 1952, and it has since become a standard drug in the treatment of tuberculosis. It is quite inactive against organisms other than the tubercle bacillus. It is readily absorbed if given by mouth, and is well distributed in the body including the C.S.F. Toxic effects in man are largely on the central nervous system and resemble those of vitamin B_6 deficiency, which is not surprising in view of the chemical nature of isoniazid. The most severe toxic effect is peripheral neuritis, which can be prevented by giving pyridoxine.

Isoniazid is bactericidal and has the advantage over streptomycin of being fully active against tubercle bacilli within macrophages, which streptomycin is not.

As in the case of streptomycin, organisms may become highly resistant to isoniazid after quite a short period of treatment, and it cannot therefore be used alone.

Pyrazinamide and *ethionamide* are other nicotinic acid derivatives which are sometimes used in the treatment of tuberculosis, but they are much more toxic than isoniazid, and are reserved for cases which have proved resistant to other drugs.

p-AMINOSALICYCLIC ACID (PAS)

This substance differs from isoniazid and streptomycin in that it is bacteriostatic rather than bactericidal, and it has rather weaker inhibitory action on the tubercle bacillus. However, it is valuable in anti-tuberculosis therapy because when given in combination with other drugs it helps to prevent the development of resistant strains during treatment. PAS is given by mouth in quite large doses, e.g. 10 g daily, and may cause gastric discomfort and nausea.

Cycloserine (oxamycin) is also occasionally used in the treatment of tuberculosis, where the infecting organisms have become resistant to other drugs. It is a broad spectrum antibiotic structurally related to the amino acid D-alanine, but is regarded as too toxic for normal use.

Because of the chronic nature of the disease, the treatment of tuberculosis needs to be continued over a period of months or perhaps years. This will inevitably lead to the development of resistant strains if a single drug is used. In practice, therefore, the standard treatment is a combination of streptomycin, isoniazid and PAS. In a few cases, where strains are already resistant to one or more of these drugs, others must be used, such as kanamycin, pyrazinamide, or cycloserine, but these latter drugs are not used in routine treatment because of their toxicity.

ANTIBIOTICS ACTIVE AGAINST FUNGI

Fungi are normally completely resistant to the antibiotics which are used to treat bacterial infections. There are, however, some antibiotic substances of which the reverse is true, i.e. they are very active against certain fungi but quite inactive against bacteria.

NYSTATIN

This substance was isolated from *Streptomyces nomsei* by workers on the staff of the New York State Department of Health, hence its name. It is only slightly soluble in water, and is given as a watery suspension by mouth or by injection. Nystatin is more active against yeasts like *Candida albicans* or *Cryptococcus neoformans* than against the filamentous fungi. It has been widely used for the local treatment of candidiases, e.g. in thrush or vaginitis, and also to prevent the overgrowth of *Candida albicans* which may accompany the use of broad spectrum antibiotics.

AMPHOTERICIN B

This is a similar type of substance to nystatin and is active against both filamentous fungi and yeasts. It is given intravenously, by slow infusion over a period of several hours, and has been effective in *Histoplasma capsulatum* and other systemic mycoses and in meningitis due to *Cryptococcus neoformans*. Toxic effects such as fever, thrombophlebitis and impairment of renal function are common and must be weighed against the probable danger from the infection itself.

GRISEOFULVIN

Unlike nystatin and amphotericin B, griseofulvin is more active against filamentous fungi than yeasts, and is especially valuable in infections due to dermatophytes since when given by mouth it becomes incorporated with the keratin of skin, nails and hair. *Trichophyton rubrum* infections of the nails, normally extremely resistant to all treatment, have been cured by griseofulvin. The drug may need to be taken over periods of up to six months, but fortunately the toxicity seems to be very low.

THE USE OF ANTIBIOTICS

SELECTION

A number of factors will govern the choice of an antibiotic in any particular case. The most important of those are of course the nature of the infecting organism and its range of sensitivity, but other matters must also be considered such as the comparative toxicity of the various drugs which might be suitable, the possibility that the patient may be allergic to them (e.g. hypersensitivity to penicillin) and their relative cost.

It is not always enough to establish the clinical diagnosis to enable a choice of antibiotic to be made, for a number of conditions (pyelitis, meningitis, pneumonia, etc.) may be caused by quite a wide variety of Gram-positive or Gram-negative organisms. In these conditions, the examination of a stained smear of suitable material, e.g. C.S.F., urine deposit, sputum and so on, will generally show the type of organism responsible and enable treatment to be started with a suitable drug. It is, however, important not to omit a full bacteriological examination on material taken before treatment.

In most cases it may be assumed that the infecting organism is sensitive. For example, penicillin resistance in pneumococci or meningococci is so rare that it can be ignored. Unfortunately, some organisms, such as staphylococci and Gram-negative bacilli, *E. coli* or *Proteus spp.* readily become resistant to a number of antibiotics. This is particularly likely to be the case in hospitals where antibiotics are used extensively, and in the case of patients who become infected with such resistant strains, the choice of an antibiotic may depend very much on the results of sensitivity tests.

ROUTE OF ADMINISTRATION

For systemic treatment, the route of administration will depend on the nature of the drug chosen, but will usually be either by mouth or by intramuscular injection. The latter method must be used in the case of substances like streptomycin which are not absorbed from the intestine. Preparations for intravenous injection are available for most of the commonly used drugs, but are not usually given except in severely ill patients in whom it is imperative to get rapid and certain absorption. However, some drugs like vancomycin are so irritating to the tissues that they must always be given intravenously.

In meningitis, antibiotics can generally enter the C.S.F. much more readily than than in health, but intrathecal injections are often given in addition to ensure a high local concentration. Quite small doses are sufficient, e.g. 10,000 units (6 mg) of benzylpenicillin or 5 mg of streptomycin. Larger doses are dangerous because they produce thecal irritation.

If necessary, antibiotics may also be injected directly into serous cavities, for example into the pleural cavity in cases of empyema, or into the joint cavities in suppurative arthritis.

For local application in skin disorders or in superficial infected wounds or burns, antibiotics are incorporated in creams or sprays. Bactericidal substances are to be preferred to bacteriostatic ones, and it is usual to use a combination of two or three together, e.g. neomycin, polymyxin and chlorhexidine, to eliminate the chance of resistance developing.

COMBINATION OF ANTIBIOTICS

The routine administration of mixtures of antibiotics is to be avoided, but in certain situations antibiotic combinations can be more effective than any single drug given alone.

Two of the more important of these situations are:

(1) In conditions where phagocytosis is relatively ineffective, e.g. in subacute bacterial endocarditis where a combination of penicillin and streptomycin is often more effective than penicillin alone.

(2) In conditions where resistance is particularly likely to develop. A good example of this is the chemotherapy of tuberculosis, where the advantages of using streptomycin, isoniazid and PAS together have already been stressed.

Mixtures of antibiotics are also sometimes recommended for mixed infections but it is often just as effective to use a single broad-spectrum antibiotic such as ampicillin or a tetracycline.

If it is decided to use a combination of two or more drugs, it is important to remember that penicillin which acts only on growing organisms, must not be combined with bacteriostatic drugs such as chloramphenicol or tetracyclines. It may, however, be used in combination with sulphonamides.

DEVELOPMENT OF RESISTANCE

An inevitable corollary of the widespread and sometimes indiscriminate use of antibiotics has been the development of resistant strains.

Organisms may become resistant to antibiotics in a number of ways. In the case of penicillin, for example, many organisms have the ability to produce an enzyme, *penicillinase*, which destroys the drug. This is the basis for resistance in the penicillin-resistant strains of staphylococci which are unfortunately so common in hospitals. With these strains the individual bacteria remain sensitive to penicillin, but the enzyme destroys the drug before it can affect the cells. Penicillinases are also produced by strains of *E. coli*, *Proteus spp.* and *Bacillus spp.* among others.

Resistance which is due to the destruction of the drug by an enzyme is uncommon and for practical purposes is confined to penicillin. With other chemotherapeutic drugs, resistance appears to depend on the development of some alternative biochemical pathway, perhaps involving the production of new enzymes which are not inactivated by the drug.

In many cases the laboratory evidence suggests that resistance arises as a spontaneous mutation in a few cells of an initially sensitive culture. In the presence of the drug only these resistant cells can grow, and the entire culture is soon composed of their descendants. With streptomycin for example, it has been shown that about 1 in 10^9 cells in a normal (sensitive) culture are resistant. This process of mutation and selection of resistant mutants may occur both with penicillinase-producing organisms and where an alternative biochemical pathway is involved.

Clinically resistant strains may cause trouble in two ways.

(1) The patient may become infected with an already-resistant organism. This is especially a problem in relation to staphylococci in hospitals, where strains of these organisms may be found which have multiple resistance, e.g. to penicillin, streptomycin and tetracyclines.

(2) The infecting strains, initially sensitive, may become resistant during treatment. This is very unlikely to occur where an acute infection is treated over a short period (5–7 days) with adequate doses of a suitable antibiotic but it is apt to be a real problem if treatment must be long continued. The example of tuberculosis has already been mentioned.

It has recently been shown that multiple resistance to several unrelated drugs can be transferred directly from one bacterial cell to another by the passage of extranuclear DNA. In this way resistance can be passed e.g. from, *E. coli* to *Salm.-typhimurium* and this may be an important general method by which resistance is transferred in intestinal strains.

THE DANGER OF CHEMOTHERAPY

It is important to realize that although chemotherapy has revolutionized the treatment of many bacterial infections and has saved millions of lives, there are dangers associated with the indiscriminate use of antibiotics. These dangers arise mainly from:

(a) Allergic and toxic reactions to the drugs used.
(b) Alterations in the normal bacterial flora.
(c) The development of resistant strains.

ALLERGIC AND TOXIC REACTIONS

The principal toxic effects of the different groups of antibacterial substances have already been described. The penicillins (+ cephalosporins) seem to have the lowest primary toxicity of any group, but unfortunately, are the most likely to give rise to allergic reactions, and there is a real danger with penicillin of severe and perhaps fatal anaphylactic shock developing in sensitive patients. It is therefore important to enquire about any previous reactions to the drug.

Allergic reactions to other antibiotics are usually confined to the development of skin rashes or urticaria. The latter may also occur during penicillin therapy.

ALTERATIONS IN NORMAL BACTERIAL FLORA

As already mentioned this is a problem associated with the use of broad-spectrum antibiotics and affects especially the alimentary tract. Suppression of the normal flora may lead to an overgrowth of the yeast *Candida albicans*, or to an invasion of the bowel with antibiotic-resistant staphylococci.

THE DEVELOPMENT OF RESISTANT STRAINS

The mechanism of this has been discussed above. It is a problem especially of infections with staphylococci and Gram-negative bacilli.

LABORATORY CONTROL OF ANTIBIOTIC THERAPY

Whenever possible, antibiotic treatment should be controlled by identifying the causative organism and determining its antibiotic spectrum. In addition, it may be necessary to estimate the amount of antibiotic present in the blood, pleural fluid, C.S.F. and so on, during treatment, to see if adequate concentrations are being obtained.

Laboratory tests therefore are necessary to determine (1) the sensitivity of a particular organism to various antibiotics, or (2) the level of an antibiotic present in the blood or other body fluid. The two methods used, use the *serial dilution technique* and *the agar diffusion test.*

SERIAL DILUTION TECHNIQUE

This is most frequently used to estimate the level of antibiotic in some body fluid or other unknown solution. Serial dilution (e.g. 1/2, 1/4, 1/8) of the unknown is made in a suitable medium in a series of small sterile tubes, and each is inoculated with an organism of known sensitivity. A similar series is set up using a standard antibiotic solution. After incubation, the end-points (highest dilution showing inhibition) are determined for each series. A comparison of the unknown with the standard enables the antibiotic content of the unknown to be calculated. The method can of course also be used to determine the sensitivity of an organism to a particular antibiotic, but it is time consuming and requires a pure culture of the organism to be tested.

AGAR DIFFUSION TESTS

These tests are less accurate, but are quicker, simpler and capable of being used in mixed primary cultures. They are, therefore, very widely used for routine purposes. The basic principle is that the organisms are grown on solid media in which are incorporated reservoirs of the antibiotics to be tested. These

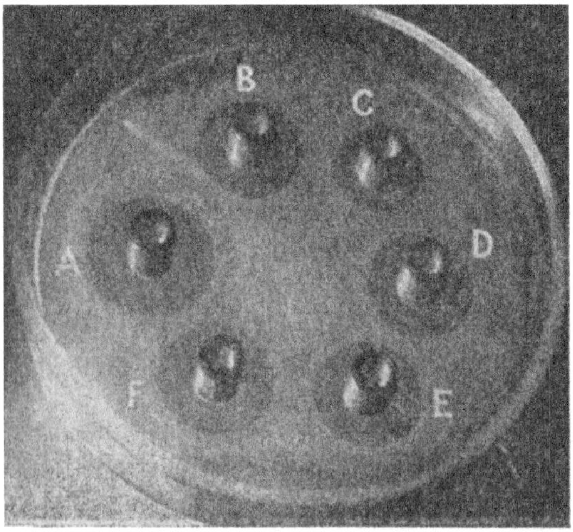

Fig. 7.VII. *Penicillin Assay.* Cylinder method, with 2, 1 and 0·5 units of penicillin per ml in cylinders A, B and C respectively and the unknown in cylinders D, E and F.

reservoirs may be glass or porcelain cylinders placed on the agar, ditches or wells cut in the agar, or filter paper discs soaked in the antibiotic, which are placed on the surface of the medium. Because of their convenience, the latter are most often used. From the reservoirs, the antibiotics diffuse out and cause zones of inhibition in the surrounding growth.

The concentrations of drugs used in the reservoirs for agar diffusion tests are important. Sensitivity is a relative term and few organisms are totally resistant to any particular drug, so that zones of inhibition of growth will almost always be seen if the

Table I

Substances Used in Antibacterial Chemotherapy

Substance	*Origin*	*Type of action*
Sulphonamides	Synthetic	Bacteriostatic
Penicillins	*Penicillium spp.*	Bacteriostatic
Cephalosporins	*Cephalosporium spp.*	Bactericidal
Streptomycin	*Streptomyces griseus*	Bactericidal
Chloramphenicol	*Streptomyces venezuelae* (now produced synthetically)	Bacteriostatic
Tetracyclines	*Streptomyces spp.*	Bacteriostatic
Erythromycin group	*Streptomyces spp.*	Bacteriostatic (bactericidal at higher concentration)
Vancomycin	*Streptomyces spp.*	Bactericidal
Novobiocin	*Streptomyces spp.*	Bacteriostatic
Bacitracins	*Bacillus spp.*	Bactericidal
Polymyxins	*Bacillus spp.*	Bactericidal
Fucidin	*Fucidium coccineum* (a filamentous fungus)	Bacteriostatic
	Antituberculous drugs	
Isoniazid	Synthetic	Bactericidal
p-Aminosalicyclic acid	Synthetic	Bacteriostatic

concentrations of drugs used are high. However, treatment based on the results may well fail because such high concentration cannot be obtained in the body. Suitable amounts for filter paper discs are (per disc), about 1–2 μg of penicillin, 5 μg of tetracyclines or 10 μg of chloramphenicol. These should give a zone of inhibition about 10 mm wide around the disc, with fully sensitive organisms.

In the treatment of urinary infections, discs with higher concentrations are used, because much higher concentrations of most drugs are found in the urine due to the concentrating effect of the kidney.

Table II

Activity of some common antibacterial drugs against different groups of micro-organisms

	Activity against							
Substance	*Gram-positive bacteria*	*Gram-negative bacteria*		*Spirochaetes*	*Rickettsia*	*Psittacosis lympho-granuloma group*	*Viruses*	*Fungi*
		Neisseria	*Others*					
Sulphonamides	+	+	+	−	−	−	−	−
Penicillins	+	+	−	+	−	+	−	−
Ampicillin	+	+	+	+	+	+	−	−
Streptomycin	+	+	+	−	−	−	−	−
Erythromycin group	+	+	−	(+)	−	(+)	−	−
Chloramphenicol	(+)	(+)	+	−	+	(+)	−	−
Tetracyclines	+	+	+	+	+	+	−	−
Bacitracins	+	−	−	−	−	−	−	−
Polymyxins	−	−	+ except *Proteus*	−	−	−	−	−

This table is designed primarily to indicate clinical use; *in vitro* activity is often much wider.
(+) = Effective, but not usually used for some reason, e.g. toxicity, or other drugs are available.

Chapter VIII

Disease Production by Bacteria

While bacteria are responsible for many diseases of man and animals, it is important to recognize that only relatively few bacteria are capable of producing disease: the majority are unable to multiply in the tissues, and are readily disposed of by the normal defence mechanisms of the body.

Bacteria, as previously indicated may be broadly divided into two rather arbitrary classes—the *parasites* which grow on living matter, and the *saprophytes*, which multiply on dead tissue or other organic substances. Another somewhat arbitrary division may be made into those forms which can produce disease—the *pathogens*, and those without this capacity—*non-pathogens*. These terms are only relative, since an organism which is pathogenic for one animal species may be non-pathogenic for another, and many normally non-pathogenic bacteria may induce pathological changes if introduced into the tissues in sufficient numbers.

NORMAL BACTERIAL FLORA OF THE BODY

A new-born infant is normally free from bacteria, but within a few hours of birth, staphylococci and other organisms can be grown from the skin, and from then on in all healthy persons the surfaces of the body, i.e. the skin and mucous membranes, normally harbour a number of bacteria which multiply without damage to the host; these are termed *commensals*. Although some of these organisms are important potential pathogens, they exist as saprophytes; their food supply is not obtained at the expense of the living cells, but is derived from local secretions, debris, excretions and foreign matter. Many varied bacterial species come into contact with the body surfaces, but only a few find the local conditions suitable for multiplication. The flora of the different regions, while not absolutely constant, tends to maintain a characteristic distribution which is largely dependent on the environmental conditions. The upper respiratory tract usually harbours various strains of streptococci, Gram-negative cocci, staphylococci, diphtheroids and pneumococci. The alimentary tract has a particularly varied flora which can be modified by dietetic changes. For example proteolytic organisms are predominant when a largely protein diet is taken, and in infants, because of their milk diet, the intestinal organisms are mainly lactobacilli. The vaginal secretions, after puberty, are acid, and here also lactobacilli are largely present.

As a general rule, therefore, those parts of the body in close contact with the external surroundings are seldom if ever free from bacteria under normal conditions. In contrast, surfaces such as those of the uterus, bronchioles, alveoli, sinuses and middle ear harbour few if any organisms in health, and serous cavities such as the joint spaces, the pleura, and the peritoneum are normally sterile. The cerebrospinal fluid and the meninges are also normally sterile.

Many commensals are also potential pathogens, and on any decrease in the resistance of the individual they may give rise to disease. Many of the pneumococci and the influenza bacillus (*Haemophilus influenzae*) are usually present as members of the normal flora of the upper respiratory tract. When, however, the resistance of the body is lowered, as during a debilitating illness such as measles, influenza or whooping cough, the pneumococcus, or the influenza bacillus may cause a secondary bronchopneumonia.

GERM-FREE ANIMALS

As mentioned above, all animals normally acquire bacteria from their surroundings soon after birth, and thereafter the body surfaces carry a distinctive bacterial flora for the rest of the animal's life. However, by performing caesarian section on the mother and subsequently raising the new-born animals under special conditions designed to ensure continued sterility, it is possible to obtain laboratory animals such as guinea pigs and rabbits in a completely bacteria-free state. Provided that a fully adequate diet is given, these *germ-free animals* will remain healthy, reach maturity and bear young, thus proving that the presence of bacteria is not necessary for the normal development of the host. It is of interest that the lymphatic system in these germ-free animals is rudimentary, thus showing the part played in the normal development of this system by bacterial antigens.

Germ-free animal colonies have been extensively used in research, since they enable the effect of injection with single selected strains of bacteria to be studied. For example, investigations into the cause of dental caries have been carried out, using germ-free animals infected with single strains of lactobacilli.

VIRULENCE

The virulence of an organism is its capacity to multiply in the body of the host and thereby produce disease, and it is usual to regard it as a combination of *invasiveness*, or the power to multiply in the tissues, and *toxigenicity*, or the ability to produce toxic substances. It is important to distinguish between these two elements of virulence since it frequently happens that the factors which allow a particular strain of bacteria to evade phagocytosis and so multiply in the body tissues are not those responsible for the toxic effects of bacterial infection. For example the presence of a capsule is essential to the virulence of pneumococci, yet the capsular substances, which are polysaccharides, are non-toxic when injected alone. Again, the presence of M-proteins on the surface of the cell wall in B-haemolytic streptococci enables this organism to grow more or less unhindered in host tissues, but the purified M-protein is not toxic when injected.

Virulence is by no means a fixed property of an organism, and a particular strain of bacterium may be highly virulent for one animal species, but of low virulence for another. For example the human type of tubercle bacillus will set up a fatal infection in a guinea pig, even if injected in very small numbers, but the same dose of organisms would cause, at most, a local non-fatal lesion in the rabbit. In some cases, particularly on the first isolation of an organism from the infected tissues, reproduction of the disease in the experimental animal is not readily obtained. By serial inoculation in animals of the same species, however, the virulence of the organism may be greatly enhanced: this is termed *exaltation of virulence by animal passage*. A striking and classical example of this is seen in the case of rabies virus: on fresh isolation the *street virus* (i.e. isolated from dogs off the street) reproduces the disease in rabbits only after a long incubation period, but on serial intracerebral passage in a succession of rabbits, the incubation period diminishes until the maximum virulence for the rabbit is reached: the virus is then termed "*fixed*" virus. While the virulence is increased by this type of procedure for the species in which the virus is being passaged, it is often lessened for other animal species, e.g. the "fixed" virus is less virulent for man than the "street" virus.

The opposite phenomenon, i.e. a decrease in virulence, is termed *attenuation*. This may result in various ways: repeated subculture on artificial media or cultivation under unsatisfactory conditions, for example by using unsuitable media or by incubation at unfavourable temperatures. Pasteur, in the case of chicken cholera (due to *Pasteurella septica*) found that when the cultures were kept at room temperature for relatively long periods they no longer produced the disease when injected into susceptible fowls. However, he found that the fowl injected with the attenuated strain were not resistant to the disease when reinoculated with freshly isolated virulent bacteria.

As a result of repeated cultivation on artificial media not only may the bacterium lose its virulence, but there may be associated changes in its cultural characteristics. The colonies on solid media may become rough and dry instead of smooth and shiny, while the growth in broth takes the form of a granular deposit instead of an even turbidity. This change from *smooth* to *rough* is usually styled *S → R variation* and is due to loss or alteration of the surface components of the organism. For example in the case of the pneumococcus, loss of ability to form capsules (associated with loss of virulence) leads to the formation of rough colonies. In the case of non-capsulated Gram-negative organisms such as strains of *Shigella* or *Salmonella*, the S → R variation is caused by the production of a different type of surface antigen, and the rough (R) strains are generally non-virulent. However, the production of rough colonies is not always an indication that an organism is avirulent. For example, in the case of the anthrax bacillus, the normal virulent form produces rough colonies.

MEASUREMENT OF VIRULENCE

The virulence of a particular strain may be expressed in terms of the *minimum lethal dose*, that is the smallest number of organisms which will regularly cause a fatal infection. In view of the considerable number of variable factors involved in such tests, this is not a very useful measure of virulence and it is more usual to calculate the *LD 50*, which is the number of organisms which will, on average, kill 50 per cent of the animals infected with it. However, as noted above, the virulence of a strain can be raised or diminished by various procedures, so that the figure for the *LD 50* is of rather limited value outside the conditions of the experiment in which it was obtained.

There is no doubt that the number of bacteria required to produce infection varies considerably with different species. In the case of anthrax and plague bacilli, infection may result from invasion of the tissues by a very small number of organisms, while with weakly pathogenic species, such as some skin staphylococci, a very considerable number of organisms is required to produce even a minor lesion. It can be seen, therefore, that the number of bacteria required to initiate infection is more or less inversely proportional to the virulence of the strain: the more virulent the organism, the smaller the infecting dose needed.

ROUTE OF INFECTION

The site of penetration and invasion by bacteria usually but not always determines the nature of the infection, and in some cases infection will not result unless the organisms are introduced by a particular route. Cholera vibrios, which are normally taken into the body by mouth with infected food and water, will not cause intestinal symptoms if injected subcutaneously, and the same is true of dysentery bacilli. In the naturally acquired diseases due to these organisms, the lesions are confined to the alimentary tract. In typhoid fever, on the other hand, the organisms are found throughout the body, and it is possible for typhoid to result from the accidental inoculation of living *Salmonella typhi* into the tissues, e.g. in the laboratory due to careless handling of a syringe, even though the normal route of infection is by mouth.

The site of invasion by anthrax bacilli determines the type of infection caused: infection through the skin gives rise to a localized malignant pustule, whereas infection by inhalation of spores will give a fulminating broncho-pneumonia. A similar state of affairs is found in the case of the plague bacillus, where infection through the skin by the bite of rat flea gives the classic bubonic plague, while infection by inhalation will cause pneumonic plague.

The intact skin offers an almost complete protection against bacterial invasion. Bacteria are usually introduced as a result of some break in its continuity such as a laceration, incision or puncture. However, some organisms such as *Leptospira icterohaemorrhagiae* or *Treponema pallidum* appear to be able to penetrate intact skin, although it is difficult to rule out the possibility that they get in through extremely minute abrasions not visible to the naked eye. The intact mucous membrane is also a definite barrier against infection: for example potential pathogens such as *E. coli* and *Cl. welchii* are normally to be found in the intestinal tract, but do not invade the tissues in health. However, if the surface is breached, as in abdominal injuries, a fatal peritonitis may result. Other organisms such as the typhoid bacillus can obviously penetrate the mucosa without too much trouble, and thereby start a generalized infection.

TOXIC SUBSTANCES PRODUCED BY BACTERIA

The injurious effects of bacteria on the body tissues are chemical in nature, and are primarily due to the action of proteins or other substances produced during bacterial growth which are inhibitory or lethal to the cells of the host. These inhibitory or lethal substances are called *toxins*.

Our present conception of bacterial toxins originated in the work of Roux and Yersin (1888–90) on the toxicity of the diphtheria bacillus. These workers found that the presence of the bacilli in the inoculum was not essential for the production of the characteristic lesions in the guinea pig. The filtrate of a broth culture, after passage through a filter which removed the organisms completely, had the same action as the whole culture. The toxic action was thus due solely to some filterable substance. This was termed an *exotoxin* and was considered to be a soluble substance secreted by the bacteria. Exotoxins were found to be formed by only a few Gram-positive bacteria, such as *C. diphtheriae*, *Cl. tetani*, *Cl. botulinum*, *Cl. welchii*, *Strep. haemolyticus* and *Staph. aureus*. In the case of other bacteria, especially Gram-negative organisms such as *N. meningitidis*, *Salm. typhi* or *V. cholerae* it was found that the toxic substances produced were largely associated with the insoluble parts of the bacterial cell, such as the cell wall, and were only liberated after death or destruction of the bacteria had occurred. These substances were called *endotoxins* to indicate their close association with the structural elements of the cell.

EXOTOXINS

As stated above, these are soluble toxic substances which are found free in filtrates of bacterial cultures. All exotoxins which have so far been purified have been found to be unconjugated proteins whose toxic properties are readily destroyed by heating. They are all highly antigenic, and it is usually easy to obtain potent antisera (antitoxic sera) against them. When the antibodies in such sera combine with toxin, the toxicity of the latter is neutralized, and serum therapy is therefore often important in the prophylaxis and treatment of diseases due to toxigenic bacteria. Another characteristic feature of exotoxins is their extremely high potency: a purified preparation of the neurotoxin of *Cl. tetani* for example may contain as many as 75 million mouse lethal doses for every milligram of nitrogen.

Although they are all similar in physical properties, the exotoxins produced by different organisms have very different, and generally highly specific, effects on body cells. For example, diphtheria toxin exerts its main lethal effect on cardiac muscle, while the toxins of *Cl. tetani* and *Cl. botulinum* are neurotoxins, and the α-toxin of *Staph. pyogenes* causes disintegration of the lysosomes of human and rabbit leucocytes. Another staphylococcal toxin, the *enterotoxin*, produced by strains which give rise to food poisoning, appears to act directly on the intestinal wall.

While the main anatomical sites of action of most bacterial toxins are known, the actual biochemical lesions which the toxins cause in host cells are generally not understood. However, in some cases, exotoxins appear to act as specific enzymes. One of the best-known examples is the α-toxin of *Cl. welchii* which is a lecithinase, and its haemolytic action is due to this property. The biochemical lesions caused by toxins do not necessarily kill the cells affected. For example in tetanus, if the patient can be successfully brought through the acute stage of the disease, he will ultimately recover without signs of residual damage to the C.N.S., indicating that the effects of tetanus toxin are reversible. More commonly, however, the actions of bacterial toxin are irreversible, and lead to death of the affected cells.

Many pathogenic organisms produce more than one toxin, and often the disease process seems to be due to several toxins acting together or in sequence. For example, anthrax "toxin", so-called, has been shown to be made up of at least three different factors, and in gas gangrene of muscle tissue caused by *Cl. welchii* the death and destruction of muscle seems to be due to a combination of the α-toxin which kills the muscle fibre, and other "toxins" (collagenases, proteases, hyaluronidases, etc.) which break up the dead tissue. It is doubtful whether the latter substances should properly be classed as toxins, since they do not appear to act on living cells.

ENDOTOXINS

Whereas the classic exotoxins are soluble proteins which are heat-labile and highly antigenic, the endotoxins are heat-stable lipopolysaccharides which are largely associated with the insoluble parts of the cell (principally the cell wall). Unlike exotoxins where each type produces its own peculiar pattern of effects in the body, all endotoxins produce an exactly similar pattern of effects in the host, irrespective of what bacterial species the toxin comes from, although the endotoxins from different species may differ in potency. In *small doses* (about $1\ \mu g = 10^{-3}$ mg per kg) the principal effects of endotoxin preparations is to cause a sharp rise in temperature. Larger doses (milligram amounts) will produce a profound degree of shock, with rapid fall in blood pressure, and in fact *endotoxic shock* is a well-recognized feature of severe infections with Gram-negative organisms, especially in elderly patients. Endotoxins do not show the extremely high toxicity which is associated with exotoxins such as those of tetanus or diphtheria, where the minimum lethal dose may be 10^{-6} or 10^{-7} mg.

Endotoxins are antigenic, and will stimulate the formation of specific antibodies in animals. However, these antibodies are very inefficient in neutralizing the toxic effects of endotoxins. For this reason, antiserum therapy is usually not effective in infections with Gram-negative organisms.

TOXOIDS

It was originally noted by Ehrlich that preparations of diphtheria toxins lost their toxicity on storage, but did not at the same time lose their ability to stimulate the formation of neutralizing antibodies in animals. These non-toxic preparations are called *toxoids*. It was discovered later that the addition of a low concentration of formaldehyde (0·1–0·2 per cent) to the toxin, followed by incubation at 37°C for some weeks, was a convenient way of inducing toxoid formation.

Although originally observed with diphtheria toxin, toxoids can be formed from all exotoxins. They are most useful agents for producing active immunity, since they are themselves non-toxic but will induce the formation of antibodies which will neutralize the effects of the original toxin.

It is a further point of distinction between exotoxins and endotoxins, that endotoxins do not form toxoids.

SIGNS AND SYMPTOMS OF INFECTION

The effects produced in patients by microbial infections may vary very widely. Some types of bacteria and viruses, such as those causing measles, smallpox, typhoid fever (*Salm. typhi*) or whooping cough (*Bordetella pertussis*) regularly produce a characteristic set of signs and symptoms, so that the presence of these organisms may be inferred from the clinical picture they produce. Diseases of this kind, with a specific clinical picture, are often referred to as *specific infectious fevers*. Pyogenic organisms, however, tend to produce a broadly similar type of lesion in whatever tissue they affect; thus meningitis may be produced by a number of different organisms, and it is not normally possible to distinguish clinically, for example, between acute meningitis due to *Haemophilus influenzae* and that caused by *Streptococcus pyogenes*. Other examples of infections due to pyogenic organisms are pneumonia, cystitis, empyema and otitis media, in all of which the clinical picture is determined largely by the site of infection and not by the infecting organism.

In the specific infectious fevers, there is a period of time, known as the *incubation period*, between the time of infection and the onset of symptoms. This may vary from a few days to as much as three weeks, the actual interval being usually fairly constant for any particular disease, for example in measles it is usually 10–14 days. During this period the infected person is usually well, but may be infectious for others, especially shortly before symptoms appear. In the incubation period the organisms are multiplying in the body, often within reticulo-endothelial cells, and symptoms do not occur until the microbial population has reached a certain level. In the case of organisms producing specific exotoxins, the incubation period is likely to be related both to the amount of toxin necessary to cause lesions, and time taken for it to reach the cells it specifically affects. In tetanus, where the neurotoxin reaches the C.N.S. by passing up the peripheral nerves, the incubation period may be over a month, whereas in diphtheria it is usually 2–3 days, since the throat lesions are caused at the site of infection.

The incubation period with pyogenic organisms is generally short, and signs of infection are usually evident within 12–24 hr.

TISSUE RESPONSE TO BACTERIAL INFECTION

The microscopic appearance of the local lesions caused by bacterial infection varies with kind of organism, but apart from the specific effects of exotoxins, there are two principal types of cellular response to bacterial invasion of the tissues. The pyogenic organisms, e.g. streptococci, staphylococci, pneumococci, gonococci, meningococci, and in many cases also Gram-negative bacilli such as *E. coli*, *H. influenzae*, *Ps. pyocyanea* and others give rise to an *acute inflammatory reaction*. This is characterized macroscopically by pain, redness and swelling of the affected part, and microscopically by a marked exudation of polymorphonuclear leucocytes, which with the remains of broken down tissue cells and dead bacteria, form *pus* (hence pyogenic = pus producing). It must be noted that bacterial infection is not the only stimulus which will produce a reaction of this kind, since the injection of irritant substances such as turpentine into the tissues will also give rise to a marked exudation of polymorphs. The other principal type of reaction to bacterial infection is exemplified especially by *M. tuberculosis*, which causes a chronic granulomatous lesion in which in the cellular reaction consists largely of mononuclear cells instead of polymorphs, and there is often much formation of fibrous tissue. Another characteristic feature of this type of response is the formation of multinucleate giant cells.

These two types of tissue response to bacterial infection are not mutually exclusive, since if an acute inflammation persists and becomes chronic, the lesions may show a granulomatous change, while an acute inflammatory type of response may occur in the early stages of tuberculous infection. The type of cellular response to inflammation is therefore at least partly determined by the length of time the infection persists: the longer this is, the more likely is there to be fibrosis and the presence of mononuclear cells.

SPREAD OF BACTERIA WITHIN THE BODY

In many instances the local tissue reaction is sufficient to keep the infection under control, but if the infecting organism is sufficiently invasive, further dissemination of the bacteria may occur in several ways:

(1) *By direct continuity* on skin or mucous surfaces, along natural channels such as the urethra, ureters, gland ducts, etc., or along tissue planes. Examples are numerous—gonococcal infection spreading from urethra to prostatic glands, streptococcal cellulitis spreading along tissue planes, bronchopneumonic infection spreading in the lung, etc.

(2) *By lymphatic spread.* An important method of spread in bacterial infections is that which occurs along the lymphatics to the regional lymph nodes. For example, with *Strep. pyogenes* a small local lesion on the hand may be followed within a few hours by lymphangitis in the lymph channels of the arm, and then lymphadenitis in the glands of the axilla. Similarly, plague bacilli which have been injected into the tissues by the bite of an infected rat flea, pass quickly to the regional lymph nodes and form the characteristic buboes of plague.

(3) *By entrance into the blood stream, and general spread throughout the body.* The persistent invasion of the blood stream by moderate to large numbers of bacteria (e.g. 100 or more per ml) implies severe infection, and is termed *septicaemia*. In some pyogenic infections a suppurative phlebitis with thrombosis may develop. On breaking down of the thrombus, infective particles are disseminated in the blood stream, and multiple abscesses form in different tissues: this condition is called *pyaemia*.

Both septicaemia and pyaemia, especially the latter, are evidence of serious general infection, and the patients in whom they occur are almost always seriously ill. However, the transient occurrence of bacteria in the blood stream in small numbers, termed *bacteraemia*, is a relatively frequent occurrence at some stage in many bacterial infections, and may also occur in the absence of disease. For example it is well known that in individuals with minor peridontal infection, mouth streptococci can be detected by blood culture immediately after tooth extraction, or even after biting on a hard object. These streptococci are removed quickly by the normal clearance mechanisms, and cause no infection, except in a few individuals with heart valves damaged by previous rheumatic fever, who may develop endocarditis (see chapter on streptococcal infections).

GENERAL MANIFESTATIONS OF BACTERIAL INFECTION

In addition to the local symptoms directly arising from bacterial activity evidence of general metabolic disturbance is usually present. One of the most general effects is *fever*, which may vary from a slight rise above normal (e.g. 99°F only in the evening), to hyperpyrexia of 106°F and over which necessitates immediate measures to reduce it. The patient will also feel generally unwell, with headache, pains in the limbs, sweating and so on. In pyogenic infection especially there is a marked increase in the blood leucocyte count, a *leucocytosis*, generally with an increase in the number of young polymorphs. The white count may rise from a normal of about 5,000 per mm^3 to 20,000 per mm^3 or over. However, in many virus diseases there is only a slight rise in the total white count, and no detectable increase in polymorphs: there may even be a decrease in white cells, a *leucopenia*.

In the case of infections with Gram-negative organisms, the fever may be explained in part by the liberation of endotoxins which are known to cause a rise in temperature even in very small doses. However, Gram-positive bacteria do not generally produce substances of this kind, and there is as yet no satisfactory general explanation for the fever associated with bacterial disease.

Chapter IX

Immunity: Innate and Acquired

All animal species are equipped to defend themselves against aggressors. They are subject to attack, not only by visible foes, but, at the cellular level, by bacterial, protozoal and viral predators and by toxic foreign proteins or polysaccharides which may gain access to the body after ingestion, inhalation or through lesions in the skin and mucous membranes. In the next three chapters the defence mechanisms of the body against such attack by foreign micro-organisms or macromolecules will be considered. The study of these mechanisms is the science of immunology.

As described in the first chapter, man's first real insight into the nature of infectious diseases came only a hundred years ago with the realization by Louis Pasteur in Paris of the role of bacteria in their causation. Among his many notable contributions to immunology, Pasteur showed that protection against a particular disease could be conferred, either by past exposure to the disease or by immunization with cultures of the causative organism which had first been rendered harmless. In thus producing immunity by active immunization, Pasteur was, in fact, using the same principle as had been used nearly a century before by the Gloucestershire physician, Edward Jenner, who induced protection against smallpox by vaccination of human subjects with material from the lesions of cowpox. These observations led many workers in Pasteur's institute and elsewhere to search for further ways to induce immunity against bacterial disease. It became obvious that certain factors in the serum of convalescent patients, known as antibodies, could, when injected, induce protection against bacterial disease and could also inactivate the causative bacteria in the test tube. Contemporaneously with Pasteur, Russian-born Elie Metchnikoff studied the intriguing capacity of phagocytic cells, even in very primitive species, to ingest and destroy all sorts of foreign material including bacteria. Thus, well before the end of the nineteenth century, it was realized that humoral factors, i.e. antibodies and cellular factors, present in phagocytic and other cells, both played a part in immunity to infection. During the next few decades, numerous methods were developed for the detection and measurement of antibody in serum. Serotherapy, i.e. passive immunization with antibody, became a popular form of treatment for many bacterial diseases. The 1914–18 war stimulated the development of prophylactic active immunization programmes using the killed bacilli of typhoid and other diseases. Effective as these measures were in reducing the mortality from certain bacterial diseases, they did not entirely solve the problem. Late in the 1930s and during the Second World War came the first effective therapeutic use of antibiotics in medicine. These drugs did what serotherapy had often failed to do completely, they produced a much more dramatic effect in many bacterial infections and a greater reduction in mortality from them. The interest of the clinician in immunology declined correspondingly, and serotherapy was, for the greater part, relegated to the history books. However, during the past 10 years or so, there has been a renaissance in immunology for several reasons. The powerful drugs which doctors use today can themselves act as immunological aggressors and may cause dangerous side effects whose nature will be discussed in a later chapter. Penicillin hypersensitivity is a common example. In the same way the wide-scale use of blood transfusion in replacement of blood loss and the avoidance of fatal transfusion reactions depended on an understanding of the immunological principles, defined by Karl Landsteiner, governing the cross reactivity of the human blood groups. More recently still, a special branch of immunology has arisen which studies the problems raised when a surgeon transplants foreign tissues such as donor skin or kidney into the human body as replacement for diseased or destroyed tissue. The related field of cancer immunology includes the study of immunological reactions against tumours which may, in many ways, be regarded as invasions by foreign cells which have originated from the normal cells of the body. During the past decade a group of disorders known as the *auto-immune* diseases has been recognized in which a misdirected immune response is mounted against the body's own tissues. Even more recently, immunology has become closely dependent on the study of protein chemistry and molecular biology in an attempt to understand the chemical structure of antibody molecules and the underlying genetic mechanisms governing the production of antibody by cells.

Thus immunology has obviously moved (in many directions) away from the original study of the response to bacterial infection. Nevertheless it is still true to say that the majority of immunological reactions constantly taking place in each of us during the course of our lives are defence reactions against the numerous bacterial and viral infections and protozoal infections as well, especially if we live in the tropics, which often do not cause any clinical disturbance but which may occasionally give rise to obvious disease.

INNATE OR NON-SPECIFIC IMMUNITY AND SPECIFIC ACQUIRED IMMUNITY

The immune response to foreign invasion may be discussed under two main headings: these are firstly *innate* or *non-specific* immune responses which may be of many types enumerated below but examples of which may be found in all animal tissues, vertebrate and invertebrate, at all stages of evolution and even in monocellular organisms such as amoeba. The term *non-specific* used in this context means that the immune response evoked is not directed solely against one particular organism. Many different bacteria and even inert particles may evoke very similar inflammatory responses upon introduction into the body. Non-specific immunity is *innate* and does not have to be learnt. Non-specific immune responses are usually as effective on the first contact with a micro-organism as on any subsequent occasion.

The second heading includes *specific, acquired* immunity. This type of immunity is seen only in vertebrates and appears phylogenetically with the evolution of lymphoid tissue, the thymus and spleen. The body mounts a *specific* immune response against a particular foreign substance. The foreign chemical substance against which such a response is directed is known as an *antigen*. Bacteria, being chemically complex, may contain numerous antigens each of which may induce a specific immune response. This type of immunity is learnt. We are not born with specific immunity to measles—except that we may sometimes acquire transient passive protection *in utero* from maternal serum antibody which crosses the placenta. With our first attack of measles, we produce a specific immune response which protects us on subsequent exposure to measles virus. Therefore second attacks of measles are rare. However, this immunity to measles is specific so that an attack of measles leaves us just as susceptible as before to chicken-pox or rubella.

NON-SPECIFIC INNATE IMMUNITY

By far the greatest number of bacterial or viral invasions of the body are dealt with by non-specific defence mechanisms probably without the need for participation of specific immune responses against the invader at all. Indeed, as mentioned above, invertebrates probably deal with foreign invasion entirely by such non-specific mechanisms. These mechanisms become operative within a very short period after invasion. On first contact with an antigen, there is a delay of several days before specific immunity begins to play a part. On later contacts, specific immune reactions are much more rapid, but it remains true that the major first lines of defence against infection are non-specific.

Non-specific immune reactions will be considered under three headings: *genetic or constitutional* innate immunity, *humoral* non-specific immunity and *cellular* non-specific immunity.

GENETIC OR CONSTITUTIONAL INNATE IMMUNITY

Species immunity. Certain organisms which cause severe and fatal disease in one species are incapable of causing disease in others. Dogs die of canine distemper; men do not develop the disease. Rats are resistant to diphtheria; guinea pigs are extremely susceptible.

Racial immunity. In laboratory animals kept under standardized conditions, differences have often been observed between different strains of the same species in their susceptibility to infectious diseases. It is not so easy to evaluate the influence of race on resistance to disease in man since so many non-standardized factors such as diet and economic and climatic conditions play a part. Racial groups most of whose members are carrying hookworm or malarial parasites may well fall prey to incidental diseases more easily than healthy groups for reasons quite unrelated to race. In central Africa, there is a high incidence of the genetically determined defect of haemoglobin synthesis which, in homozygotes, leads to sickle-cell anaemia. Interestingly enough, persons carrying this defect have an increased resistance to infection with the malarial parasite *Plasmodium falciparum*. It might be argued that the sickle-cell defect survives in these racial groups simply because it protects the population from malaria, as individuals who do not have the defect are weeded out by *P. falciparum*. Genetic selection probably, therefore, plays a part in resistance. Racial groups exposed to diseases which they have not experienced in their previous history tend to suffer from more fulminating infections than other racial groups who have lived with these diseases for many years. Hence the virulence of syphilis in a previously unexposed population in renaissance Europe and the relative virulence of tuberculous infection in negro populations first exposed to contact with the disease carried by white men, themselves exposed for many centuries. In negroes, fulminating, haematogenous dissemination of tuberculosis was common, whereas the same disease underwent a chronic course in Europeans with a tendency to formation of circumscribed fibrotic lesions.

AGE AND NON-SPECIFIC IMMUNITY

The immunology of old age is not well understood. Clearly, in advanced age, we fall prey to many infections which would be trivial in a young adult. Degenerative changes may lead to increased susceptibility of tissues to bacterial toxins and obstructive changes in, e.g. the urinary tract, may lead to the stasis which favours infection. At the cellular level, however, the influence of ageing of tissues on their capacity to mount defence reactions has not been well evaluated.

At the opposite pole, rubella, which causes a mild exanthem in normal children and adults, causes severe organizational defects when it attacks the foetus *in utero* in the first trimester of pregnancy. Toxoplasmosis, a protozoal infection, causes a fatal encephalitis when acquired *in utero*, whereas the adult form of the same disease is much milder, usually without C.N.S. lesions.

HUMORAL NON-SPECIFIC IMMUNITY

External defences. An intact skin or mucous surface prevents infection simply by acting as a physical barrier and preventing ingress of microorganisms. Other physical barrier mechanisms include the microscopic trap operated by the bronchial ciliated epithelium which prevents any particle larger than 4 μ in size from reaching the alveoli. Apart from their function as barriers, body surfaces also act by secreting various chemical substances inhibitory to bacterial invaders. The sebaceous secretions of the skin contain fatty acids with bactericidal properties. If bacteria such as *E. coli* are spread over the skin surface, it is quite readily shown that they are destroyed within a quarter of an hour or so by the action of these secretions. Another external bactericidal secretion is the enzyme, lysozyme, powerfully lytic for Gram-positive cocci which, as well as being present in serum, is secreted in high concentration in saliva, nasal mucus and lacrimal fluid. The highly acidic gastric juice also causes death of most bacteria taken by mouth unless these are protected from adverse pH by being taken in a large and indigestible meal. Lower in the gastro-intestinal tract, bile salts inhibit growth of most organisms other than Gram-negative bacilli. These latter proliferate as commensals and may themselves act as metabolic competitors to keep out potentially pathogenic invaders.

INTERNAL HUMORAL DEFENCE MECHANISMS

If the skin is injured, even quite trivially, a local area of reddening is seen at the site of injury within seconds and this is quickly followed by the appearance of a central area of oedema. These events, fundamental in the initial response to trauma, comprise the weal and flare response of Thomas Lewis, who showed that they were caused by local dilatation of small vessels due to the action of histamine. Histamine release causes immediate vasodilatation and increases the permeability of small vessels. The weal and flare form the visible manifestations of these vascular events. Circulating cells and bactericidal chemical substances are thus enabled to reach the injured or infected site much more rapidly than they would normally. Histamine is a small molecule and other low molecular weight substances such as serotonin (5-hydroxytryptamine) may also play a part in the Lewis response. In addition, following tissue damage and particularly contact of tissues with foreign surfaces, inactive enzyme precursors may be activated and initiate a complex series of chemical events of importance in the inflammatory reaction. Such an event is the formation of substances known as "kinins" which are peptides produced by the action of enzymes named "kallikreins" on an inactive substrate. Kinins have a slow acting but very powerful effect in increasing vascular permeability and in vasodilatation. Among the other events which occur following tissue injury are the initiation of blood coagulation and the conversion of fibrinogen to fibrin. The Hageman factor, a factor activated by contact with foreign surfaces is probably of importance in the early stages of both coagulation and kinin production. Apart from its essential haemostatic effect, the formation of fibrin may, by localizing the injured site, play some part in preventing rapid spread of microorganisms from the initial focus of infection.

Among the chemical substances present in inflammatory exudates are *chemotactic factors* which influence the migration of leucocytes and other phagocytic cells. Probably, once such cells have passed through the capillary walls, or even earlier, they come under the influence of chemotactic substances and are attracted through the tissue spaces to the site of infection or injury. Bacteria, when incubated in the presence of normal serum produce such chemotactic substances and indeed polymorphs themselves produce chemotactic factors capable of attracting other polymorphs. Complexes of antigen and antibody (with serum complement) stimulate the appearance of powerful chemotactic factors in serum.

Cells in inflammatory lesions utilize the sugars in their environment and convert them to lactic acid. The pH of inflammatory exudates is thus usually acid and is, in itself, inhibitory to the growth of many pathogenic bacteria.

An important humoral factor in the mechanism of resistance to virus infections is *interferon*. This is a protein released from cells in response to virus infection which, when taken up by other cells renders them resistant not only to the original virus but to a wide range of other viruses. Interferon is produced within a few hours of infection, much more rapidly therefore than specific antibody, but it also disappears more rapidly than antibody.

INTERNAL CELLULAR DEFENCE MECHANISMS

The cellular inflammatory response. During the vasodilatation associated with injury, *polymorphonuclear leucocytes* become adherent to the vessel walls within minutes and then emigrate through them into the tissues. These cells are actively phagocytic and, under the influence of chemotactic factors, accumulate at the site of infection where they take up bacteria and necrotic tissue. Later in the inflammatory response, larger amoeboid, mobile, mononucleate cells known as *macrophages* accumulate in the site of infection. These cells are also phagocytic and ingest bacteria and necrotic leucocytes. Acute bacterial infections such as staphylococcal boils and abscesses are characterized by a predominantly polymorphonuclear inflammatory response. However, in many more chronic infections such as tuberculosis, macrophages are the predominant inflammatory phagocytic cell. Phagocytes are of fundamental importance in immune reactions in all species. Firstly, they are able to destroy ingested bacteria or foreign particles. This probably happens as follows. The phagocyte surrounds the foreign particle with pseudopodia. The particle is thus invaginated and then taken up into a capsule surrounded by phagocyte cell membrane. This capsule is known as a *phagosome*. Soon the phagosome becomes surrounded by the *lysosomes* of the phagocyte which attach themselves to the phagosomal wall and finally rupture into it releasing powerful hydrolytic enzymes which digest the contents of the phagosome. This process is not always effective in destroying bacteria. Certain bacteria such as brucella and mycobacteria may continue to live for long periods and to multiply inside macrophages and bacteria may themselves destroy the cell containing them and be released. This brings us to the second important role of phagocytic cells. Macrophages are able to carry antigenic material away from the site of infection to the local lymph glands, which contain those cells capable of initiating specific immune responses. This transport mechanism may therefore provide direct access of antigen to these cells.

Hormones play an important part in determining the effectiveness of the body's cellular immune responses. Adrenal cortical steroid hormones depress the inflammatory response and inhibit phagocytosis. This may be due to their property of stabilizing lysosomal membranes thus making release of enzymes into the phagosome more difficult. Steroids may also depress specific immune responses such as antibody production and delayed hypersensitivity. The well-known increased incidence of infection in diabetes mellitus may be related to the fact that the lowered pH of the site of inflammation is maintained by keto-acids rather than by lactic acid. Keto-acids are known to have a far less bacteriostatic effect than lactic acid.

The lymphoid tissue. Lymphatic channels drain most of the tissues of the body with certain exceptions such as the bones and the central nervous system. Foreign material taken up by phagocytes at any site can therefore be transported *via* these channels to the draining lymph nodes. In addition, foreign material which is disseminated haemotogenously is likely to become trapped in the large central reservoirs of lymphoid tissue, the spleen and bone marrow, both of which are provided with a good blood supply. Other important sites of lymphoid tissue are the Peyer's patches of the small intestine and the dense lymphoid tissue in other areas of the alimentary tract including the tonsils, adenoids and the wall of the appendix. Therefore, almost all the possible sites of entry of infectious agents into the body are supplied by a plexus of lymphatics. Finally, there is the thymus in the anterior thorax, the role of which in the control of the function of other lymphoid tissues will be referred to later.

Commonly, at the time of an infection such as a boil, tender swelling of the draining lymph nodes is observed. Bacteria pass rapidly up the lymphatics to the node. This is followed by migration of polymorphs into the lymph node both from the site of infection and from the blood stream. In addition to the mobile phagocytes already mentioned, the polymorphs and macrophages, lymphoid tissues contain a network of fixed phagocytic cells capable of trapping and ingesting foreign material. These cells are metalophilic and have a stellate appearance on staining with silver. They are present in large numbers in the lymph nodes and spleen and probably similar cells in other tissues include the Kupffer cells of the liver, the histiocytes of connective tissue and the microglial cells of the central nervous system. The mobile and fixed phagocytic cells of the body are collectively known as the *reticulo-endothelial system*. Besides the phagocytic cells described above, the lymphoid tissues contain those cells responsible for mediating immune responses of the specific type. These cells include lymphocytes, plasma cells and their precursors. These cells are found in close anatomical relationship with reticulo-endothelial cells. Their function in specific acquired immune responses is closely dependent on this relationship and cannot be considered apart from it. Detailed consideration of the function of lymphocytes and plasma cells introduces the next broad class of defence mechanisms, those concerned in specific acquired immunity.

SPECIFIC ACQUIRED IMMUNITY

The immunity mechanisms discussed so far are basic first-line defences many of which are brought into operation within a very short time after infection or injury of tissues. But in the vertebrates a sophisticated immunity system has developed by means of which the tissues of the body recognize not only that any material that has been introduced is foreign but recognize it as a specific chemical entity and mount a defence reaction which is itself specific for this particular foreign substance. Such a sophisticated defence system entails the reception and processing of detailed information about the chemical nature of the invader before a specific immune response can be mounted. This mechanism takes some time to come into operation. Several days elapse after the first contact with an invader before a specific immune response begins to be produced. On later contact, a much more rapid immune response is mounted within the first 48 hr.

Specific acquired immune responses can be divided into two types. The cells which mediate specific immunity are known collectively as *immunologically competent cells* and, as mentioned above, these include lymphocytes and plasma cells and their precursors. In the first type of specific immunity, the *humoral* type, the plasma cells produce *antibody* which is secreted and circulates in the blood stream and tissue fluids and which reacts specifically with antigens. In the second type, *cellular* specific immunity, the immune response is mediated directly by immunologically competent lymphocytes. In discussing these immunity mechanisms, it is first necessary to consider the nature of the antigens against which such specific immune responses are directed.

ANTIGENS

An antigen is a substance which when introduced into the body is capable of stimulating an immune response which is specific for that injected antigen.

For a substance to be antigenic, the following criteria must be fulfilled.

(1) The substance must be a *foreign* substance, i.e. it must not be a normal constituent of the tissues or fluids of the body of the injected person or animal. The specific immune mechanism is able to recognize its own tissue constituents as *self* and does not mount an immune response against them. If foreign substances are introduced, these are recognized as *not-self* and an immune response follows. Thus, injection of rabbit serum albumin into a genetically similar rabbit leads to no immune response. If serum albumin from a foreign species such as the guinea pig is used instead, a good immune response is usually elicited in the rabbit. Apparent exceptions to this rule of foreignness will be discussed later on. The fact that a substance is foreign does not invariably mean that it is antigenic. Any given substance may be antigenic in one species but not in another. Some substances, which chemically might be expected to do so, fail to act as antigens at all.

(2) The substance must be a *macro-molecule.* Inorganic salts and small organic molecules are not antigens. The smallest known naturally occurring antigen is *glucagon* (mol. wt. 3,800) although synthetic antigens have been made with molecular weights down to about 1,000. While low molecular weight dextrans are poor antigens, if the molecular weight of dextran is increased by polymerization to over 100,000, it becomes much more strongly antigenic. In this case, the larger the antigen, the better its antigenicity.

There are some apparent exceptions to this rule that antigens must be large molecules. Certain drugs and other substances of low molecular weight such as penicillin, sulphonamides or picryl chloride, may induce specific immunity or hypersensitivity when introduced into the body. These substances are known as *haptens.* They are not antigenic in themselves and only become so because, on introduction into the body, they become attached to a large molecule such as serum albumin. This attachment confers a new specificity on the hapten–protein complex so that the two in combination act like a foreign macromolecule and an immune response directed against the haptenic groups is produced. Haptens may be combined with foreign proteins *in vitro* and then injected. An immune response is then formed against both the carrier protein and the hapten group. Haptens are of considerable clinical importance as many instances of drug hypersensitivity are examples of immune reactions against haptens.

Antigens may be either soluble or particulate. Many antigens exist normally as chemical groupings on the surfaces of bacteria or tissue cells but they may be detached from their parent cell and circulate in soluble form. In this form they remain strongly antigenic. Bacterial exotoxins are examples of antigens which are found both within the bacterial cell and free in solution in the tissues of the infected host.

(3) The *chemical nature* of antigens is important. Proteins and polysaccharides are usually, but not invariably, good antigens. Lipids are not antigenic unless they are combined with proteins or polysaccharides. Lipoproteins, glycolipids and lipopolysaccharides are usually strongly antigenic. Nucleic acids are antigenic but rather poorly so.

Antigenic determinants. A protein antigen is made up from a number of polypeptide chains each of which consists of a hundred or more amino-acids linked in sequence by peptide bonds. If the protein is examined structurally to determine where in the sequence of amino-acids its antigenicity lies, it is found that certain areas consisting of about 10 linked amino-acids are antigenically active but that the remainder of the sequence does not react as antigen. There may be several of these active sites, each made up of a few linked amino-acids on the whole molecule but the remainder is antigenically inert. The active sites are known as antigenic *determinants* and the molecule reacts specifically at these sites with complementary *combining sites* on the antibody molecule which, again, consist probably of about 10 linked amino-acids. Similarly, many polysaccharide antigens are made up of a backbone chain of polysaccharide with repeating series of sugar-containing side groups branching out on each side like a fishbone. It is the side groups which determine the antigenic specificity as the central chain is inaccessible and so does not normally react with antibody. Minor chemical differences at the determinant sites of antigen molecules may therefore cause clear differences in antigenic specificity. On the other hand, substantial structural differences at inert sites of the molecule do not lead to differences in antigenic specificity. Haptens may be regarded as artificial determinants which, by modifying the structure of their carriers, are able to change an antigenically inert protein into a specifically antigenic hapten–protein conjugate. An antigen may be compared with a key, one end of which is specially shaped to fit a particular lock, its specific antibody. The shank and handle of the key do not fit the lock directly and can be altered without destroying the fit, but they are still essential to get the key into the lock. However, antigens differ from keys inasmuch as each molecule has not just one but several different active sites capable of fitting together with antibody.

Antigens of clinical importance include the antigens of the bacterial cell wall, which may be polysaccharide or lipopolysaccharide in character. Many are linked complexes of, for instance, lipopolysaccharides with peptides, glycopeptides or lipid. Examples of very powerful bacterial lipopolysaccharide antigens are provided by the somatic or (O-) antigens of the coli-typhoid-dysentery group of Gram-negative bacteria. There are many serotypes of Salmonella and each has its specific O-antigen distinguishable from that of other types. This is because, in each organism, the determinant side-chain sugars are slightly different and each one stimulates formation of specific antibody. This distinction forms the basis of the agglutination tests used for Salmonella typing in the identification of food poisoning organisms. The flagellar (H-) antigens which are proteins also differ in different Salmonella types and they can be classified serologically in the same way. Other important bacterial polysaccharide antigens include the capsular polysaccharide of the pneumococcus. Again, there are many antigenic types of pneumococci. They vary in the determinant groups on the capsular polysaccharide and these variations can be detected by specific antisera. Detection of pneumococcal serotypes was very important when patients with pneumococcal infection were treated by passive serotherapy. It has largely lost clinical importance since the introduction of antibiotics. Other streptococci also contain numerous surface antigens (e.g. "C" group specific carbohydrate antigen, streptolysin-O, streptokinase) and an intracellular M-protein antigen which is of considerable importance in some of the delayed immunological effects of streptococcal infection referred to later. Bacterial exotoxins such as diphtheria and tetanus toxins are proteins. An important chemical manipulation of the antigenic properties of bacterial exotoxins is their conversion to toxoids for use in active immunization. The toxic property and the antigenic property of these toxins depend on different chemical linkages within the molecule. If the molecule is treated, usually with formalin, so that the toxic property is destroyed while the antigenic property is retained, the toxin can then be safely used for immunization. Toxins modified in this way are known as *toxoids*.

Many tissue cells and erythrocytes also contain surface antigens which are lipopolysaccharides structurally similar to the O-antigens of Gram-negative bacilli. Erythrocyte blood group antigens are of this type and many of the antigens of tissue cell membranes which may play a part in transplantation immunity and cancer immunity may be chemically similar. ABO blood group antigens on the surface of red cells have a polysaccharide backbone with sugar (oligosaccharide) side chains. The same antigens present in solution in ovarian cyst fluid have a polypeptide backbone but identical oligosaccharide side chains. Both types of blood group substance give identical reactions with antisera because blood group-specific antisera are directed against the side chain determinants and the unreactive parts of the molecule can be modified markedly without altering this antigenic specificity so long as the antigenically reactive groups are identical in structure and sequence.

THE ANTIBODY RESPONSE

Antibodies are proteins produced by the body as a specific response to stimulation by antigens. They contain active *combining sites* capable of bonding chemically with the antigen which caused their formation but unreactive with unrelated antigens. This reaction of antibody with antigen usually results in neutralization and inactivation of the antigen. It also leads to accelerated elimination of antigen from the body, as can be shown by following the elimination of antigenic substances which have been labelled with a radioactive isotope. Antibodies are present in serum and form that fraction of the serum proteins which migrate in the gamma-globulin region on electrophoresis. They have a distinctive chemical structure adapted to their biological requirements as antibodies. This unique structure is discussed in detail in the following chapter.

Antibody production. The mechanism of production of specific antibodies is still one of the most controversial fields in immunology. Antibodies are produced by plasma cells which are typically oval cells containing a small eccentric nucleus whose chromatin is arranged in clumps like the spokes of a wheel and a deeply basophilic cytoplasm which occupies most of the cell. The cytoplasm contains a profuse rough endoplasmic reticulum, thus indicating active protein synthesis and secretion. Plasma cells are found, after immunization, in the draining lymph nodes, in the spleen and bone marrow, and at the site of injection of antigen. Not all antibody-producing cells are typical plasma cells. Some are immature forms and some are morphologically similar to lymphocytes although the classical lymphocyte does not contain the characteristic endoplasmic reticulum necessary for the production of protein and its secretion from the cell.

For specific antibody to be produced by plasma cells, precise information about the nature of the antigen must in some way reach the sites of antibody production. The details of this process are still not fully understood, but antigen at the site of injection or infection is taken up by many phagocytic cells and transported, either by these cells or directly in the lymph, to the local lymph nodes. Antigen which is disseminated in the blood stream may similarly be trapped by reticulo-endothelial cells in the spleen. Therefore, the local concentration of antigen in lymphoid tissue is clearly important for antibody production and is followed by division and maturation of cells within the lymphoid tissue. *Plasmablasts*, primitive nucleolated cells with large nuclei and pale cytoplasms which contain only a small amount of endoplasmic reticulum develop into the mature plasma cells described above. When maturation is complete, maximal production of antibody by plasma cells follows. The essential link which would explain how these cells receive the information which triggers off their division and maturation is still missing as antigen, although present in the sites where these cells occur, cannot be demonstrated actually to enter the antibody-producing cells at any stage.

Primary and secondary response. The process described above, involving as it probably does several cell divisions, takes some days to complete. After the *first* contact with an antigen, this process does not lead to appreciable antibody production until about 10 days after inoculation. A small amount of antibody is produced but this antibody response is short-lived, tailing off after a few days. This is known as the *primary response*. Although, in the primary response antibody production is poor and short-lived, as a result of the contact with antigen, cells are now present in the lymphoid tissue which are *informed* about the specific antigen, are committed to production of specific antibody against it and are primed to produce large amounts of antibody at short notice. These cells have what is called *immunological memory* so that they are immediately able to recognize and respond again to their specific antigen on further contact with it. Therefore, the *secondary response*, which occurs on injection of a second dose of antigen, is characterized by production of high levels of antibody within 48 hr of injection and these levels are sustained over a period of many weeks (Fig. 1.IX). The level of antibody does eventually tail off but usually immunological memory is retained and further injections of antigen lead to responses similar to, and often even better than, the secondary response. This sequence of events explains clearly why most immunization programmes entail the use of two or more injections of antigen separated by periods of a month or more and why a single antigen injection alone is not an efficient way of producing immunity.

Factors affecting the antibody response. The antibody response can be abolished or diminished by the use of any agent which is toxic to antibody-forming cells or which interferes with cell division. Thus *cytotoxic drugs*, anti-mitotic agents such as 6-mercaptopurine and X-rays, when given at or a day or two

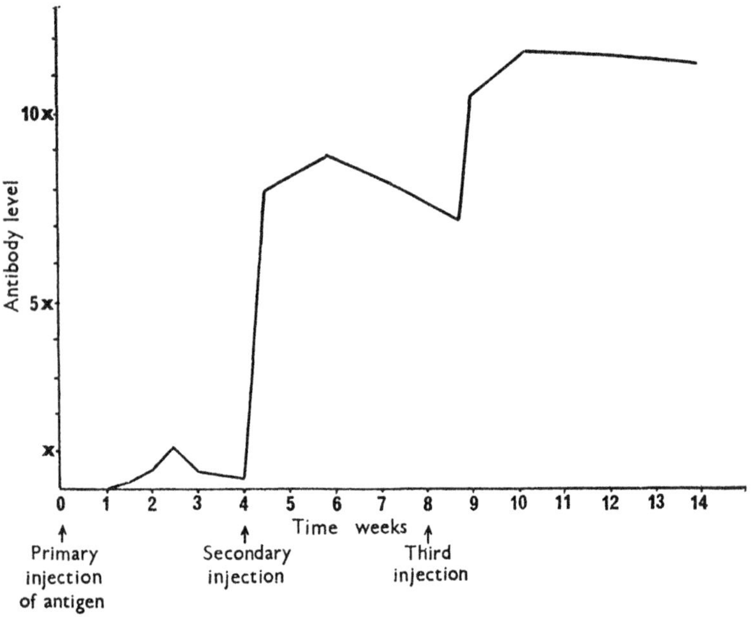

Fig. 1.IX

before the time of injection of antigen will inhibit the production of antibody. This effect is pronounced when these agents are given at the time of the primary response in which cell division is important. They have far less effect on the secondary response which involves antibody production by mature cells which are relatively radio resistant. Other agents which inhibit antibody production include drugs which interfere with the genetic control of protein synthesis. Drugs such as actinomycin and puromycin act in this way. As mentioned earlier, steroid hormones may also act to decrease antibody production. The use of such *immuno-*

suppressive agents plays an important part in increasing the chance of successful surgical transfer of incompatible tissue grafts but, at the same time, increases the chances of intercurrent bacterial infections.

The antibody response may be enhanced by the use of *adjuvants*. An adjuvant is a substance which, when injected together with an antigen, increases the level of antibody production or cellular immunity against that antigen, without necessarily itself playing any direct part in the immune response. Thus adjuvants act as catalysts for immune responses. The most powerful adjuvant yet discovered is Freund's complete adjuvant which is a water-in-oil emulsion containing heat-killed *Mycobacterium tuberculosis*. Freund's adjuvant gives very much enhanced levels of antibody but, unfortunately, cannot be used in man because it produces large necrotic local granulomata. Other bacteria such as *Bordetella pertussis* also act as adjuvants but again are obviously unsuitable for use in man. Many inorganic substances such as silica, kaolin and calcium salts also act as adjuvants but the substances which have gained most general acceptance for human immunization have been the aluminium salts, potassium alum, aluminium phosphate and aluminium hydroxide. These cause mild granuloma formation and give acceptable levels of antibody while being very safe. Most adjuvants cause retention of antigen at the site of injection, alum because it causes precipitation and adsorption of the antigen, and Freund's adjuvant partly because the antigen is retained in water droplets in an oily suspension. Antigen is therefore not disseminated rapidly through the blood stream and destroyed as would otherwise be the case. Macrophages accumulate at the site of injection and ingest the antigen which they then transport to the local lymph nodes. It may well be this slow, continued stimulation of lymphoid cells by macrophage-borne antigen which leads to the effect of adjuvants in enhancing immune responses.

Sites of antibody production. After subcutaneous injection of antigen, large numbers of antibody-producing plasma cells can be demonstrated in the draining lymph nodes. These lymph nodes make more antibody than more distant lymphoid tissues and some antibody is also made by plasma cells which have migrated to the site of injection of antigen. However, when the antigen is injected with Freund's adjuvant, more distant lymph nodes are stimulated and these also make antibody. The bone marrow also contains plasma cells and this tissue is an important site of antibody formation against antigen which has been dispersed from the injection site via the blood stream rather than via the lymphatics.

After intravenous injection of antigens, the major site of antibody production is the spleen in which haematogenously disseminated antigen becomes trapped by reticulo-endothelial cells.

Theories of antibody production. Any theory of antibody production must try to explain (1) the high specificity of antibody produced against a given antigen, and (2) the fact that the body recognizes and responds only to foreign antigens. If unhindered immune responses against the body's own tissue took place, we would rapidly destroy ourselves completely. An early theoretical explanation of antibody production was put forward by Haurowitz who suggested that antigen entered the antibody-forming cell and acted as a *template* on which the antibody was modelled. Thus antigen *instructed* the cell how to make antibody. Haurowitz's view can no longer be held in its original form, not only because antigen has never been demonstrated inside plasma cells but because the genetic control of protein synthesis by DNA must rule out antigen as a direct model for antibody production. However, modified "instructive" theories are still popular. These theories presuppose that, before antigenic stimulation, immunologically competent cells are uncommitted and that, in the presence of antigen, information is transmitted which directs the cell into a particular line of antibody production. This need not happen as a direct result of contact with unaltered antigen. Antigen contained in macrophages may cause modifications in the macrophage RNA which is then itself transmitted to lymphocytes or plasma cells. Modified antigenic material could also act directly in the way that viruses act, to alter the genome of the antibody-forming cell. The cell would then be committed to production of specific antibody.

The second group of theories of antibody production are *selective*. According to Sir Macfarlane Burnet's *clonal selection theory*, the precursors of immunologically competent cells are committed, each to production of a single specific antibody from the earliest foetal stages. The *clone* or family of cells derived from the original parent cell will thereafter produce antibody of one type alone. During foetal life, clones directed against antigens which are themselves present in the foetus, i.e. "self"-antigens will be suppressed because the normal immunological response at this stage is not proliferation and antibody production but instead destruction or suppression of immunologically competent cells. However, as the individual matures, at or near the time of birth, the immune response changes to the adult type in which antigenic challenge leads to proliferation and antibody production. By this time all the "anti-self" clones have been suppressed and antigens now met with should be exogenous or "not-self". Injection of an antigen after this stage leads to proliferation of cells belonging to its own precommitted clone.

Other theories of antibody production suggest that cell clones are not irrevocably committed to production of one antibody only but that "recognition factors" for antigen—possibly natural antibodies—exist in serum. Once "recognition factor" has combined with a particular antigen, information is passed to immunologically competent cells which thereafter are committed to production of antibody against that antigen.

"Natural antibodies". Certain antibodies present in normal serum cannot be accounted for by direct stimulation by the antigen against which they are directed. The best known of these antibodies are the ABO blood group agglutinins which are found in high titre in the sera of individuals who have certainly never been exposed to incompatible blood group antigens. Another antibody of this type is the antibody, present in many species, and in low titre in normal human sera, which reacts with a tissue antigen known as Forssman antigen present in tissue cells of other species such as sheep and horse. These antibodies are present from a very early stage of life. The probable explanation of their presence is that they are, in fact, produced in response to antigen but not the antigen against which they are demonstrably directed. Bacterial polysaccharide antigens contain many similar determinant groups to those found on the surfaces of mammalian tissue cells and erythrocytes. These antigenic determinants are widespread throughout nature and possibly natural antibodies represent an immune response, in neo-natal life, to bacterial or other antigens of this type ingested through the gastrointestinal tract.

Maternal–foetal transmission of antibody. The new-born baby is not capable of making very much antibody itself but during the first months of life, on the earliest contacts with exogenous antigens the baby rapidly develops the capacity to produce its own antibodies. The serum of new-born babies does nevertheless contain antibody against quite a number of bacteria and viruses. This antibody is not home-made but derived from the mother by placental transfer. The levels of such maternally-derived antibodies decline quite rapidly as the foetus begins to manufacture its own antibody.

CELL-MEDIATED IMMUNITY

Three types of cell are fundamental to specific immune processes. These are the plasma cell, the lymphocyte and the various types of phagocytic or reticulo-endothelial cell. Plasma cells are *par excellence* the antibody-producing cells and so have been considered earlier with humoral immunity. Phagocytic cells have the important function of trapping and destroying toxic exogenous substances and mobile phagocytes carry antigens to the site where specific immune responses are initiated. Macrophages may also possibly play an important role in the recognition of foreign antigens and in processing them in a way which enhances their immunogenicity. The third type of cell, the *lymphocyte*, is of crucial importance in specific cell-mediated immune processes. Lymphocytes are present in large numbers in all lymphoid tissues. Circulating lymphocytes are also found in the lymphatic channels and in the blood stream and they are constantly found at the sites of inflammatory lesions throughout the body. In lymphoid tissues, lymphocytes at all stages of development from primitive nucleolated cells to maturer forms may be seen. The development of lymphocytes is still a source of debate. It used to be thought that small lymphocytes were end-cells, the final result of maturation, but it is now clear that, upon appropriate stimulation, these small cells can develop into large nucleolated blast cells. Whether plasma cells develop from the same precursors as lymphocytes or follow a separate line of development is still unknown. The lymphocytes are not static cells. Gowans has shown that there is a circulation of lymphocytes through the lymphatics and a constant exchange of lymphocytes between the lymphoid tissue and the blood stream.

Following inflammation there is proliferation and maturation of cells of the lymphocyte and plasma cell series within the lymph node draining the inflamed site. At the same time, in the cortex, discrete spherical collections of large pale immature cells surrounded by a ring of lymphocytes appear. These discrete round foci are known as *germinal centres*. They appear in response to stimulation by circulating antigens and antigen which reaches the lymph node from outside can be shown to become localized in them. They therefore play a role in specific immunity but this role is not yet clearly understood. They may be responsible for the maintenance of immunological memory. Similar germinal centres are seen in the spleen and in the submucosal lymphoid tissue of the intestine after intravenous administration of antigen. Adjacent to the germinal centres are areas known as "paracortical" or "thymus-dependent" areas. These are of special importance in cell-mediated immunity as the cells which mediate phenomena such as delayed hypersensitivity and graft rejection appear to proliferate there following antigenic challenge.

Lymphocyte function. Typical small lymphocytes contain no endoplasmic reticulum and therefore probably do not secrete antibody globulin. However, not all antibody is produced by typical plasma cells. Antibody production can be demonstrated in cells morphologically similar to lymphocytes. Until the precise line of development of the lymphocyte and plasma cell series is established, it will be difficult to be certain to which of the two lines these cells belong. Leaving aside the question of antibody production, the possible functions of lymphocytes are as follows: (1) They may be stem cells from which the other immunologically competent cells derive. Small lymphocytes will, when stimulated, develop into large immature pyroninophilic blast cells which may mature into other types of lymphoid cell. It has never been proved, however, that plasma cells are derived from lymphocytes in this way. (2) Lymphocytes may carry "immunological memory" after a first contact with antigen, thus ensuring that a good secondary response will occur. (3) "Sensitized" lymphocytes certainly react with antigen in cell-mediated responses. Thus for instance, lymphocytes can, like antibody, cause lysis of cells bearing foreign antigens. Under the heading of *cell-mediated immunity* are classified those specific responses to antigen which are unrelated to the presence of circulating antibody. They include delayed hypersensitivity and the rejection of grafts of foreign tissue and they are, probably, also of importance in immunological reactions against tumours and in auto-immune reactions against tissue antigens. Cell-mediated immune processes are also important in bacterial infections particularly of the more chronic type such as tuberculosis, syphilis or brucellosis. It is known that in certain virus infections, such as smallpox, effective immunity depends more on cellular responses than on the presence of circulating antibody. If an animal is stimulated by injection of antigen to produce a cellular immune response such as, for example, delayed hypersensitivity, and serum antibody is then taken from this immune animal and injected into a second non-immune animal, the cellular immunity is *not* transferred to the second animal. If, however, instead of serum, the primed *lymphocytes* from the first animal are taken and injected into the second animal, then cellular immunity *is* transferred. This power of lymphocytes to transfer immunity has been shown in the case of graft tolerance, delayed hypersensitivity and auto-immune disease. Lymphocytes are therefore essential for the development of these phenomena.

The thymus. Following injection of antigen into peripheral tissues, the lymphoid tissues, lymph nodes, spleen and bone marrow show proliferative changes indicative of an immune response. Although densely populated with lymphoid cells, the same is not true of the thymus. Until recently nothing was known about thymus function. It was known to be a lymphoid organ, derived from the third branchial cleft, prominent at birth but undergoing involution during childhood. Other than being a refuge for destitute pathologists who ascribed the cause of otherwise unexplained sudden deaths to status thymo-lymphaticus, the thymus remained a mystery until a very few years ago.

The importance of the thymus was uncovered by observing the effects of its surgical removal. Thymectomy has been performed for myasthenia gravis for many years in more or less mature individuals without apparent harmful effects. However, when experimental thymectomy was carried out in *new-born* mice, it was shown that the mice failed to thrive, became wasted and failed to develop normal immune responses. These thymectomized mice could be grafted with incompatible skin grafts which they accepted and which their normal siblings rejected. The blood, lymph and certain cortical areas of the lymph nodes of thymectomized mice were depleted of lymphocytes. Neonatally thymectomized mice therefore showed deficiencies of lymphocytes and of the cellular types of immunity mediated by them. They also showed some deficiencies of humoral antibody production but to nothing like the same extent. Similar, but much milder, deficiencies also occur in mice thymectomized in adult life. The thymus is therefore an essential organ in the development of normal immune responses. Possibly cells migrate from the thymus to "seed" the other lymphoid tissues. Possibly, the thymus elaborates a circulating substance which can control the immunological capacity of lymphoid cells in distant tissues. Although the thymus is itself packed with lymphocytes, these thymic lymphocytes, or "thymocytes" are themselves poor mediators of immune responses. The importance of the thymus resides much more in its influence on other lymphoid tissues than in its own capacity directly to mount specific immune responses. The thymus is of paramount importance in the development of immunity in early life. Later its importance declines but is probably never completely lost.

IMMUNIZATION PROCEDURES

The object of immunization is to establish effective immunity against pathogenic invaders without at the same time subjecting the immunized individual to the harmful lesions which those invaders would normally cause. Immunization may be *prophylactic*, in which the object is to induce protective immunity in a healthy individual who might, at some time in the future, be exposed to infection. It may be *therapeutic* in which artificial methods are used to boost the immune state of an individual who is already suffering from an infectious disease.

Immunization is considered under two sub-headings. These are (1) *active immunization*, in which immunity is induced by injecting bacterial or viral *antigens* into the subject in order to stimulate the individual's own tissues to production of antibody and immunologically committed lymphocytes, and (2) *passive immunization* in which pre-formed *antibody* derived from human or animal serum is injected into the individual. Thus in passive immunization, the tissues of the individual are not themselves called upon to produce any immune response.

Active immunization. Under normal circumstances, most of our active immunity is acquired as a result of the various overt or sub-clinical infections from which we suffer during our lives. Thus, serum gamma-globulin consists of a mixture of a large number of antibodies directed against the antigens with which we have come into contact at various times. Sometimes this immunity is long-lasting as, for instance, in the case of mumps or rubella. Against other diseases such as the common cold or influenza, immunity is only transitory. Active immunity could be induced by injecting whole unaltered organisms but this would obviously be highly undesirable as it would involve exposing the individual to the disease against which we wished to protect him. Therefore, active immunity is in practice achieved by injecting organisms or extracts of organisms which have, in some way, been rendered harmless while retaining their normal antigenicity. By dissociating antigenicity from pathogenicity, good immune responses could be achieved in safety.

Active immunization may be achieved safely in three ways:

(1) In the case of diseases caused by organisms whose main pathogenic effects are due to their production of exotoxins, e.g. diphtheria or tetanus, active immunity is induced by the injection of *toxoid*, i.e. toxin which has been treated with formalin in such a way that its toxicity is lost while its antigenicity is retained.

(2) Organisms may be killed by treatment with heat, formalin merthiolate or other chemical substances. If such treatment does not destroy the antigenic properties of the organism and the killed vaccine is relatively non-toxic, such killed organisms may be used to induce active immunity.

(3) Living organisms may be injected in an *attenuated* form. Such organisms have been treated or cultured so that while they have retained antigenicity, they have lost virulence. They are therefore unable to cause disease but still stimulate good immune responses. An extension of this principle is to use an avirulent

organism antigenically related to, but not identical with, the one against which immunity is desired. This is the principle in smallpox vaccination in which vaccinia virus is used to produce immunity against the closely related variola virus.

In order to induce satisfactory active immunization, the principles governing antibody production put forward in earlier sections must be understood. Immunization may be used in prophylaxis either for the protection of the individual or, by producing a highly immune population, in whom epidemics are unlikely to start, for the protection of the community as a whole. As the period of life at which we are most vulnerable to many endemic infections is childhood, active prophylactic immunization against these infections is begun early in life. As placentally transmitted maternal antibody is present in foetal serum during the first two or three months of life, any antigen which might be injected during this period may be neutralized by such maternal antibody and thus not become available to stimulate active antibody production. Therefore immunization programmes are not usually begun until about three months after birth. A single injection of antigen will induce only a small primary antibody response. A second injection is therefore usually given one month later to stimulate a much greater secondary response, and a third a month to six weeks later. A sustained high level of serum antibody is thus achieved. Booster injections are usually given at the time of entry to school and, if required, thereafter at intervals of a few years. Active immunization against non-endemic diseases such as cholera is obviously only necessary in populations at risk. Routine immunization against such diseases is therefore not practical except for persons travelling to endemic areas.

The uses of active immunization are principally prophylactic in healthy individuals. As the primary antibody response has a delay of about three weeks, active immunization is of no immediate therapeutic benefit in unimmunized subjects who are already suffering from serious infectious diseases. However, in previously immunized persons who are exposed, say, to tetanus through a dirty wound, a booster dose of tetanus toxoid is able to stimulate high levels of antibody before clinical lesions are set up. Similarly, in diseases such as rabies which have very long incubation periods, active immunization after exposure may lead to a good immune response before the clinical lesions of the disease begin to appear. Active immunization usually gives long-lasting immunity which can be maintained by booster injections as required.

The clinical uses of active immunization. (1) Toxoids. (a) Diphtheria toxoid is prepared by formalin treatment and is injected with an aluminium adjuvant, either aluminium hydroxide (APT) or hydrated aluminium phosphate (PTAP). Aluminium adjuvants are important in the early injections but probably not so in later injections. Diphtheria toxoid is given to children following a programme similar to the one described above. It is usually given as a "triple vaccine" in a single injection together with pertussis and tetanus vaccines (see below). When used in adults who sometimes show severe reactions to APT, diphtheria toxoid may be given in the form of toxoid-antitoxin floccules (TAF). Active immunization against diphtheria is safe and mass immunization has virtually eradicated this disease as a killer in childhood.

(b) Tetanus toxoid is similarly prepared by treatment of purified toxin with formalin. It is often given in the "triple vaccine" in childhood. As everyone is potentially at risk from tetanus, and active immunization against it is very safe while complications following passive immunization with antitoxin are common, it is desirable that the population should be actively immunized against this disease. It is then simple to boost with toxoid persons exposed to *Clostridium tetani* following injury, etc., thus ensuring a vigorous immune response within 48 hr or so. Because tetanus toxoid is safe, it is replacing tetanus antitoxin as a routine prophylactic in the management of dirty wounds in casualty departments.

(2) Killed vaccine. (a) Pertussis. Whooping cough may be a serious disease in young infants although the mortality has dropped markedly in recent years. Pertussis vaccine is prepared by formalin or merthiolate treatment of virulent strains of *Bordetella pertussis*. Killed *Bord. pertussis* organisms themselves have adjuvant activity and, when given as a triple vaccine, the presence of *Bord. pertussis* probably enhances the level of antibody against diphtheria and tetanus. Severe encephalitis has been reported occasionally after pertussis vaccination.

(b) Enteric fever. The vaccine used for protection against enteric fevers is generally known as TAB vaccine. It contains heat-killed, phenol-preserved *S. typhi* and *S. paratyphi* A, B and C in the smooth phase containing the O- and Vi-antigens. It is given to persons travelling to endemic areas or otherwise at risk, as a series of three injections. Such vaccines contain endotoxin and may cause considerable local reaction and systemic febrile illness for about 48 hr after injection. Debate about the effectiveness of TAB has been prolonged but its value in prophylaxis among exposed populations is now clear.

(c) Cholera, plague, rabies. Cholera and plague vaccines are given in several doses to persons travelling to endemic areas. The first dose of these vaccines should be given as early before departure as possible, so that antibody production may be boosted quickly if necessary on arrival. Rabies vaccination is dangerous and is not used for routine prophylaxis. The vaccine is prepared from infected nervous tissue (homogenized sheep or rabbit brain) in which the virus has been killed with phenol. The vaccine is only given after a bite by a suspected rabid animal (usually but not always a dog). It involves multiple injections of foreign brain tissue, usually into the abdominal wall and carries a real risk of allergic encephalitis. As rabies is uniformly fatal, the risk of vaccination is preferable to the disease. Vaccination is given only after bites from an infected dog, not after licks. It is usually given synchronously with anti-rabies serum, which reduces the risk of post-vaccination encephalitis.

(d) Poliomyelitis. Salk-type poliomyelitis vaccine is prepared in monkey kidney cells and formalinized. It contains antibiotics and is given by subcutaneous injection in three doses as outlined above. A history of antibiotic sensitivity should be sought. This killed polio vaccine has now largely given place to oral vaccine (see below).

Attenuated or live vaccines. (a) Tuberculosis. BCG (Bacille Calmette-Guérin) is an attenuated bovine strain of *M. tuberculosis* of very low virulence. BCG vaccination has had a chequered career chiefly as a result of a disaster in the attractive medieval Baltic town of Lübeck where, in 1930, 72 children died after vaccination with BCG which had become contaminated with virulent *M. tuberculosis.* Only comparatively recently has BCG re-established itself as a safe protective measure against tuberculosis. It is offered to all tuberculin-negative individuals at risk, i.e. nurses, medical students and staff and contacts of patients with tuberculosis and to tuberculin-negative pre-adolescent school children. In some parts of the country it is given routinely to new-born babies. It should not be given to tuberculin-positive subjects in whom it is not only useless but may cause severe reactions and even re-activation of primary lesions. A single intradermal injection of 0·1 mg over the deltoid is all that is required. The bacilli remain viable *in vivo* long enough to provide a secondary-type response without further injection.

(b) Smallpox. Vaccination against smallpox, one of the scourges of the world throughout history, is the oldest of all immunization procedures. Once attempted by injection of smallpox virus itself, since the days of Jenner, it has been achieved by injection of vaccinia virus, the causative organism of cowpox. Virus suitably prepared from the skin lesions of cowpox or from chick embryo cultures is scarified into a small area of the epidermis on the upper arm by the multiple pressure method, taking care not to penetrate the dermis. In a primary vaccination, a vesicle surrounded by a circle of erythema develops after about seven days if the vaccine has taken. If the vaccine has not taken, the vaccination must be repeated. Thereafter, revaccination is performed every three years. In a successful revaccination, an accelerated response is seen. After two or three days, a small papule is seen which may vesiculate. This is much milder than the primary response and heals rapidly. Smallpox vaccination is obligatory for travel to or from the many areas where smallpox is endemic. There is also no doubt that doctors and nurses should be vaccinated, particularly as, in cosmopolitan populations, sporadic epidemics of smallpox occasionally break out. It has, for many years, been standard practice to vaccinate infants in the first years of life despite persistent opposition. Nevertheless, smallpox vaccination, particularly when carried out after the first two years of life, carries a small risk of encephalomyelitis (about 20 per million vaccinations in Great Britain) and recently it has been questioned whether the risk of post-vaccinial encephalomyelitis has not now, in Great Britain, become greater than the risk from smallpox. It must be remembered, however, that it is only strict control of the vaccination status of visitors to and from endemic areas which maintains this state of affairs. This risk of encephalomyelitis is least if vaccination is carried out in the second year of life, so vaccination is best carried out after the first birthday but before the second. Vaccination should be avoided in eczematous children in whom there is a risk of generalized vaccinia.

(c) Poliomyelitis. Oral poliomyelitis vaccination with attenuated strains of the three main types of poliomyelitis virus has, within a few years, become standard practice for the routine immunization of children. Poliomyelitis, once a feared disease in the prosperous nations of the world, has almost disappeared within the lifetime of a first year medical student. The oral vaccines have now largely replaced intramuscular injection of killed poliovirus. These attenuated viruses multiply in the gastro-intestinal tract and stimulate formation of high levels of antibody. They have undergone very extensive trials and are now considered to be extremely safe. Multiple doses are required to provide full immunity against all strains of poliovirus.

(d) Yellow fever. A strain of yellow fever virus known as 17D is used for vaccination against yellow fever. This strain is avirulent for man and is used for immunization in the form of a single injection given to persons travelling to those tropical areas in which yellow fever still occurs. Long lasting immunity (at least six years) follows injection.

(e) Measles. Extensive clinical trials have shown that live, attenuated measles vaccine is the best and safest of the recently introduced measles vaccines. Immunization is achieved with a single injection given to children more than 12 months but less than 10–15 years old, and preferably before school entry. It is effective and safe but should not be given to children with histories of allergy or convulsions.

Passive immunization with antibody globulin gives an immunity which lasts only as long as the injected globulin. As gamma-globulin has a half-life measured in days or weeks only, this form of immunization is obviously incapable of producing long-lasting immunity. On the other hand, it confers immunity immediately from the moment of injection. To produce effective levels of immunity, fairly large doses of protein must be injected. As such protein is often derived from immunized animals, it is very prone to act as a foreign antigen and to give rise to hypersensitivity reactions of varying severity. For this reason, injections of antitoxins should generally be preceded by a test intradermal injection to see if any hypersensitivity reaction may take place. It there is no reaction, treatment may proceed. If there is a weal and flare, further injections of antitoxin should be avoided. A history of asthma, hay fever or drug sensitivity should be sought and adrenaline should always be available for injections in case of serious anaphylactic reactions.

Passive immunization is not used to induce long-lasting prophylaxis. It may be used for short-term prophylaxis in persons exposed to infections who may be expected to contract the disease in the immediate future. It is also used (in much higher doses) in treatment of certain bacterial diseases.

Diphtheria antitoxin is prepared from the purified globulin fraction of serum from horses immunized against diphtheria, and is used in the treatment of that disease. Tetanus antitoxin, again prepared in the horse, is also used in the form of purified globulin, a procedure which cuts down the possibility of side reactions. Antitetanus serum has been used for many years as a prophylactic measure in the management of dirty wounds, but it is safer to use tetanus toxoid for this purpose, particularly in previously immunized individuals. Tetanus antitoxin in large doses is used in the treatment of tetanus. Antitoxins are also used in the treatment of other clostridial infections. Anti-gas gangrene serum contains antibodies against *Clostridium welchii*, *oedematiens* and *septicum* and is used together with antibiotics and surgical treatment in gas gangrene. An antitoxin is also available against *Cl. botulinum.*

Probably the most important use of passive immunization at present is the injection of pooled human γ-globulin for the treatment of hypogammaglobulinaemia and for the prophylaxis of rubella in early pregnancy. This form of therapy is free from the risk, associated with antitoxins, of hypersensitivity to the proteins of foreign species, but should, nevertheless, be given by deep intramuscular injection and never intravenously. In prophylaxis against the severe lesions of rubella in the foetus during the first trimester of pregnancy, pooled human γ-globulin is given because such a pool will invariably contain antibodies from individuals who have recovered from rubella. Convalescent serum from patients with rubella is even better but is not easily come by.

The dangers of immunization. The dangers of immunization, some of which have already been mentioned, may be classified under several headings. Firstly, faulty preparation of vaccines: "killed" organisms may not be dead, attenuated organisms may still be virulent and toxoids may still be toxic. Vaccines may be contaminated with extraneous bacteria, viruses or foreign toxins. After many years of experience, vaccines produced by reputable firms are now strictly controlled and these complications ought not to happen. Secondly, the operative technique of vaccination may be faulty and lacking in proper aseptic precautions. This is most likely to happen under panic conditions such as, for instance, in a mass immunization campaign during an epidemic scare. This is another argument for routine prophylactic vaccination carried out under non-stressful conditions when there is no epidemic. Thirdly, and the most difficult group to avoid, are the numerous hypersensitivity reactions. These include anaphylaxis and serum sickness after passive immunization with foreign proteins as well as the encephalomyelitic complications which occur sporadically after smallpox vaccination, more frequently after rabies vaccination and more rarely after other immunization procedures. Generalized vaccinia may be classified here. These dangers may again be reduced by careful enquiry about previous allergic episodes and present therapy with drugs such as steroids (again, slips are more likely in mass vaccination programmes). Finally, there is the possibility that active immunization in someone incubating a disease, particularly typhoid or poliomyelitis, may actually provoke an acute attack of that disease. This risk may be rare but should not be forgotten. Vaccines and antisera thus, like other drugs, have been and will be of great benefit to mankind but must be used with an intelligent awareness of their limitations and complications.

Chapter X

Immunoglobulins: The Reaction of Antibody with Antigen

In the preceding chapter, it was shown that antibodies are proteins which migrate in the electrophoretic *gamma-globulin* fraction of serum. These antibody globulins belong to a single chemical family of proteins known as *immunoglobulins* whose structure has only recently been discovered. Structural studies of these antibody globulins reveal a heterogeneity unusual among the proteins of biological fluids. This diversity is closely linked to the biological properties of immunoglobulins and to their capacity to react specifically with the enormous number of antigenic substances which the body may encounter.

THE FAMILY OF IMMUNOGLOBULINS

Five immunoglobulins have been discovered in normal human serum up to the present time. These five proteins have many features in common with one another and other features which allow them to be distinguished from one another. Immunoglobulins are known either by the family initials Ig (immunoglobulin) followed by a capital letter to indicate which particular class is being referred to, thus IgA, or by the Greek γ followed by the same capital, thus γA. IgA and γA are therefore two names for the same protein.

The immunoglobulins of human serum are as follows: (1) IgG (γG). The major immunoglobulin of serum forming about 80 per cent of the total immunoglobulin (serum concentration about 1,000 mg per 100 ml). Molecular weight, 150,000. Electrophoretically it forms a wide band spreading from the slowest γ-globulin fraction (γ_2) through the fast γ-globulin fraction (γ_1) into the β_2 and β_1-globulin regions. It is known as a 7S globulin because when spun in an ultracentrifuge, it has a sedimentation coefficient of 7. (The sedimentation coefficient is derived by a formula from the speed of centrifugation. It depends largely on molecular weight so that the bigger the protein, the higher the sedimentation coefficient.)

(2) IgA (γA). About 15 per cent of the total serum immunoglobulin (serum concentration 140–400 mg per 100 ml). Molecular weight 160,000 but it has a tendency to form polymers of higher molecular weight. Sedimentation coefficient 7S with polymers of 11S and 13S. Electrophoretically a fast γ-globulin (γ_1).

(3) IgM (γM). About 5 per cent of the total serum immunoglobulins (serum concentration 50–200 mg per 100 ml). Otherwise known as γ-macroglobulin, it has a molecular weight of 900,000 to 1 million. Sedimentation coefficient, 19. Electrophoretically migrates as a fast γ-globulin (γ_1).

(4) IgD (γD). Present in quantities of only a few mg per 100 ml in serum. Molecular weight 150,000 and sedimentation coefficient, 7.

(5) IgE (γE). Recently discovered. Present in minute amounts in normal serum (under 1 mg per 100 ml) but it may be very important functionally (see below). Isolated too recently for definitive chemical data to be available at the time of writing.

The concentrations of immunoglobulins given above refer to serum but immunoglobulins are present in other body fluids as well. IgG distributes itself between the intravascular and extravascular compartments of the body fluids according to the protein concentration of those fluids but, under normal circumstances, IgM is mainly intravascular and does not cross the capillary wall. When the external secretions of the body are considered, a rather different situation is present. Very little IgG or IgM is present there. On the other hand, high concentrations of IgA are present in secretions such as milk and colostrum, saliva, lacrimal and bronchial secretions and gastric and intestinal juices. The evidence suggests that IgA is not merely filtered, but actively secreted into these juices possibly by plasma cells or epithelial cells in the lining walls. IgA in secretions is of a higher molecular weight than serum IgA as it has an additional "transport piece" attached to it.

The placental transfer of immunoglobulins varies with the species and the type of placenta which it possesses but it has already been mentioned that human foetal serum contains maternally-transmitted antibody. This antibody is virtually entirely of the IgG type, little or no IgM or IgA being present in cord blood. Again IgG is actively transmitted rather than passively filtered across the placental barrier.

THE STRUCTURE OF IMMUNOGLOBULINS

Immunoglobulins, like other proteins, are built up of polypeptide chains of amino-acids linked together by peptide bonds.

The present knowledge of the structure of immunoglobulins has resulted from chemical methods which were devised to split the molecule in various ways and then to examine the chemical make-up and biological behaviour of the fragments thus obtained. Two such fractionation methods yielded the most interesting information. These were (1) splitting the molecule across certain peptide linkages using the proteolytic enzyme papain, and (2) chemical reduction to separate linked polypeptide chains from one another. Porter, using papain hydrolysis, was able to split immunoglobulin molecules into three large fragments. After these fragments were separated from one another, he found that two of the fragments still retained antibody activity. These are named the Fab fragments (ab for antibody). The third fragment did not act as

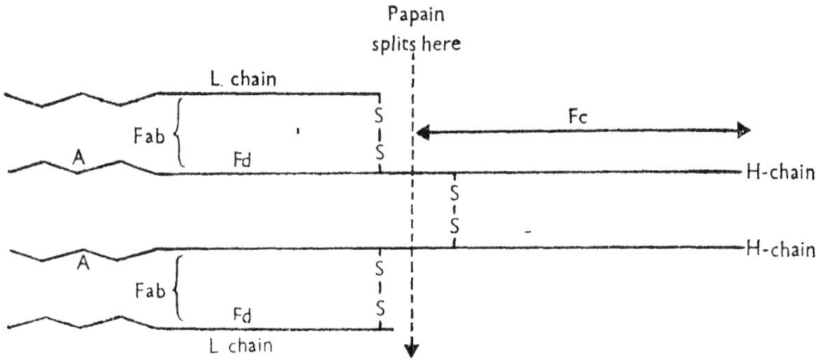

Fig. 1.X

antibody but was still active as antigen. In other words, if this fragment was injected into a rabbit, the rabbit made an antibody against it. This fragment had a tendency to crystallize, so it was called the Fc fragment.

The polypeptide chains of immunoglobulins are linked to one another by disulphide bonds. Using mercaptoethanol, these —S—S— bonds could be reduced to –SH groups and the chains thus separated from one another. It was thus found that the IgG molecule is made up of four polypeptide chains. Two of these are similar to each other and each has a molecular weight of 20,000. These are known as the light (L-) chains. The other two chains are again similar to each other with a molecular weight of 50,000 each. These are known as the heavy (H-) chains. By relating the properties of fragments obtained by enzyme hydrolysis and the chains obtained by reduction, Porter devised the schematic model for the structure of an immunoglobulin molecule shown in Fig. 1.X.

A further refinement of these procedures has been to split the Fab fragment further into two pieces by reduction. This yields one light chain and half a heavy chain (Fig. 1.X). The half-heavy chain thus obtained is known as the Fd fragment. Thus the heavy chain has an Fd half and an Fc half. The light chains are entirely contained in the Fab fragment.

BIOLOGICAL PROPERTIES OF IMMUNOGLOBULIN FRAGMENTS

The *antibody combining site* (i.e. the part of the molecule which combines with antigen) is contained in the Fab fragments (A, Fig. 1.X). The whole IgG molecule has two combining sites and each Fab fragment one combining site. The IgG molecule is therefore *divalent* as it can combine with two antigen molecules and each Fab fragment is *univalent* combining with only one antigen molecule. The heavy chain (Fd fragment) certainly has antibody activity. It is less likely that the light chains have antibody activity. The antibody combining site is probably at the N-terminal end (i.e. the left end of Fig. 1.X) of the Fd fragment (and possibly of the L-chain). This end of the chain has a variable amino-acid sequence so that serum IgG consists of many molecules which vary at this site (represented by irregular line in Fig. 1.X). The other half of the Fd fragment and of the light chain is constant in composition. Variable amino-acid composition at the N-terminal site would, of course, make sense inasmuch as serum γ-globulin consists of many antibodies, each of which must be different at its combining site in order to combine with its specific antigen.

Immunoglobulins, like most other proteins, are themselves antigenic when injected into a foreign species. The light chains have their own antigenic determinants and heavy chains also have their own quite different determinants. If these chains are injected into rabbits, the rabbits will then make anti-human light chain antibody and anti-human heavy chain antibody. Using this method, it is found that if *light* chains from the different human immunoglobulins (IgG, IgA, etc.) are injected into different rabbits, the rabbits make the same antibody against all of them. In other words, the light chains of all immunoglobulins are the same. The family likeness of the immunoglobulins resides in their sharing the same light chains. On the other hand, different rabbits injected with *heavy* chains from different human immunoglobulins make quite different antibodies against each. Each immunoglobulin therefore has its own unique heavy chain which gives it its individuality by which it is characterized as IgG or IgM and so on. The heavy chains are known by Greek letters corresponding to the roman capital characterizing the immunoglobulin. Thus IgG has γ-heavy chains, IgA, α-chains and IgM, μ-chains.

The light chains of all immunoglobulins are common. There are, however, two different types of light chain called kappa (κ) and lambda (λ). A single immunoglobulin molecule has either κ or λ light chains, never both, but serum IgG always consists of a mixture of molecules some with κ and some with λ light chains. Serum IgM and the other immunoglobulins consist of a similar mixture. Most immunoglobulins have molecular weights of about 150,000. IgM or γ-macroglobulin, however, is much bigger with a molecular weight of 900,000. It shares the basic heavy and light chain structure with the other immunoglobulins but is, in fact, a pentamer made up of five of the four-chain units linked in the shape of a five-pointed star by disulphide bonds. It therefore has a valency of 10 (two combining sites for each four-chain unit).

From the foregoing description, it can be seen that the immunoglobulins represent a very heterogeneous population of molecules. Still more is this so when it is considered that sub-classes of each immunoglobulin class are being discovered. This is so for two reasons, (1) to allow for a large number of different specific antibodies capable of reacting with all possible antigens, and (2) because the different immunoglobulins have specialized functions localized at sites other than the antibody combining site. These functions are described in further detail below.

BIOLOGICAL PROPERTIES OF IMMUNOGLOBULINS

All antibodies against a single determinant on an antigen such as diphtheria toxoid are identical or nearly so at their combining sites. All are therefore capable of reacting equally well with diphtheria toxoid. Nevertheless, anti-diphtheria antibodies of IgG, IgA and IgM types do differ from each other quite markedly at other sites of the molecule. The way they handle the antigen depends very much on these chemical differences. Some antibodies react with antigen to fix complement, others may become fixed to tissue cells, others cross membranes to appear in biological fluids other than serum. Different immunoglobulins differ from each other in these biological properties as outlined below. Such differences are not of purely academic interest. One type of antigen–antibody reaction may result in classical immunity. Another may result in cell damage and a harmful hypersensitivity reaction.

IgG is the major immunoglobulin of serum and after most natural infections with bacteria or viruses, or after artificial immunization, the majority of the serum antibody produced is of the IgG type. It was mentioned earlier that IgG alone is actively transported across the placenta and so maternal protective antibody of this type appears in foetal serum. This is because, on the Fc portion of the IgG molecule, there is a "transport site" which allows placental transfer and which is not present on immunoglobulin molecules other than IgG. There is another active site on the Fc portion of the IgG molecule which enables it to fix complement. Complement consists of a complex series of serum proteins whose participation is essential for certain antigen–antibody reactions to be carried to completion. These reactions include the lysis of erythrocytes or bacteria by antibody. In order for this to happen, two adjacent molecules of IgG must become attached to adjacent antigenic determinants on the cell surface. Complement becomes fixed to the Fc fragment sites on the two adjacent antibody molecules. A complex series of events (described later) then takes place with the end result that a small hole is punched in the cell which then bursts due to loss of its osmotic properties. IgG molecules are also ideal precipitating antibodies. Each of their two combining sites becomes attached to a different antigen molecule. The antigen molecule becomes attached by further determinant sites to other antibody molecules and an antigen–antibody "lattice" is formed which, being a macromolecular complex, is insoluble and forms a precipitate. When not reacting with antigen, the antibody molecule has the rough shape shown in Fig. 1.X. When it reacts optimally with antigen, the Fab fragments spring open upon the —S—S— bond joining the two heavy chains so that the molecule has the shape of a Y, the two top branches being the Fab fragments, with their combining sites at the top, and the stem, the Fc fragment. It is then in the optimal position for maximum antigen–antibody combination. IgG antibodies probably also possess, on their Fc portions, combining sites which allow them to become attached to tissue cells. They can be shown to become attached in this way to the skin cells of the guinea pig and to the mast cells of the mouse. In this way, antigen–antibody reactions may occur at the surfaces of cells, a process which may damage the cell or bring into play the various physiological activities of the cell as a result.

IgM. There have been many reports that the earliest antibodies produced in the primary response are IgM and that IgG antibodies are produced later. After two or three weeks the predominant antibody against most antigens is of the IgG type. However, against certain antigens, IgM antibody remains predominant. Naturally occurring antibodies against red cells such as the ABO isohaemagglutinins and the Forssman antibodies are almost entirely IgM. Other strong red cell agglutinins such as the

Paul-Bunnell antibody in glandular fever and saline anti-D in maternal-foetal Rhesus incompatibility are also IgM. Strong bacterial agglutinins such as, for instance, the antibodies to the O-antigens of Gram-negative bacteria are also of the IgM type. IgM antibodies are much more efficient at agglutinating large particles such as bacteria and red cells than IgG. This is because they have 10 combining sites which are able to combine like cogs on a wheel with multiple determinants on the surface of adjacent erythrocytes or bacteria and thus bind the particles very tightly together so that a clump of agglutinated cells is formed. IgM antibodies are also more efficient in promoting lysis of red cells and bacteria than IgG. Whereas two IgG molecules are required to fix complement and punch a hole in a cell, only one molecule of IgM is sufficient to do the same job. IgM antibodies can therefore be seen to be extremely efficient in dealing with large particulate antigens. Against smaller particles such as viruses or against antigens in solution, such as bacterial toxin, IgG is probably the more efficient immunoglobulin.

IgA antibodies are *par excellence* the antibodies of external secretions. Many serum antibodies may be of the IgA type, e.g. anti-brucella, anti-diphtheria toxoid, but IgA antibodies against these antigens are never predominant in serum. For example, larger amounts of IgM or IgG anti-brucella than IgA are almost always found. In gastric juice, saliva, tears and other secretions, however, IgA is present in large quantities. IgA antibody against *Vibrio cholerae* is secreted into the gut in large amounts in cases of cholera. IgA antibody in the oral and nasal cavities may also be of importance in preventing recurrent herpes simplex infections. This immunoglobulin is probably, therefore, an important first line of defence in the secretions in respiratory and gastro-intestinal disorders. It is also present in colostrum and natural IgA may therefore be of importance in preventing gastro-intestinal infections in the new-born child exposed to a new and potentially hostile bacterial environment. IgA does not fix complement.

IgE antibodies have only recently been discovered. They are present in minute amounts in serum but there is good evidence that they are the so-called "reaginic" antibodies which react with pollens and other antigens in asthma and hay fever. These antibodies become attached to skin cells and, presumably, the cells of the upper respiratory mucosa or sub-mucosa. They react with antigen on the surface of these cells, cause cell damage and subsequent histamine release. This results in the lesions of urticaria in the skin and those of asthma and hay fever in the respiratory tract (see Immediate Hypersensitivity).

IgD. Antibody activity has not yet been discovered in the IgD fraction, although it is reasonable to presume that IgD must eventually be found to possess such activity.

IMMUNOGLOBULIN ABNORMALITIES

The level of serum immunoglobulins rises after any long continued antigenic stimulation. Thus high levels of γ-globulin are found in long-standing bacterial diseases such as chronic abscesses, syphilis or sub-acute bacterial endocarditis. Similarly in many diseases in which auto-immunity may play a part or in which protein breakdown by the liver is impaired, γ-globulin levels may be high. Thus, rheumatoid arthritis, systemic lupus erythematosus, auto-immune haemolytic anaemia, Hashimoto's thyroiditis and hepatic cirrhosis are all diseases in which high γ-globulin levels are found. Populations from areas in which malaria and hookworm are endemic have higher γ-globulin levels than European populations. Hypergammaglobulinaemia is less common in acute bacterial or viral diseases in which the levels of antibody produced over a short period are not usually sufficient to affect grossly the total serum γ-globulin level.

Paraproteinaemias: myeloma. There is a group of diseases known collectively as paraproteinaemias which are usually regarded as neoplasms of the antibody-forming tissues. This group includes *multiple myelomatosis* and *Waldenström's macroglobulinaemia* and patients with reticuloses such as *lymphosarcoma* or *chronic lymphatic leukaemia* may show similar abnormalities. Essentially, in these disorders, a clone of antibody-forming cells becomes neoplastic and multiplies aberrently eventually replacing the normal antibody-forming tissue. This neoplastic clone of cells produces large amounts of γ-globulin and the serum γ-globulin levels in these diseases are usually very high. However, this γ-globulin, derived as it is from a single clone of cells is not heterogeneous like the normal immunoglobulin population, but shows a remarkable homogeneity, being always of the same immunoglobulin type in any particular patient and always being of either the K or λ light-chain type, never both.

Myelomatosis occurs in the older age groups and is characterized by invasion of the bone marrow with abnormal plasma cells. Bone marrow replacement leads to anaemia and to cystic changes in the bones which may spontaneously fracture and there is usually a grossly elevated erythrocyte sedimentation rate and serum γ-globulin level. The disease is fatal within a few years. Myeloma globulins are electrophoretically homogeneous forming a single dense band in contrast to the diffuse γ-globulin band of normal serum. Myeloma globulins are always either IgG or IgA in type, never both. IgD and IgE myelomas are extremely rare. There is over-production of light chains in the bone marrow and these light chains, unattached to heavy chains, are of low enough molecular weight to be excreted free in the urine. About 40 per cent of myeloma patients therefore excrete free light chains, otherwise known as Bence–Jones protein, in their urine. Replacement of normal antibody-forming tissue may render patients with myelomatosis and other paraproteinaemias increasingly susceptible to intercurrent bacterial infection.

Waldenström's macroglobulinaemia is the IgM analogue of myeloma. It occurs in elderly men and there is infiltration not only of the bone marrow, but of the lymph nodes and spleen with large basophil cells which produce abnormally large quantities of electrophoretically homogeneous IgM. The same systemic changes are seen as in myeloma except that there is no rarefaction of bone and there may be splenomegaly and enlargement of lymph glands. The disease is milder than myeloma and may persist for many years. There is no Bence–Jones protein in the urine. Macroglobulins may precipitate in the cold thus causing peripheral vascular abnormalities.

Secondary paraproteinaemias involving any one of the three major immunoglobulins may occur in reticuloses, lymphosarcoma or reticulosarcoma or in chronic lymphatic leukaemia. They may also be seen rarely in cases of primary carcinoma, particularly in the alimentary tract and, occasionally, in otherwise healthy elderly patients. These latter have a perfectly good prognosis.

Paraproteins occasionally have activity as antibodies. More often than not, however, no antibody activity can be ascribed to them. The diagnosis of paraproteinaemia can be made by the striking clinical and serological abnormalities outlined above together with the search for increased numbers of neoplastic plasma cells by sternal marrow puncture. The paraproteinaemias may be classified by immunoelectrophoretic analysis of the serum immunoglobulins (*vide infra*) which gives a positive indication of the particular immunoglobulin type affected in any given patient.

HYPOGAMMAGLOBULINAEMIA: ANTIBODY DEFICIENCY SYNDROME

Hypogammaglobulinaemia is physiological during the first months of life after which the γ-globulin levels of a normal baby should rise rapidly. Low serum levels of γ-globulin after this time may be due to a genetically determined primary agenesis of antibody-producing cells or may be secondary to other diseases which depress antibody production or which cause destruction or replacement of lymphoid tissue.

Primary hypogammaglobulinaemia may be (1) due to agenesis of all immunologically competent tissues. This severe form (Swiss type) is familial but not sex linked and is fatal in neonatal life. Such infants have thymic atrophy and produce neither lymphocytes nor plasma cells and are not able to mount either cell-mediated immune responses (lymphocytes) nor to manufacture antibody (plasma cells). They are therefore very defenceless and invariably succumb in early life to intercurrent infections. (2) There may be agenesis of plasma cells but perfectly normal lymphocyte function. This is a sex-linked form of hypogammaglobulinaemia occurring in male children with a family history of similar disturbances. Serum levels of all immunoglobulins are low in some of these children who, therefore, have virtually no γ-globulin in the serum. Few plasma cells are seen in the lymphoid tissues but normal numbers of lymphocytes are present. These children are very prone to bacterial infections but show normal cell-mediated immunity. They take smallpox vaccinations normally and show resistance to many viral infections and they give positive tuberculin reactions (delayed hypersensitivity). Once diagnosed, they are maintained on regular doses of parenterally administered human γ-globulin. A milder familial form of hypogammaglobulinaemia occurs in which deficiencies of only one immunoglobulin are observed. Depending on the severity of the defect, these patients may be detected in childhood or may grow to adult life without symptoms. There may be compensatory increases in the level of other immunoglobulins. The families of such patients usually show similar abnormalities. (3) Di George's syndrome is not strictly an antibody deficiency syndrome but should be grouped with them. These babies have normal immunoglobulin levels but have aplastic thymuses and deficiencies of cell-mediated immunity with normal humoral immunity. They are very susceptible to virus infections. They may, for instance, die of generalized vaccinia following smallpox vaccination.

Secondary hypogammaglobulinaemia. Low serum γ-globulin levels may be found in any disease associated with gross protein loss such as chronic nephritis. Diseases in which antibody-forming tissues are infiltrated by neoplastic cells may also result in low serum γ-globulin levels. This may particularly occur in reticuloses, leukaemias and carcinomatosis. Although in myelomatosis serum γ-globulin levels are high, there is an antibody deficiency syndrome, as the normal antibody is replaced by globulin formed by the neoplastic plasma cells. Patients with all these disorders have an increased susceptibility to bacterial infection.

ANTIGEN-ANTIBODY REACTIONS

Some of the ways in which antigens and antibodies react together, e.g. precipitation, agglutination and complement fixation have already been mentioned. These reactions can very often be reproduced in the test-tube and so can be of value in the diagnosis and management of bacterial and viral diseases as well as other diseases in which immunological mechanisms play a part. Diagnosis of disease by these means is usually called *serological* diagnosis as it characteristically involves taking serum from a patient and allowing the serum antibody to react in a test-tube with its appropriate antigen.

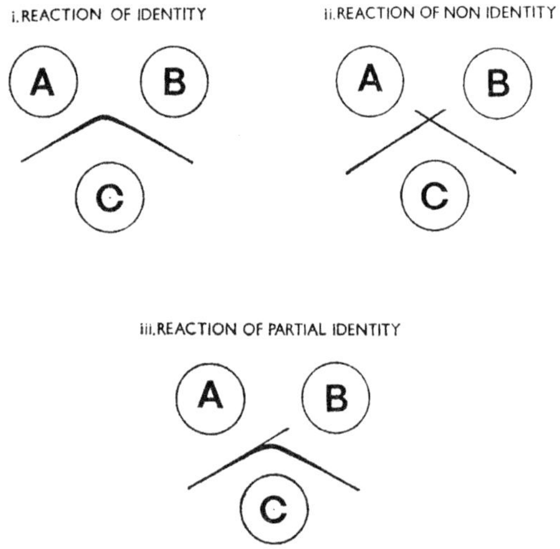

Fig. 2.X. Gel Diffusion Tests.
A and B are solutions of antigens; C is an antibody against them.

Serological tests may be qualitative. Antibody is incubated with antigen and an effect is observed but not measured. They are more usually quantitative. It is often of importance in diagnosis and management to have an idea how much antibody is present in a patient's serum. To do this, a row of test-tubes is set up. Standard amounts of antigen are added to each tube. Serum is put into the first tube at a dilution of, say, 1/5. In the next tube the serum is diluted 1/10 and in the following tubes, 1/20, 1/40, 1/80 and so on. The antibody level in each tube is therefore diminishing by *doubling dilutions* down the row. A strongly reactive serum may still be reactive with antigen as judged by the presence of precipitation, agglutination, etc., depending on which test is being used, at a dilution of, say, 1/640, whereas a weaker serum is reactive only at 1/20. The *titre* of serum antibody is the reciprocal of the final dilution at which it still reacts with antigen. Thus a serum which reacts at 1/80 but not at 1/160 has a *titre* of 80. Changes in *titre* of sera taken from a patient over a period of time are often valuable indicators of the progress of the patient.

Precipitation. Precipitation tests can only be used to detect reactions between antibodies and *soluble* antigens. In the presence of antigen, divalent antibody molecules open out so that each combining site reacts with a separate antigen molecule. Most antigens are *polyvalent,* having a variable but quite large number of determinant sites. Each antigen molecule is therefore capable of reacting with several different antibody molecules. An insoluble macromolecular lattice is thus formed of antigen and antibody molecules firmly bound together. This insoluble complex forms a visible precipitate usually within a few minutes after mixing. The importance of the valency of antigen and antibody molecules for maximal precipitation is shown in the following example. If an antibody has a valency of two and its corresponding antigen has a valency of 10, then maximal precipitation will occur when a mixture of 10 antibody molecules is present to every two antigen molecules. At these proportions, all the combining sites of the antibody molecules and all the determinant sites of the antigen molecules will react with one another. In other words, the antigen and antibody are combining at *optimal proportions.* If the number of antigen molecules is decreased, antibody molecules will be present which have free combining sites unable to react with antigen. There is, therefore, *antibody excess.* If the number of antigen molecules is increased over optimal proportions, free antigenic determinants are present. There is, therefore, *antigen excess.* In antigen excess, antibody reacts with antigen to form *soluble* complexes rather than a precipitate. The principle of optimal proportions is used for the estimation of antibody in serum in absolute terms. If constant volumes of antibody-containing serum are incubated in a series of tubes each with increasing doses of antigen, little precipitation will occur in the tubes containing little antigen. As the region of optimal proportions is approached, the amount of precipitation increases. Where there is antigen excess, the precipitate again decreases. In the tube with maximal precipitation, at optimal proportions, all the antibody is precipitated. If the amount of protein in this precipitate is measured and the known amount of antigen subtracted, the resultant figure is an absolute value for antibody in the test serum in terms of milligrams of protein.

Antigen–antibody reactions are chemically firm but are reversible. Antigen can be dissociated from antibody more or less easily depending on the firmness of union. Dissociation is accelerated if the temperature is raised. For antigen–antibody precipitates to form, the presence of electrolytes in the solution is also essential.

Precipitin tests may be done in test-tubes in saline solutions. (Precipitin is the name sometimes used for precipitating antibody.) Examples of such tests are the titration of antibody and the determination of optimal proportions described above. Another common form of precipitin test is the *ring* test. This is used for streptococcal grouping. An appropriate antiserum is layered into the bottom of a tube. An extract of the test streptococcus is then layered over it without mixing. If the antiserum reacts with the streptococcal antigen, a ring of precipitate appears at the interface between the two layers. As antisera specific for each streptococcal group can be prepared, the serological group of any given streptococcus can be determined by this method.

Precipitin tests may also be performed in agar gels. Purified agar is layered into a Petri dish and two circular holes are punched into the agar at appropriate distances from one another. One hole is filled with antibody, the other with antigen. The antibody and antigen will diffuse radially from the holes through the agar. In the agar between the holes, antigen will eventually meet antibody. At this spot a line of precipitate will form. If the antigen solution is a mixture of two substances and the antibody is able to react with both antigens, then two parallel lines will be seen. This is called the Ouchterlony technique and can be used for analysis of antigens and as a check on their purity. This technique can also be used to see if two unknown solutions of antigen are identical with each other as follows. Three holes are punched in the agar at the angles of an equilateral triangle. Holes A and B are filled with the unknown antigens, hole C is filled with antibody (Fig. 3.X). If the precipitin line of A and B against C is continuous, then antigens A and B are identical. However, if A gives a separate line from B and the two lines cross, antigens A and B are not identical. There is a third possibility. A may have one determinant in common with B but another determinant that B does not share. Then A will give a precipitin line continuous with that of B because of the shared determinant but with a spur coming off it representing the unshared determinant. These three reactions are known respectively as the reaction of *identity*, the reaction of *non-identity* and the reaction of *partial identity*.

A further refinement of the agar gel precipitation technique is the technique of *immunoelectrophoresis*. By this method, serum

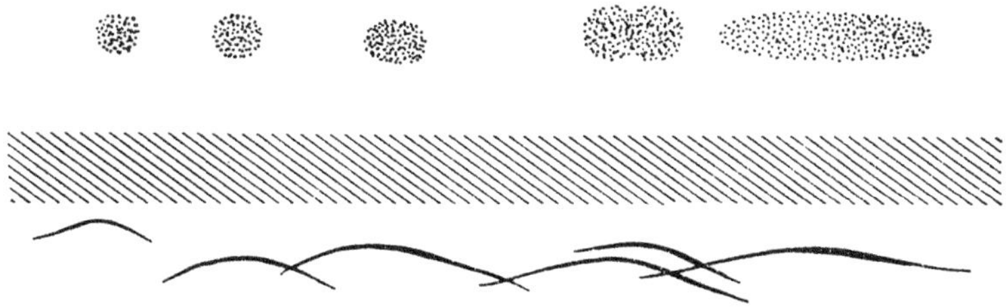

Fig. 3.X. Immunoelectrophoresis of antigen in agar gel. Upper row: Spots of protein separated from complex mixture by electrophoresis. Central row: Trough containing polyvalent antiserum. Lower row: Arcs of precipitate formed by antiserum diffusing from trough meeting parts of antigen complex separated by electrophoresis.

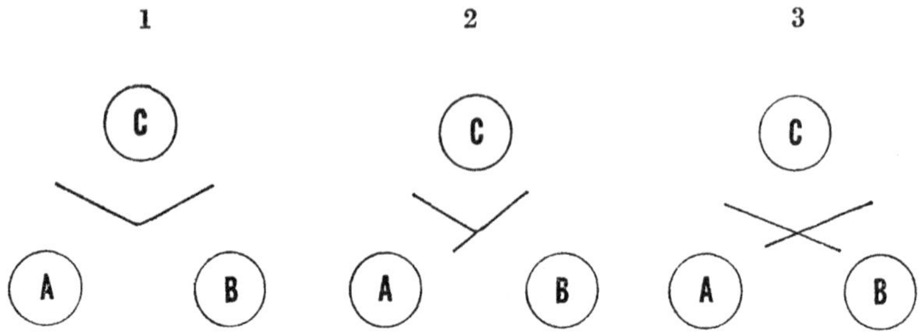

Fig. 3.X.1 Reaction of Identity. A & B = unknown antigens. C = antiserum. Continuous Line of Precipitate.
2 Reaction of Partial Identity. A & B = unknown antigens. C = antiserum. Continuous Line of Precipitate with Spur showing a separate antigen in B.
3 Reaction of Non-Identity. A & B = unknown antigens. C = antiserum. Lines of Precipitation cross showing Non-Identity.

proteins or other antigenic mixtures are first electrophoresed in the agar gel. After the current is stopped when the proteins have migrated through the gel, a trough is cut in the gel close to and parallel to the plane of migration. This trough is filled with antibody and the antibody and the electrophoresed antigens are allowed to diffuse towards one another through the gel. A series of precipitin arcs will form corresponding to the various antigens in the mixture (Fig. 3.X). By this means, detailed information can be obtained about very complex mixtures of proteins, not only about their antigenic behaviour but about their electrophoretic mobility. The method is very suitable for analysis of mixtures of proteins such as serum or other body fluids.

Agglutination. Agglutination tests detect antibodies which react with *particulate* antigens. These may be the surface antigens of bacteria or red cells. Soluble protein antigens can also be made to stick to the surface of artificial particles of materials such as polystyrene latex. They then behave like particulate antigens and can be detected by agglutination tests. Other soluble antigens can be attached to the surface of red cells. This is made easier if the red cells are first treated with tannic acid. This is the basis of the *tanned cell haemagglutination test* which can be used as a very sensitive detector of many protein antigens. In agglutination tests, antibody (agglutinin) binds to antigenic determinants on contiguous particles or cells. It thus holds the particles tightly together causing them to agglutinate. Visible clumps of bacteria or red cells are then seen at the bottom of the tube instead of the even suspension where no antibody is present. It was mentioned earlier that IgM antibodies are better at doing this, because of the multiplicity of their combining sites. It is also probable that some antigenic determinants are recessed deep in the red cell or bacterial surface. Where this is so, a big antibody molecule has a better chance of reaching down to the determinant than a smaller one.

The clinical uses of agglutination tests are legion. Some of the more important ones are enumerated below.

Against bacterial antigens: The Widal test for antibody against the somatic (O) and flagellar (H) antigens of *Salmonella typhi* and *Salm. paratyphi* A and B in enteric fevers; similar tests for the serological grouping of Salmonella species in food poisoning or of Shigella species in bacillary dysentery; tests for antibodies against *Brucella abortus* and *melitensis* in undulant fever and Malta fever; serological grouping of coliform organisms (especially useful in infantile gastroenteritis).

Against erythrocytes: Blood grouping (ABO and Rhesus groups and other rarer groups) and the cross-matching of blood for transfusion. Investigation of auto-immune haemolytic anaemias in which patients develop antibodies against their own red cells. Investigation of maternal–foetal Rhesus incompatibility. The Paul–Bunnell test for antibody against sheep erythrocytes found in infectious mononucleosis (glandular fever) and in serum sickness.

ABO blood grouping depends on the presence of "natural" agglutinins in normal human sera. Normal human erythrocytes are of either group A (43 per cent of a British population), i.e. they have the A antigen on their surface; group B (8 per cent) with the B antigen, group AB (3 per cent) with both antigens, or group O (46 per cent) with neither A nor B antigens on their surface. Sera of normal individuals carry antibodies against the ABO antigens which are *not* present on their cells, i.e. serum of a group A individual contains anti-B but *not* anti-A; serum from group B, anti-A but *not* anti-B. Serum from someone of group O will therefore contain both anti-A and anti-B but serum from someone of group AB will contain neither antibody. By incubating sera with appropriate cells or vice versa and observing agglutination, the blood group of any given individual may be determined. In the case of Rhesus (D-) groups, natural antibodies are not present in serum, but a Rhesus negative mother may make anti-D against the erythrocytes of her Rhesus (D-) positive child.

In glandular fever, patients produce high levels of antibody active against the erythrocytes of the sheep and other species. In serum sickness, the normally low level of Forssman antibody—also active against sheep cells—is grossly elevated. The two diseases both show high levels of sheep cell agglutinins but may be distinguished in the Paul–Bunnell test because glandular fever antibody is removed by pre-treatment of serum with ox cells but not with guinea pig kidney, whereas Forssman antibody is removed by pre-treatment with guinea pig kidney but not with ox cells.

Agglutination tests based on the attachment of soluble antigens to a particle followed by the agglutination of that coated particle by antibody are also commonly used in clinical practice. In rheumatoid arthritis, tests are used in which human γ-globulin is coated either on to sheep cells (Rose–Waaler test) or on to latex particles. Rheumatoid factor, present in sera from cases of rheumatoid arthritis is an antibody against human γ-globulin and can be revealed by the agglutination of such coated particles because the rheumatoid factor reacts with the γ-globulin attached to them. Among other uses, the tanned cell haemagglutination test is used for titration of antibodies to a variety of protein antigens such as diphtheria toxoid and for detection of antibodies to thyroglobulin in auto-immune thyroiditis.

"Incomplete" antibodies. Many antibodies combine with the antigenic determinants on the surface of a single particle such as a bacterium or red cell, but are not able at the same time to react with the determinants on a contiguous cell. They therefore coat the cells but do not agglutinate them. 7S antibodies are particularly likely to do this. They are called "incomplete" or "blocking" antibodies. As no visible reaction occurs when such antibodies coat a cell, special methods have to be used for their detection. The methods used are as follows:

(1) To treat the cell surface with an enzyme such as papain. This strips off the outer coat of the cell and reveals deeply recessed antigenic determinants. Small antibodies may thus be able to "bridge the gap" between two adjacent cells. This method is very useful in blood transfusion serology.

(2) To treat the cell with antiglobulin. To cells coated with human globulin antibody is added an anti-human globulin serum prepared in a rabbit. The anti-globulin reacts with the globulin molecules coating the cells. The anti-globulin molecule itself is thus able to act as a bridge between the globulin molecules attached to the surfaces of the two adjacent cells. The cells then agglutinate. This test is known as the Coombs' test and is of great value in transfusion serology, in the diagnosis of Rhesus incompatibility and haemolytic anaemia and is now being successfully applied to the detection of incomplete antibodies in brucellosis.

(3) By observation of the inhibiting effect of "incomplete" antibodies on the agglutinating ability of "complete" antibodies. If cells are first coated with "incomplete" antibody and then exposed to agglutinating antibody, all the determinant sites will have been occupied by incomplete antibody and so the agglutinating antibody will be unable to react. Incomplete antibodies are sometimes known as blocking antibodies because they are able to "block" agglutination. This happens normally in long-standing cases of brucellosis whose sera contain a mixture of blocking and agglutinating antibody. In the first few tubes of an agglutination titration, the blocking antibody is so strong as to prevent agglutination. As it becomes diluted out, agglutination begins to appear. This phenomenon, where there is no agglutination in the tubes containing the highest level of antibody is known as a "prozone".

Agglutination inhibition is used in investigations of antibody production against viruses of the myxovirus group. These viruses have the property of adsorbing to erythrocytes and agglutinating them. This agglutination is *not* immunological as it is a property of the viruses themselves. Serum antiviral antibody from patients suffering from influenza, mumps, etc., will prevent the viruses from becoming attached to erythrocytes and can be measured by its ability to inhibit the agglutination caused by its specific virus.

Complement fixation. The next two types of serological test to be described are lysin and complement fixation tests. As both depend on the reaction of antibody with complement, it is now necessary to discuss the nature of complement.

Bordet discovered in 1898 that antibodies from fresh rabbit sera, which were able to lyse sheep cells, no longer did so if the serum were heated to 56°C or if it were allowed to age. The antibody could still combine with the cells but not lyse them. If fresh normal serum was added to the aged or heated sera, the lytic ability of the antibodies was restored. A heat-labile factor was therefore present in serum which could not lyse cells by itself but which was essential in order for antibody to do so. This heat-labile activity of normal serum is known as complement and much biochemical research has gone into investigating its nature. It is now known to consist of at least nine serum globulins which form a complex enzyme system which reacts in a clearly defined sequence. So far, inhibitor enzymes have also been discovered for at least three of the components of complement. Complement is known by the symbol C' and its various constituents are known as C'_1, C'_2, and so on. This enzyme system is activated when antibody reacts with antigen. One of its end results is to cause lysis of bacteria or red cells but it has other important biological properties. Let us say that an antibody reacts with an antigen on the surface of an erythrocyte. In the presence of Ca^{++} ions, the first component of complement will be activated and become attached to the Fc fragment of the antibody. The resulting complex may be represented thus; EAC'_1 (E = erythrocyte, A = antibody and C'_1 = the first component of complement). The reaction proceeds further in the presence of Mg^{++} ions, so that two more complement components are attached, thus EAC'_{142}. If another component, C'_3, now becomes attached (EAC'_{1423}) we have a biologically active complex which is capable of enhancing phagocytosis of red cells and increasing vascular permeability by activating a substance called *anaphylatoxin* which induces histamine release. Add three more components ($EAC'_{1423567}$) and we have another complex capable of activating a serum factor which is chemotactic for granulocytes. Two more components complete the story ($EAC'_{142356789}$). This impressive chain reaction now at last ends in the lysis of the erythrocyte. All nine components are essential for cell lysis. It can be seen that complement activity is important, not only for lysis of red cells and bacteria but in activating several mechanisms of basic importance in the inflammatory response.

DISORDERS OF COMPLEMENT

Serum complement levels may be low in any disease in which complement is being used up in large amounts by antigen–antibody reactions. In addition, in the disease *hereditary angioneurotic oedema*, there is a specific genetic enzyme deficiency so that the inhibitor enzyme for C_1' is missing. These patients also have a deficiency of an inhibitor of kinin activation, possibly the same enzyme. They are liable to recurrent attacks during which vascular permeability is increased. During the attacks, they suffer from non-irritant oedema which is harmless when it is cutaneous but which may occur in the larynx causing dangerous respiratory obstruction.

BACTERIOLYSIS AND HAEMOLYSIS

The mechanism of lysis of red cells and bacteria has been outlined above. Tests for lytic antibodies can be carried out in the same way as agglutination tests except that fresh normal serum *must* be added as a source of complement.

Complement fixation. Certain antigen–antibody reactions occur without causing noticeable precipitation or agglutination. These reactions can very conveniently be detected if they involve the uptake or *fixation* of complement. The best-known clinical uses of the complement fixation test are the Wasserman reaction for the diagnosis of syphilis and the gonococcal complement fixation test for the diagnosis of gonorrhoea. Antibodies against many viruses and auto-antibodies against tissues such as thyroid and gastric mucosa can also be detected by complement fixation.

The gonococcal complement fixation test can be taken as an example. Serum from a suspected case of gonorrhoea is taken and any complement in it is *inactivated* by heating at 56°C. This is to ensure that complement from the patient's own serum does not interfere with the test. The patient's serum is then incubated with an antigenic extract made from whole gonococci. To this mixture is added a measured amount of fresh guinea pig serum as a source of complement. The mixture is allowed to incubate. If the patient's serum contains antibody, it will react with the antigen and fix the guinea pig complement. If the patient's serum contains no antibody, the guinea pig complement will remain free. However, complement fixation does not cause any visible change in the test tube. An indicator has next to be added to detect the presence or absence of free complement. This indicator consists of sheep erythrocytes which have been coated with a haemolytic anti-sheep cell antibody made in the rabbit. When these coated sheep cells are added, if the patient's serum has already reacted with the gonococcal antigen and fixed complement, there will be no complement left to lyse the coated sheep cells. If the patient's serum contained no anti-gonococcal antibody, however, complement will still be free. This complement will be available to lyse the sheep red cells. So lysis of the sheep red cells indicates a negative test, no lysis indicates a positive test. This is shown diagrammatically in Fig. 4.X.

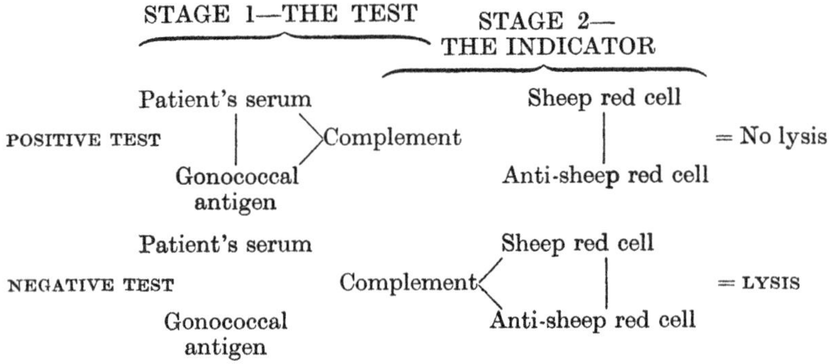

Fig. 4.X. The complement fixation test.

Neutralization tests. (1) Bacterial toxins. The ultimate test of the strength of a bacterial toxin is its lethal effect for an animal *in vivo*. Antitoxins may be measured by their effects in protecting animals against the lethal effects of the toxin. Alternatively, if toxins are injected intradermally, they give rise to a local inflammatory lesion. Addition of antitoxin will again protect against such local lesions.

(2) Virus neutralization. The growth of viruses may be demonstrated in various ways including their cytopathogenic effect in tissue culture or, in the case of bacteriophage, their lytic effect on cultures of their target organism. By the addition of anti-viral antibody to such systems, the viruses are neutralized and are unable to exert the expected effect on their target tissue. Thus no cytopathogenic effect is seen in the presence of high levels of antibody, but when the antibody is diluted out, it is again seen. Such tests can therefore be carried out quantitatively by the addition of progressive dilutions of the anti-viral serum to the virus under test.

Opsonization. *Opsonins* are antibodies which become attached to bacteria in such a way that the coated bacteria become increasingly susceptible to phagocytosis. While such a mechanism is obviously of considerable importance in the defence reaction against infection, the *in vitro* estimation of opsonizing antibody is difficult and has, at present, little place in the laboratory diagnosis of bacterial diseases.

Detection of labelled antibodies. (1) Immunofluorescence techniques. Proteins such as γ-globulin or many other antigenic proteins can be combined chemically with a group of dyes known as fluorochromes. Such a combination does not interfere with the immunological properties of the protein. Thus antigens in tissues can be localized by their reaction with fluorochrome-labelled antibody. A section of tissue containing an antigen such as, for example, thyroglobulin is prepared and is then coated with a solution containing a fluorochrome-labelled antibody against thyroglobulin. The antibody will combine with thyroglobulin wherever it is present in the section whether in cells or tissue spaces. The section is then washed to free it of uncombined protein and is examined in a microscope in a dark field. Wherever antibody has combined with antigen, bright fluorescence will be seen against the dark background. The most popular fluorochromes for attachment to antibody are fluorescein and rhodamine. Antibodies conjugated with fluorescein emit an apple-green fluorescence; those conjugated to rhodamine emit a reddish fluorescence. The examination of tissue sections treated in this way is achieved by exciting fluorescence in the tissue with an ultra-violet or violet light source. The labelled protein then emits fluorescence at its own characteristic wavelength. Above the slide a barrier filter is put into the microscope so that harmful ultra-violet light is excluded but light of the wavelength emitted by the fluorochrome is allowed through. There are three main variations on immunofluorescence techniques. These are (a) The single layer technique: in a search for antigens in tissue, e.g. thyroglobulin, tissue sections are treated with fluorescein-labelled anti-thyroglobulin. The labelled antibody will home on to its target antigen wherever it occurs in tissue so that, by examination of the section, all areas containing thyroglobulin will show bright green fluorescence. (b) The double layer technique: if the single layer technique were used to examine all sera from cases of suspected auto-immune thyroiditis, it would involve labelling every single patient's serum for test with fluorescein. This would be a very lengthy business. A simpler way to do this is as follows. Tissue sections containing antigen are treated with *unlabelled* antibody, i.e. patient's serum.

The antibody becomes fixed wherever antigen is present. Then fluorescein-conjugated rabbit anti-human γ-globulin is taken. This in turn becomes fixed wherever human γ-globulin, i.e. unlabelled antibody, has become fixed to the tissue antigen. The end result is again that fluorescence is localized at the sites of antigen localization. The technique is more convenient and also more sensitive than the single layer technique. (c) The *sandwich* method is used for the localization of antibody-producing cells in lymphoid tissue. This time, in a search for plasma cells making, for instance, anti-diphtheria toxoid, unlabelled diphtheria toxoid is layered on to the lymph node section first and becomes attached to the sites of antibody production in plasma cells. Then fluorescein-labelled rabbit anti-diphtheria toxoid is layered on and combines with the diphtheria toxoid attached to antibody-producing cells. In this method the sequence is (i) tissue (plasma cell) antibody, (ii) antigen and (iii) fluorescent antibody. There is therefore a sandwich of antigen between two layers of antibody. A single layer method can also be used for antibody localization. Fluorescein-labelled anti-γ-globulin is applied directly to a section of lymphoid tissue. All plasma cells producing γ-globulin will then show fluorescence. A fourth variation, used for localization of antigen–antibody reactions which take up complement, is to use a fluorescein-labelled anti-complement. As guinea pig serum is one of the best sources of complement, this is generally an antiserum against guinea pig globulin.

(2) **Radioactive tracers.** Antigens or antibodies can be labelled with radioactive tracers, generally by incorporating radioactive iodine (^{131}I or ^{125}I) or radioactive carbon (^{14}C) into their amino-acids. This technique has multidinous uses. Radioactivity can be counted in antigen–antibody precipitates (Farr test) or in agar gel precipitates (radioimmunoelectrophoresis) or antigen and antibody can be localized in tissue sections by autoradiography. The fate and distribution of isotope-labelled antigens or antibodies can be followed *in vivo* after their injection into experimental animals. By the use of such tracer techniques, antigen–antibody reactions can be detected with a sensitivity not approached by conventional methods but, as they are expensive and difficult to carry out, their use is still largely restricted to research laboratories.

Chapter XI

Hypersensitivity

The term *hypersensitivity* is used to describe a number of very different immunological responses to antigenic contact. While conventional immunity characteristically results in increased resistance to the antigen in question, hypersensitivity responses typically lead to greater susceptibility. Thus, whereas classical immunity is followed by neutralization and elimination of antigen without significant damage to the host's own tissues, in hypersensitivity states, antigen–antibody reactions or cellular immune responses frequently take place in such a way that cell damage occurs and harmful pathological lesions may ensue. Thus the antigen–antibody reaction itself initiates a sequence of events which result in tissue lesions. Hypersensitivity states are therefore associated with clinical disturbances provoked by antigenic contact. It would, nevertheless, not be true to say that hypersensitivity reactions are inevitably detrimental. There is indeed evidence that such tissue reactions can play a defensive role in certain bacterial and protozoal diseases. Suffice it to say that, in hypersensitivity, the immunologically competent cells show a heightened reactivity to antigen and that this altered reactivity may manifest itself in various tissue changes whose end result may be either harmful or beneficial. The term *allergy* is frequently used synonymously with hypersensitivity but has fallen into some disfavour as it has been used very vaguely, often in completely non-immunological contexts. When used precisely to indicate a heightened reactivity to antigen, it is a perfectly useful descriptive term.

Like classical immune responses, hypersensitivity states can be divided into two broad groups: those dependent on the presence of circulating antibody and those mediated by immunologically competent cells in which antibody production is of secondary importance. In subjects previously sensitized with antigen, antibody-mediated hypersensitivity is frequently manifested by a rapid response within minutes or seconds of further contact with antigen. This form of hypersensitivity is therefore often known as *immediate hypersensitivity*. In other forms of antibody-mediated hypersensitivity, the reaction may be slower, but it is characteristic of cell-mediated hypersensitivity that, again in previously sensitized persons, a delay of between 24 and 72 hr occurs after contact with antigen before a reaction develops. Cell-mediated hypersensitivity is therefore usually known as *delayed hypersensitivity*. The relationship of the different types of hypersensitivity described below to classical immune responses are represented schematically in Table I.

ANTIBODY-MEDIATED HYPERSENSITIVITY

(1) IMMEDIATE HYPERSENSITIVITY (TYPE 1 HYPERSENSITIVITY OF COOMBS AND GELL)

In immediate hypersensitivity, a reaction takes place between antigen and a specialized class of antibodies which, when found in human sera, are known as *reagins*. Human reaginic antibodies are unusual inasmuch as their activity is heat-labile, they do not form precipitates when they react with antigen and they probably belong to the special immunoglobulin class, IgE. They have an affinity for the surfaces of certain cells to which they become fixed. Antigen subsequently becomes bound to reaginic antibody at the cell surface and the antigen–antibody reaction causes cell damage. The presence of complement is not necessary for this type of cell damage to take place. In some species, reagins are particularly likely to become fixed to the surface of *mast cells*, the granular tissue basophil cells which contain large amounts of histamine, serotonin and heparin. As a result of the antigen–antibody reaction on the cell surface, the mast cells are disrupted and liberate histamine. The result is a weal and flare reaction with increased vascular permeability, erythema and oedema. Histamine release also leads to smooth muscle contraction and therefore to bronchoconstriction and, if generalized, to *anaphylactic shock*. Antibodies with biological properties similar to those of human reagins exist in many species. Many of the details of the immediate hypersensitivity reaction outlined above have been studied in species such as the guinea pig and it is not clear whether the exact process is identical in man. It is, however, quite certain that the reaction of human reagins with antigen does lead to the release of pharmacologically active substances such as histamine. Other substances also released are bradykinin and SRS (slow reacting substance) which causes a delayed but prolonged contraction of smooth muscle not susceptible to inhibition by anti-histamines.

LOCAL IMMEDIATE HYPERSENSITIVITY REACTIONS

If antigen is introduced into a particular site, reaginic antibodies may combine locally causing signs of histamine release restricted to the site of entry. This may happen frequently on contact not only with foreign proteins but also with haptenic low molecular weight substances which become attached to proteins *in vivo*. There is a particular group of hypersensitivity states, which

Table I

The Relationship of Hypersensitivity Reactions to Classical Immunity

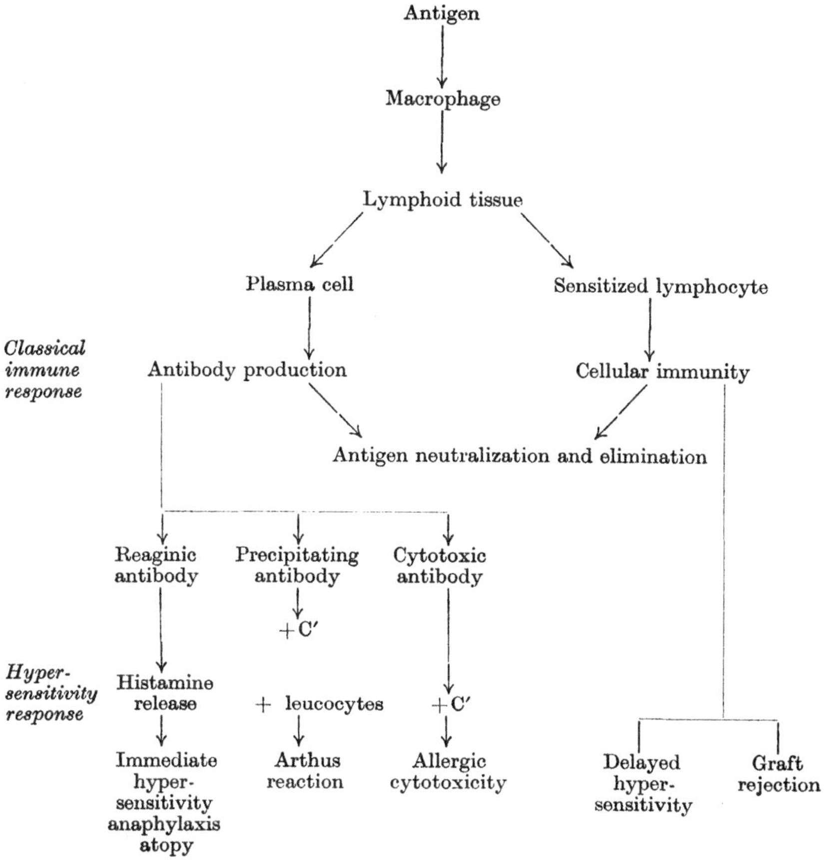

occurs most commonly in individuals who have a genetic predisposition to react to contact with extrinsic antigens by the formation of reaginic antibodies in circumstances where normal persons would either not react or would produce classical non-reaginic antibody. This type of hypersensitivity has been named *atopy* or *atopic hypersensitivity*. The manifestations of atopy vary according to the site of contact with antigen. Thus *urticaria* may occur in the skin either locally after direct contact with, for instance, animal fur or, in a more generalized form in persons who are hypersensitive to fish, strawberries or other foods and who are unwise enough to eat them. Probably the most common form of atopic hypersensitivity is *hay fever*, which occurs when the nasal or conjunctival mucosae of an atopic subject come into contact with the antigen (allergen) to which he is sensitive. Usually this is a pollen but it may be a constituent of house dust, possibly the spores of certain fungi, or animal fur or numerous other proteins. Inhalation of the allergen leads to immediate mucosal hyperaemia and copious exudation of nasal and lacrimal secretions. Lower in the respiratory tract, atopic subjects may suffer from *asthma* due to the histamine-induced bronchoconstriction which may follow inhalation of allergens and their fixation to the bronchial lining cells. These allergens are often similar to those which cause hay fever, i.e. pollens, furs, dusts. Asthma and hay fever are often associated with the presence in the blood of large numbers of eosinophils. However, asthma is a complex disease in which many aetiological factors play a part other than immediate hypersensitivity. In some asthmatic individuals clear evidence of an allergic process is present but in many patients such evidence is not obtained. Asthmatic patients often give a family history of atopic disorders and themselves may be subject to hay fever, urticaria, etc., as well as asthma.

The demonstration of the presence of reaginic antibodies in serum from patients with immediate hypersensitivity states is not easy. These antibodies do not form precipitates with antigen and there is no simple way of detecting them in the test tube. The surest way to demonstrate them is the Prausnitz–Küstner reaction. Küstner was a German research worker who was hypersensitive to fish. He took a little of his own serum and injected it into the arm of his colleague Prausnitz. A day or two later he took an extract of the fish tissue antigen to which he was sensitive and injected it into the same spot on Prausnitz's arm. Prausnitz got a local weal and flare reaction because Küstner's reaginic antibody had become fixed to his skin cells and was able to react with the allergen and cause histamine release. The Prausnitz–Küstner reaction is, however, not used clinically because it requires willing normal volunteers for skin testing and because it involves a risk of transferring serum hepatitis. The more usual clinical tests for atopy are the intradermal or prick skin tests. In these tests a small quantity of the suspected allergen is injected intradermally or is pricked into the epidermis of the allergic subject. In sensitive persons, a weal and flare reaction develops within the next quarter of an hour. These tests must be carefully controlled as many persons react to a simple needle prick without allergen by a weal and flare reaction. These tests do not detect reaginic antibodies directly but may give an indication which of several allergens the subject is sensitive to. This information can be used if it is decided to treat the patient by densensitization, a procedure which involves the repeated injection of gradually increasing doses of the allergen in the hope of stimulating the formation of "blocking" antibodies which inhibit the skin sensitizing activity of reaginic antibody. Patients may thus be relieved of asthma or hay fever but desensitization is usually only of temporary benefit and the injections may, themselves, give rise to unpleasant hypersensitivity reactions.

Generalized immediate hypersensitivity reactions—anaphylactic shock. The term anaphylaxis, which implies an increased reactivity to antigen, is often used synonymously with immediate hypersensitivity. Thus, the phenomena described above are manifestations of local anaphylaxis. However, if antigen is introduced in such a way as to be disseminated rapidly, systemic anaphylaxis occurs. There is considerable species variation in the manifestations of systemic anaphylaxis. In man, itching of the skin and flushing or urticaria may be associated with bronchospasm and with symptoms and signs of hypotension due to the effects of histamine-like substances on the peripheral circulation. In more severe reactions, *anaphylactic shock* may follow immediately upon injection of antigen. There is profound peripheral circulatory failure with hypotension, pallor, collapse, dyspnoea and fever. Death may occur quite rapidly, particularly after intravenous administration of allergens. Adrenaline should be given immediately followed by antihistamines. Anaphylactic shock is, unfortunately, an iatrogenic disease. It occurs following therapy with antisera as mentioned earlier and with many drugs which act as haptens. Penicillin hypersensitivity is one of the most carefully investigated examples but the list of drugs is almost endless and includes sulphonamides, iodides, arsenicals, benzene derivatives, plant extracts and many others. A careful history should therefore always be sought before giving antisera or the commoner allergenic drugs, particularly in atopic persons.

A form of general anaphylaxis which has only recently been recognized is *cot death*. This occurs in bottle-fed babies and is probably an acute hypersensitivity reaction to cow's milk proteins which are capable of crossing the infant's gut wall undigested, and are therefore absorbed intact. They retain their antigenicity and are capable of stimulating an immediate hypersensitivity response.

It would be curious if the body had evolved such a potentially dangerous way of dealing with antigens if it had no biological advantage. An explanation of the evolution of immediate hypersensitivity may be sought in the immune reaction against intestinal parasites, ubiquitous pathogens in most lower vertebrate species. In many animals, high levels of "anaphylactic" antibody are produced after parasitic infections, associated with marked eosinophilia; sheep infected with certain intestinal parasites show a dramatic "self cure" phenomenon in which, after feeding with the larvae of the parasite with which they are infected, increased intestinal mobility and oedema is observed associated with a rise in blood histamine and expulsion of the adult worms adherent to the intestinal wall. It is probable that this "self cure" is due to a protective anaphylactic reaction in the intestinal wall directed against the antigens of the parasites and leading to their expulsion.

(2) Allergic cytotoxicity (Type 2 hypersensitivity of Coombs and Gell). This is a hypersensitivity reaction against either (a) cell membrane antigens or (b) circulating antigens which become attached to cells. Antibody then becomes fixed to antigens at the cell surface. If complement is involved, the antigen–antibody reaction leads to cell lysis. If not, the cells may be damaged in other ways and then phagocytosed. The process in which *antigen* first becomes fixed to cells and then combines with *antibody* must be distinguished from immediate hypersensitivity in which *antibody* first becomes fixed to cells and later combines with antigen. Mechanisms of this type may play some part in certain auto-immune diseases and in graft rejection. Cytotoxic antibodies have been shown to destroy thyroid cells in auto-immune thyroiditis. An interesting disease in which such a mechanism plays a part is sedormid purpura. Sedormid was a drug which was formerly used as a sedative, until it was discovered that patients treated with it developed thrombocytopenic purpura. The reason for this was that sedormid was a hapten and, *in vivo*, it became attached to platelets. The platelet–sedormid complex behaved like an antigen and antibodies were formed against it which subsequently destroyed the platelets. Auto-immune destruction of erythrocytes by antibody in, for instance, acquired haemolytic anaemia in which antibodies become attached to the patient's own erythrocyte antigens and destroy the cells, is also essentially of this type.

(3) The Arthus reaction (Type 3 hypersensitivity of Coombs and Gell). The Arthus reaction occurs when a soluble antigen is injected into animals whose serum already contains precipitating antibody against that antigen following previous sensitization. Antigen combines with precipitating antibody in the circulation and the antigen–antibody reaction binds complement. Circulating antigen–antibody complement complexes become adherent to the endothelial lining cells of small capillaries. These complexes have, as described earlier, an effect in increasing capillary permeability, perhaps by production of the histamine-releasing agent, anaphylatoxin. They also liberate a chemotactic agent powerfully attractive for polymorphonuclear leucocytes. At the site of an Arthus reaction, a diffuse oedematous swelling appears accompanied by skin reddening. This lesion is more diffuse than the immediate hypersensitivity reaction and appears only after several hours. It may be accompanied by petechial haemorrhages. Histologically, clumps of platelets and granulocytes adhere to the capillary and venular endothelium and these may completely block the circulation through the small vessels. Some capillaries may rupture causing local small haemorrhages and, in severe reactions, local necrosis ensues. Heavy infiltration of the whole lesion with polymorphonuclear leucocytes is seen. Later, lymphocytes and macrophages may also invade the lesion.

Clinical Arthus-like phenomena. Local lesions very similar to the experimental Arthus reaction are seen in several important diseases. (a) The occupational disease *Farmer's lung* is an allergic reaction to the spores of fungi in mouldy hay, It is characterized by acute attacks of respiratory distress following exposure to mouldy hay which are associated with pulmonary alveolitis with exudation of polymorphonuclear leucocytes. Local Arthus reactions can readily be demonstrated on injection of the test allergen into the forearm of a sensitive subject and the essential pulmonary lesion is probably predominantly an Arthus reaction, following inhalation of the spores of the fungus *Micropolyspora polyspora*. Immediate hypersensitivity reactions also play a role in this and other similar pulmonary diseases. (b) Acute *glomerulonephritis* can be produced experimentally by injection of anti-kidney antibodies. This antibody globulin localizes to its corresponding antigen in the glomerular basement membrane and becomes bound there. The body then makes its own antibody against the foreign globulin bound in the kidney, so that anti-globulin–globulin complexes can be demonstrated on the basement membrane and these complexes are capable of binding complement and initiating the lesions described above. Antigen–antibody complexes of other types can also be shown to bind to the glomerular basement membrane and to set up the lesions of acute glomerulonephritis.

What relevance have these experimental results to human glomerulonephritis? One pointer may be the fact that in many patients, acute nephritis appears two or three weeks after a streptococcal throat infection. Furthermore, nephritis is associated chiefly with infection with streptococci of Type XII bearing a type specific M-protein antigen. Possibly, such cases represent hypersensitivity reactions against either glomerular antigens which share determinants with streptococcal M-antigens or against streptococcal antigens which become fixed in antigen–antibody complexes on the glomerular basement membrane and there bind complement. Serum complement levels in acute nephritis are usually low. (c) *Rheumatic fever* is another disease indirectly associated with recent infection with Group A streptococci. Streptococci have never been demonstrated in the cardiac or arthritic lesions but it has recently been shown that the infecting streptococci have an antigen similar to one in human heart tissue. Antibodies made against streptococci could therefore be capable of initiating an auto-immune carditis.

Serum sickness is the classical example of a generalized Arthus-type reaction in man. It derives its name from the fact that it commonly occurs after passive immunization with antiserum. About a week after a primary immunization, or more quickly after a second injection, the immunized subject becomes febrile and develops generalized lymphadenopathy, splenomegaly, joint swelling and urticaria. There is a low serum complement level and precipitating antibodies against the injected serum can be demonstrated as well as a high level of Forssman-type antibodies against sheep erythrocytes. In serum sickness, antigen–antibody precipitates are present in the circulation and can set up vascular Arthus-like lesions in many sites. In experimental serum sickness in rabbits, necrotizing arteriolitis is seen. The lesions of *polyarteritis nodosa* in man are very similar and many cases of polyarteritis occur following treatment with sulphonamides, thrioureas and other potentially haptenic drugs. It has therefore been argued that polyarteritis nodosa is a generalized manifestation of Arthus-like hypersensitivity to such drugs but direct evidence in favour of this hypothesis, i.e. the demonstration of circulating antigen–antibody complexes is, as yet, lacking. It is possible also that Arthus-type phenomena play a role in the pathogenesis of systemic lupus erythematosus (SLE) and other auto-immune diseases.

DELAYED OR CELL-MEDIATED HYPERSENSITIVITY (TYPE 4 HYPERSENSITIVITY OF COOMBS AND GELL)

The classical example of the delayed hypersensitivity reaction is the tuberculin or Mantoux test. In a person not previously exposed to tubercle bacilli, intradermal injection of a small quantity of tuberculo-protein into the forearm leads to no response. In a person who is suffering from, or who has been previously infected with tuberculosis or who has had previous sub-clinical contact with tubercle bacilli, a dull red, indurated, firm papule about 1 cm in diameter appears at the injection site between 24 and 72 hr after injection. This is quite different from the immediate hypersensitivity or Arthus response, both of which are essentially vascular lesions characterized by oedema and erythema. Histological examination of the Mantoux reaction reveals little vascular change but a dense infiltration of the site of antigen injection with lymphocytes and macrophages is seen. The characteristics of delayed hypersensitivity reactions are as follows: (a) slow onset, 24 to 72 hr; (b) the vascular component is slight and circulating antibody is not essential to the development of the reaction; (c) lymphocytes are essential. Tuberculin hypersensitivity can be transferred from a hypersensitive animal to a normal animal by the passive transfer of lymphocytes but never by transfer of cell-free serum.

DELAYED HYPERSENSITIVITY IN BACTERIAL DISEASES

Delayed hypersensitivity plays an important role in many bacterial diseases and has, in fact, been named bacterial allergy. This name is only partly correct because delayed hypersensitivity plays a part in many phenomena such as auto-immune disease, graft rejection, etc., which are unrelated to bacterial antigens. Among the bacterial diseases, the role of delayed hypersensitivity has been most intensively studied in tuberculosis. The altered reactivity of tuberculous animals to tuberculoprotein was first described by Robert Koch. Normal guinea pigs responded to the injection of tubercle bacilli by the delayed formation after 14 days of a permanent, ulcerating lesion whereas, in previously sensitized animals, an indurated lesion appeared after two days which ulcerated but eventually healed. If, however, tuberculoprotein from heat-killed bacilli is substituted for live bacilli in Koch's experiment as is done in the Mantoux test, normal animals show no reaction, whereas the previously sensitized animals give the typical delayed hypersensitivity response. In other words, Koch's normal animals had lesions directly due to tuberculous infection in the absence of hypersensitivity. The sensitized animals showed the lesions of delayed hypersensitivity against tubercle bacilli 48 hr after injection. The clinical lesions of tuberculosis also vary considerably in unexposed and in sensitized individuals. In unexposed persons, primary tuberculosis develops. If this is progressive, it leads to massive haematogenous dissemination and death from miliary tuberculosis or tuberculous meningitis. More often, the disease remains confined to the primary complex, a small peripheral pulmonary focus with associated granuloma in the draining lymph node, both of which eventually calcify. On the other hand, post-primary tuberculosis in previously exposed subjects is characterized by gradually expanding caseous granulomata with marked fibrosis which spread locally but only rarely disseminate haematogenously. The very different behaviour of the two lesions raises the questions; are these differences a reflection of delayed hypersensitivity in the post-primary lesion and to what extent is delayed hypersensitivity protective against tuberculosis and to what extent is it harmful? There is, unfortunately, no simple answer. It is probable that the delayed hypersensitivity state leads to a modification of the body's cellular reaction against *Mycobacterium tuberculosis* which may be reflected in the much larger necrotic macrophage epithelioid cell granuloma seen in post-primary lesions. Whether such a cellular reaction is harmful or beneficial is still doubtful. In Koch's experiment the second injection of tubercle bacilli in sensitized guinea pigs produced a local lesion which healed before death. Delayed hypersensitivity and immunity are probably but not necessarily associated. Circulating antibodies against various mycobacterial antigens are present in the sera of tuberculous patients but it is not clear that these antibodies play any important part in checking the progress of the disease.

Note that a positive Mantoux test does not necessarily indicate active tuberculous infection. It may only indicate previous exposure. The Mantoux reaction converts to positive several weeks after clinical or sub-clinical primary tuberculous infection. A negative Mantoux test in persons at risk from tuberculosis, e.g. nurses, contacts, denotes susceptibility and in the presence of a normal chest X-ray is an indication for B.C.G. vaccination.

The disease, *sarcoidosis*, is characterized by small epithelioid-cell tubercles without caseation or the tendency to coalesce. Persons with sarcoidosis have persistently negative Mantoux tests and it has been suggested that this disease is associated with an impairment of the capacity to mount delayed hypersensitivity reactions. Sarcoid-like syndromes are associated with exposure to several chemical agents such as beryllium and it has been suggested, without substantial evidence, that some cases represent atypical tuberculous infections.

Delayed hypersensitivity is present in many infectious diseases other than tuberculosis, particularly those which run a chronic course, but in none of them has it been so well studied. In many of these diseases, diagnostic skin tests similar to the Mantoux test are used and show the same delayed appearance of a dull red indurated nodule. Some of the most important of these are listed as follows: the *brucellin* test in *brucellosis*, in which a culture filtrate of *Br. abortus* or *Br. melitensis* is injected intradermally; the *lepromin* test in leprosy; the *Casoni* test in hydatid disease due to *Echinococcus granulosus*, a metazoan parasite of sheep which may infect man (this test frequently gives both immediate and delayed reactions); the *Frei* test in lymphogranuloma venereum and a similar test in psittacosis. Delayed hypersensitivity to staphylococci, streptococci and pneumococci can also be demonstrated experimentally and probably plays a part in the evolution of the lesions in long-standing or recurrent infections with these organisms.

Drug allergies have already been mentioned in connection with antibody-mediated hypersensitivity. As well as anaphylactic and urticarial lesions, *contact dermatitis* may frequently follow direct exposure of the skin to many chemical substances. This may develop a day or two after exposure as an eczematous eruption at first papular, then vesicular with exudation, later leading to crusting and skin thickening. Penicillin, arsenicals, hair dyes, nickel and extracts of many plants, common and rare, are among the numerous substances which have given rise to such lesions. Experimentally, dinitrochlorbenzene and picryl chloride are haptens which provoke powerful delayed hypersensitivity reactions. They do so by combining with proteins but the mechanism by which they do this in a local site on the skin is not clear. Individuals vary widely in their sensitivity to such allergenic agents. Frequently the clinical lesions following exposure are not simple but show elements of both antibody-mediated and delayed-type hypersensitivities.

The phenomena of graft rejection, auto-immune disease and tumour immunity bear a close relationship to delayed hypersensitivity inasmuch as, in general, they can be transferred by injection of lymphocytes from a sensitized animal to a normal one, but not commonly by transfer of serum antibody. Furthermore, the lesions in these conditions are predominantly of the lymphocyte-macrophage type as they are in delayed hypersensitivity. These phenomena are not simple and probably several immunological mechanisms are of importance in them, but due to the features which they have in common with delayed hypersensitivity, it is appropriate to consider them here.

AUTO-IMMUNE DISEASE

Paul Ehrlich, who studied the chemistry of antigen–antibody reactions during the 1890s used the phase *horror autoxicus* to emphasize the fact that antibodies normally combine with and reject foreign antigens only and never under normal circumstances react against the body's own tissues. In the last two decades, a group of diseases, the "auto-immune" diseases, has been recognized in which an abnormal immune response against "self" antigens is possibly the primary pathogenetic event and is certainly important in the evolution of the lesions. Such an abnormal immunological response could arise in the following ways. (1) The primary abnormality may be in the immunologically competent cells themselves. Self antigens are tolerated normally because they are recognized as "self" in foetal life. This tolerance may fail because the suppressor mechanism preventing anti-self immune responses later breaks down in adult life, leading to the reappearance of "forbidden clones" of "anti-self" immunologically competent cells. Random somatic mutation of immunologically competent clones has been suggested as a mechanism which may lead to the development of "anti-self" clones in adult life. (2) The immunologically competent cells may be normal and the abnormality may lie in the way they are presented with antigen. (a) Certain foreign antigens may be very similar to self antigens. The similarity between certain streptococcal antigens and heart and glomerular antigens has already been mentioned. Antibodies against colonic mucosa extracts which cross-react with certain *E. coli* strains have been found in cases of ulcerative colitis. Antibodies against the I antigen of human erythrocytes are found in pneumonia due to *Mycoplasma pneumoniae*. (b) Certain self antigens may never meet immunologically competent cells in foetal life either because, due to their sequestration in specialized and protected sites, they are not accessible to these cells, e.g. brain, lens antigens, or because they are themselves not formed until adult life, e.g. spermatozoa. If these antigens escape into the circulation later in life due to some non-immunological process, they may not be recognized as "self" because their first contact with immune tissues is so late and they may therefore stimulate an immune response. (c) "Self" antigens may be altered enough by infection, trauma or other pathological processes to become recognized as foreign and set up an immune response.

The immunological lesions in auto-immune diseases may be due to the action of circulating antibody on tissue antigens or to the direct action of immunologically competent cells such as lymphocytes. Antibody may be directly cytotoxic for tissue cells, e.g. thyroid (type 2 hypersensitivity) and cause their lysis or phagocytosis. Antibody-mediated Arthus-type reactions probably also play a part in tissue damage in diseases such as systemic lupus erythematosus and rheumatoid arthritis as well as in those auto-immune diseases related to bacterial infection mentioned above. Certainly, it is not difficult to detect circulating anti-tissue antibodies in many auto-immune diseases using complement-fixation, fluorescent antibody, tanned cell haemagglutination or other techniques. It is not always so easy to evaluate the significance of these antibodies. Probably, cell-mediated hypersensitivity plays a greater role in many auto-immune diseases than the direct action of antibody. These diseases are characterized histologically by intense lymphocyte proliferation. In the experimental auto-immune disease, *allergic encephalomyelitis*, which follows the injection of extracts of brain into laboratory animals, circulating antibodies against myelinated brain tissue are often found. However, the disease cannot be transmitted by transfer of this serum antibody but only by transfer of lymphocytes, just as is the case for delayed hypersensitivity to tuberculin. Cellular hypersensitivity is therefore the essential mechanism of allergic encephalomyelitis and serum antibodies have been found, not only to be unable to cause this disease, but even possibly to be protective against it.

(1) Organ specific auto-immune diseases. In one group of auto-immune diseases exemplified by allergic encephalomyelitis as described above the target antigen or antigens are confined to a single organ which therefore bears the brunt of the immunological attack and may eventually become largely replaced by lymphoid tissue. The major diseases of this group in man are listed below.
(a) The thyroid: In auto-immune thyroiditis, there is lymphoid replacement of the thyroid and destruction of active thyroid acinar tissue. Auto-antibodies against thyroid tissue antigens may be detected by complement-fixation, tanned red cell haemagglutination, immunofluorescence and by tests for cytotoxic antibody in tissue culture. There may be a family history of similar disease and the patient may suffer concurrently from auto-immune diseases of other tissues. The patient is most likely to be a woman. Hashimoto's thyroiditis is the classical auto-immune thyroiditis. Cell

damage and replacement by lymphoid tissue may lead to hypothyroidism but auto-antibodies against thyroid are also found in many cases of classical thyrotoxicosis and are associated with focal accumulation of lymphocytes in the thyrotoxic gland. (b) The gastric mucosa: In pernicious anaemia, auto-antibodies against intrinsic factor are present. In the absence of intrinsic factor, vitamin B_{12} cannot be absorbed across the gastric mucosa and the lesions of vitamin B_{12} deficiency develop. By immunofluorescence, antibodies against gastric parietal cells can be detected in sera from the majority of patients with pernicious anaemia. (c) The red cell: Although red cells do not form a discrete anatomical organ, they are functionally distinct and specific auto-antibodies may be directed against their surface antigens. This happens in the various forms of acquired haemolytic anaemia which may occur in association with other diseases such as reticuloses or leukaemias or which may not be associated with disease of other tissues. There is increased red cell breakdown due to coating by antibody which may cause lysis in the presence of complement or increased phagocytosis and cell destruction in the spleen. Antibodies against erythrocytes may be detected by Coombs' antiglobulin test, either coated on to the patient's red cells or free in his serum. In certain patients, particularly those with haemolytic anaemia associated with other reticulo-endothelial diseases, "cold agglutinins" are present which are detected by doing agglutination tests at refrigerator temperature but which are much less easy to detect at body temperature. In idiopathic thrombocytopenic purpura, antibodies against platelets are probably present and some cases of leucopenia may be associated with anti-granulocyte antibodies. (d) Auto-antibodies against tissue-specific antigens may be present in certain cases of nephritis, rheumatic fever and other diseases following bacterial infection. These diseases have been discussed in an earlier section.

(2) **Auto-immune diseases in which many tissues are involved.** In these diseases, auto-antibodies active against antigens common to many tissues are present. The classical example is *systemic lupus erythematosus* (S.L.E.). S.L.E. is one of the group of collagen diseases in which, clinically, involvement of many tissues is obvious. Lesions in skin, kidneys, peripheral vessels, pleura, joints and peripheral nerves are common. It is most often seen in women and is usually fatal. Serologically, most patients have antibodies against cell-nuclei, detected by immunofluorescence. They also have antibodies reactive against the cytoplasm of cells of many tissues and there may be antibodies against red cells, platelets, etc. L.E. factor is an antibody against leucocyte

nuclei which, after incubation with L.E. factor, are phagocytosed and can be demonstrated as "nuclear ghosts" in the cytoplasm of polymorphs. Despite the presence of these antibodies, the pathogenesis of S.L.E. is still unclear as anti-nuclear factor is unable to damage normal cells in which the nucleus is sequestrated from the action of serum antibodies. S.L.E. occurs in genetically predisposed individuals and may represent a complex hypersensitivity reaction to immunological stimuli such as bacterial or viral infections, drugs or even ultra-violet light which would not cause any allergic reaction in normal individuals.

Related to S.L.E. and sometimes almost indistinguishable from it is *rheumatoid arthritis* which, although characterized by chronic inflammatory polyarthritis, is far more than a simple arthritis as there is usually a varying degree of systemic disturbance. The serum contains *rheumatoid factor* which is an antibody against altered human γ-globulin. Rheumatoid factor is detected by the latex test or the Rose–Waaler (sheep-cell agglutination) test in which particles of latex or sheep cells coated with γ-globulin are agglutinated when incubated with serum from patients with rheumatoid arthritis. The joints in rheumatoid arthritis show lymphocyte, macrophage and plasma cell infiltration and similar lesions have been reproduced experimentally in animals by the injection of antigens such as streptolysin-S. Other diseases related to the above are dermatomyositis and Sjögren's disease, a syndrome of salivary gland atrophy, keratoconjunctivitis sicca and arthritis in which non-organ specific auto-antibodies against nuclear and cytoplasmic factors are found. Cases of hepatic cirrhosis following infective hepatitis may also show positive tests for antinuclear factors. In several of these diseases, there is also an increased incidence of organ-specific antibodies against thyroid or other tissues.

(3) **Myasthenia gravis** has been treated for many years empirically by thymectomy with benefit. Recently, it has been shown that patients with myasthenia have auto-antibodies, not only against striped muscle but also against thyroid and gastric mucosa and certain cases may also have anti-nuclear factor. The thymus may be enlarged and contain germinal centres which are not present normally in this organ. In myasthenia gravis, there is therefore clearly a general auto-immune disturbance, related to abnormalities in thymic function and, indeed, in certain cases to thymic tumour, but the essential link between the immunological abnormality and the neuromuscular block which causes the symptoms of the disease is still not fully understood.

(4) **Auto-immunity and cancer:** Patients with tumours of

immunologically competent tissue, reticuloses and leukaemias may show various manifestations of auto-immunity such as acquired haemolytic anaemia or the presence of anti-nuclear factor. In addition, it is possible that in neoplastic disease, immunological reactions may be mounted against tumour antigens which help to limit the spread of such tumours. In the sensory type of carcinomatous neuropathy which occurs in oat-cell bronchial carcinoma, antibodies against neurones are present. These antibodies may represent, in the first place, an immune response against a bronchial tumour antigen which shares determinants with neurones. The tumours in patients with this neuropathy are usually small and do not metastasize widely.

IMMUNOLOGICAL TOLERANCE

The hypothesis of Burnet and Fenner, discussed earlier on, that antigens encountered by foetal immune tissues are recognized as "self" whereas those encountered later are regarded as foreign, forms the basis of present concepts of immunological tolerance. Antigens are tolerated *in utero*, i.e. no immune response is formed against them, whereas later in life there is no tolerance and antibody production follows. Medawar and his colleages put this hypothesis to experimental test. They took skin cells from one inbred strain of adult mice (A) and grafted them on to a second strain (B). The graft was rejected because strain B formed an immune response against the A skin. However, if they took A spleen cells and injected them *at birth* into B mice, the B mice regarded the antigenic stimulus appearing at that stage as "self". They therefore formed no immune response against A. Then, if A skin were grafted on to the same B mice later in life the B mice failed to reject the graft. The information had been incorporated into their lymphocytes that A was self, and so no immune response was formed against the graft. Medawar had thus artificially produced tolerance at a critical point at about the time of birth. Such tolerance could not be produced if the A cells were grafted later, even a few days after birth.

Immunological tolerance has the following characteristics. (a) It is specific. Induction of tolerance to one antigen does not lead to tolerance of unrelated antigens. (b) Tolerance is easiest to induce using antigens biologically related to those of the host, e.g. A mouse skin in B mice, but very difficult to induce, e.g. in mammals against very dissimilar antigens such as bacterial or invertebrate antigens. (c) Tolerance, when artificially induced, does not necessarily last throughout an animal's lifetime. It tails off unless repeated doses of antigen are given. (d) High doses of antigen may lead to tolerance where lower doses fail to do so. (e) Animals tolerant of grafts can be made to reject those grafts by transfer of lymphocytes from a grafted but non-tolerant animal but not by serum transfer. The mechanism of tolerance (and of graft rejection) therefore depends, like that of delayed hypersensitivity, on the immunological activity of lymphocytes rather than that of specific antibody.

Immunological *paralysis* is induced by injecting very large doses of antigens into adult animals. Whereas an animal given a normal dose of antigen makes antibody, an animal given a large dose makes no antibody and is unable to do so for some time afterwards. It is possible that all the antibody produced is mopped up by the large amount of circulating antigen and so never appears free but it is more likely that no antibody is in fact made and that the animal has been temporarily rendered tolerant as a result of depression of function of lymphocytes specific for that antigen possibly because antigen reaches the immunologically competent cells in the wrong form, e.g. without having been modified by passage through macrophages.

Tolerance and tissue grafts. Grafts from one part of an animal to another or from one genetically identical twin to another will not be rejected immunologically because they are antigenically identical to the host tissue. *Homografts* are grafts from an animal of the same species but with a different genetic constitution. These are the ones which concern the surgeon and they will be rejected immunologically to a degree depending on the antigenic dissimilarity between donor and recipient. Heterografts are grafts from one species to another, and will be rejected, but do not much concern the surgeon. The clinical problem in tissue graft surgery, therefore, apart from the technical difficulties, is the problem of homograft rejection. Skin homografts from a donor to an incompatible recipient may look healthy for a variable period and begin to become vascularized but later the graft becomes necrotic and sloughs. At this stage, the graft is shown histologically to have become infiltrated with lymphocytes and macrophages. Antibodies against the grafted tissue may be present and immunologically competent lymphocytes are present which, if transferred to an individual previously rendered tolerant of the grafted tissue, would be capable of causing graft rejection in that recipient.

In the reaction described above, the lymphoid tissues of the host reject the graft. A second possibility is the occurrence of a graft-versus-host reaction. This happens when the graft contains immunologically competent cells. In this case the graft may take, but a few weeks later the graft cells may populate the host and form an immunological reaction against his antigens leading to a generalized and fatal illness known as "secondary disease". Graft-versus-host reactions are usually delayed because it takes a variable but protracted period for grafted lymphocytes to populate the host's tissues.

The problem for the surgeon is to overcome immunological graft rejection without at the same time leaving the patient completely defenceless against intercurrent infection. The approaches to this are as follows. (1) To use graft tissue from a donor who is genetically close to the recipient. This usually means grafting tissue from parent to child or vice versa or, best of all, from identical twin to identical twin. The survival of kidney grafts between close relatives is much better than that of those between unrelated individuals. Graft rejection depends on recognition of *transplantation antigens* on the surface of tissue cells. As yet transplantation antigens are poorly understood. It is now becoming possible to type an individual's transplantation antigens just as one does for blood groups and so to select suitable donors for graft surgery. (2) Immunosuppressive therapy. This is a method of depressing the recipient's immune response during and after the graft in the hope that the graft will thus not be rejected. It usually involves the use of cytotoxic drugs, heavy X-irradiation or steroids. Immunosuppressive therapy is undiscriminating. X-rays and cytotoxic drugs may destroy many cells other than those responsible for graft rejection. There is depression of haemopoiesis and generalized depression of immune responses. The recipient is therefore very unwell after such therapy and very susceptible to infection. Such patients are treated with antibiotics and blood and fluid replacement and are nursed in a sterile environment but the difficulties associated with this type of surgery are still great. Patients in whom renal transplantation is carried out are no longer treated with whole body X-irradiation but with the anti-metabolic drug azathiopurine and with prednisone in large doses. In an attempt to provide a more discriminating tool, anti-lymphocyte sera have recently been introduced. By injecting lymphocytes into experimental animals, it is hoped to produce antisera which will specifically depress lymphocyte function and not affect erythropoiesis, granulopoiesis, etc. At present, such therapy is still at the experimental stage and has not yet been used in humans.

Tumour immunity. In many ways, tumours may be regarded as grafts of antigenically foreign cells which have arisen from the body's own normal cells and which act as successful grafts to a greater or lesser extent depending upon how they are tolerated by their host tissue. Many tumours can be looked upon as foreign because there is good evidence that they contain new antigens not present in the tissue from which they arose. Other tumours may lack antigens present in the parent tissue. In experimental carcinogenesis using tumour viruses, the tumours caused by a given virus always have new antigens not present either in their parent tissue or in the oncogenic virus and these new antigens are always the same even in different types of tumour caused by the same virus. The virus modifies the genetic material of the cancer cell in a constant way so that the new cell antigens remain constant. Chemical carcinogens, on the other hand, cause a more sporadic change in the cancer cell so that many tumours induced by the same carcinogen may have different antigenic structures. It has been postulated that neoplastic change is constantly occurring but that most of these changes are eliminated instantly by surveillance mechanisms. These mechanisms may be immunological, i.e. new tumour antigens are recognized as foreign and eliminated by lymphocytes or by antibody, or they may be non-immunological in that the cells are recognized as foreign in a non-specific way, possibly because of their new surface configurations, and are rejected by their neighbour cells and phagocytosed or destroyed by enzymes. Successful tumours may be said to have escaped from surveillance but their growth and metastasis may be modified considerably by immunological or non-immunological rejection mechanisms. At present, much of the theory of the control of tumour growth by such mechanisms is still speculative but the possibility arises that the behaviour of neoplasms in man may be altered by immunological manipulation. The use of such methods in detection and treatment of cancer is therefore a hope and a challenge for the future.

Chapter XII

The Classification of Bacteria

Many attempts have been made to evolve a satisfactory classification of the bacteria. A simple but useful working classification is that based on a combination of overall shape (coccus or bacillus) and reaction to Gram stain. This gives four primary subgroups, viz. *Gram-positive cocci*, *Gram-positive bacilli*, *Gram-negative cocci* and *Gram-negative bacilli*. In Table I some organisms important to the medical bacteriologist are arranged according to this scheme. It is important to remember that under some conditions, for example in old cultures, Gram-positive organisms may decolourize very readily and appear Gram-negative; moreover it may sometimes be difficult to decide from a single examination whether an organism should be regarded as a coccus or a short bacillus, although this can generally be established without difficulty if cells are examined at different stages of growth.

Table I

Reaction to Gram Stain

GRAM POSITIVE	*GRAM NEGATIVE*
Cocci	*Cocci*
Staphylococcus	Gonococcus (*N. gonorrheae*)
Streptococcus	Meningococcus (*N. meningitidis*)
Pneumococcus (*Strep. pneumoniae*)	*N. catarrhalis*
Bacilli	*Bacilli*
Anthrax bacillus (*B. anthracis*)	Influenza bacillus (*H. influenzae*)
Diphtheria bacillus (*C. diphtheriae*) and other corynebacteria	*Bord. pertussis*
Tetanus bacillus (*Cl. tetani*)	*Br. abortus* and *melitensis*
Cl. welchii and other clostridia	*V. cholerae*
Actinomyces	Members of the coli-typhoid-dysentery group
Nocardia	Friedländers bacillus (*Kl. pneumoniae*)
Streptomyces	Plague bacillus (*Past. pestis*)
Tubercle bacillus (*Myco. tuberculosis*) and other mycobacteria	*Proteus* and *Pseudomonas*

In a general classification of living matter, bacteria are regarded as being part of the plant kingdom and are given the name *Schizomycetes* or "fission fungi". The blue-green algae, thought to be the nearest relatives of the bacteria, are called *Schizophyceae* or "fission algae". All schemes of bacterial classification are similar to those used for plants and animals, being based on the system proposed by the Swedish botanist Linnaeus in 1753. The basic unit in this system is the *species*, and related species are grouped into *genera*, genera into *families*, families into *orders*, orders into *classes* and so on. A widely used system of bacterial classification is that proposed in Bergey's "Manual of Determinative Bacteriology", and this has been followed here in considerably simplified form (Table I): many genera which are not of medical interest have been omitted from the table. It must be realized that such classifications will inevitably need to be changed as further properties of the organisms are discovered.

There are certain difficulties in applying the Linnean system of classification to organisms as simple as bacteria. Morphological characters, which are so useful in distinguishing species in larger organisms, are of much less value in bacteria and it is often impossible to separate related organisms on purely morphological grounds. For example it is not usually possible by examining stained films to distinguish between members of different genera of the family *Enterobacteriaceae* (Gram-negative intestinal bacilli) some of which are important causes of intestinal disease, while others are normal inhabitants of the bowel. Furthermore in animals and higher plants the species is to a large extent a group within which interbreeding can occur, and this concept cannot yet be applied to bacteria. In fact bacterial species are defined in very different ways in different genera.

Despite these and other difficulties, however, there is so far no sign of the present system being abandoned.

NOMENCLATURE

Under the Linnean scheme the name of an organism is a binomial derived by giving the generic name first with a capital, and the specific name afterwards with a small letter, e.g. *Clostridium welchii*, *Mycobacterium tuberculosis*, *Streptococcus pyogenes*, etc. It is not permissible to have a specific name composed of three elements, e.g. *Bacillus aerogenes capsulatus*, although such an organism could be described as a particular variety of a species, e.g. *Bacillus aerogenes* var. *capsulatus*. The names of organisms are italicized in print, or underlined in typescript.

The rules governing the naming of bacteria were agreed at the International Congress of Microbiology in Copenhagen in 1947, and are contained in the International Bacteriological Code of Nomenclature published in 1948. However, the existence of these rules has not abolished differences of opinion as to how organisms should be classified, and it must be realized that in some cases organisms of medical importance may have more than one scientific name in more or less common use. These synonyms are indicated where necessary in the text.

As well as these more formally derived names, many pathogenic organisms have common or trivial names which are well understood and frequently used in clinical bacteriology, e.g. the tubercle bacillus, the diphtheria bacillus, the gonococcus, the pneumococcus, etc. In most cases also the generic name can be used in slightly altered form as a collective noun without a capital letter to describe a group of organisms, e.g. the corynebacteria, the salmonellas, etc.

BACTERIAL CHARACTERS USED IN CLASSIFICATION

The properties most commonly used in the classification of bacteria of medical interest can be grouped under the following broad headings:

1. Morphology and staining reactions.
2. Cultural characteristics.
3. Biochemical tests.
4. Antigenic properties.
5. Pathogenicity and toxin production.

MORPHOLOGY AND STAINING REACTIONS

The principal factors involved are the size, shape and arrangement of the individual bacteria, the reaction to staining by Gram's method, and the presence or absence of flagella, capsules, spores and so on. In some cases the reaction with special stains such as acid-fast stains are of importance.

CULTURAL CHARACTERISTICS

The appearances of the colonies on solid media, changes in the surrounding medium as a result of growth, the type of growth in liquid media, and so on, are all characters to be noted. It is usual also to include under this heading oxygen requirements (aerobe or anaerobe), need for extra CO_2, rate of growth, optimum temperature, temperature range, requirement for special growth factors, etc., since these directly affect the ability of the organism to grow in any particular culture medium.

BIOCHEMICAL TESTS

This heading includes broadly both fermentation tests with a range of substrates and also tests for the identification of specific breakdown products of sugars, amino acids and occasionally other compounds. A very wide range of such tests has been used, but only a few are normally applicable in the case of any particular strain or group of strains.

ANTIGENIC PROPERTIES

The great specificity of antigen–antibody reactions has been much used in bacteriology as a means of identifying different strains of bacteria. Most of the chemical components of capsules, flagella, cell walls and other anatomical elements of bacteria are good antigens, that is, they will cause the formation of antibodies when injected into man or other animals (see Chap. IX). The antisera so produced are specific for the components concerned, and can be used to identify these components in other strains. This enables bacteria to be classified into groups whose members have a similar *antigenic composition*. Antigenic composition is often related to the ability to produce disease, and so its determination may be important in identifying pathogenic strains.

PATHOGENICITY AND TOXIN PRODUCTION

This is obviously an important character as far as human and animal pathogens are concerned, although of limited value in other fields of bacteriology. Most organisms pathogenic to man are also pathogenic to laboratory animals and can be recognized by the type of infection produced in them, although this is not necessarily the same as that found in natural infection in man. Some organisms produce well-defined toxins, and the presence of such toxins can be used to identify these strains. This is usually done either by animal tests or by identifying the toxin serologically *in vitro*.

IDENTIFICATION OF UNKNOWN STRAINS

In the identification of an unknown organism it is generally convenient to work through the list of properties more or less in the order given above, i.e. to start with morphology and staining reaction. It is generally possible to make a preliminary identification of an organism by the time its morphology and cultural characteristics have been studied, and this preliminary process will indicate which biochemical and antigenic tests are likely to be of most value. However, the process can often be shortened by omitting one group of tests altogether. For example in the case of many Gram negative intestinal bacilli, very specific antisera are available and it is often possible to apply serological tests directly to organisms from the initial culture without waiting for the results of fermentation and other biochemical tests which may take 48 hr or longer.

It is relatively easy in most cases to identify pathogenic bacteria isolated from diseased tissue, because the number of types which commonly cause disease in any particular anatomical situation is limited. For example there are only four or five different kinds of bacteria which are likely to be found in acute urinary infection, and therefore media can be used which are known to be suitable for their identification. Similarly the great majority of cases of acute leptomeningitis are caused by meningococci, pneumococci, streptococci or *Haemophilus influenzae*, although a great variety of other organisms have been less frequently found in this condition.

The following are the main features of the more important genera of medically important bacteria.

PSEUDOMONAS

Gram-negative aerobic rods: motile by polar flagella: ferment sugars without gas production: usually produce water soluble diffusible yellow and/or green pigments: grow readily on ordinary media: one species, *Ps. pyocyanea* (*Ps. aeruginosa*) found in intestinal tract of man and may cause pyogenic infections.

SPIRILLUM

Short rigid spiral organisms, motile by polar flagella: Gram-negative but easier to stain by Romanowsky stains (e.g. Giemsa): only one species pathogenic, *Spirillum minus*, normally found in oral cavity of rats and causes rat-bite fever in man.

VIBRIO

Gram-negative curved rods: motile by single polar flagellum: aerobic and facultatively anaerobic, ferment sugars without gas production: most species saprophytes, important pathogenic species is *V. cholerae*, cause of asiatic cholera.

ALCALIGENES

Gram-negative rods, motile by peritrichate flagella: do not produce acid from sugars in peptone-containing media. Non-pathogenic: found in intestinal tract of man.

ESCHERICHIA

Gram-negative rods, motile by peritrichate flagella: grow readily on all ordinary media: ferment lactose and other sugars with production of gas: indole positive: cannot use citrate as sole carbon source. Constantly present in intestinal tract of man and animals, e.g. *E. coli:* some strains pathogenic.

CITROBACTER

Gram-negative motile rods, grow readily on ordinary media: usually ferment lactose and other sugars with production of gas: indole negative: can use citrate as sole source of carbon: non-pathogenic inhabitants of soil and water.

SALMONELLA

Gram-negative motile rods: grow readily on ordinary media: ferment many carbohydrates, but not lactose or saccharose, usually with gas production: do not form indole: intestinal pathogens of man and animals causing enteritis, e.g. *Salm. typhi*, the cause of typhoid fever.

SHIGELLA

Gram-negative non-motile rods: ferment sugars (generally not lactose) with production of acid only: intestinal parasites of man causing dysentery or gastroenteritis, e.g. *Sh. sonnei.*

PROTEUS

Gram-negative motile rods, typically showing a spreading growth on moist agar ("swarming"): ferment various sugars (but not lactose) with production of gas: decompose urea very rapidly to ammonia. Normal inhabitant of intestine of man and animals: may be associated with chronic infections, e.g. *Proteus mirabilis* isolated from cystitis, bed sores, etc.

KLEBSIELLA

Gram-negative non-motile rods: usually capsulated, producing large viscid colonies on solid media: ferment a variety of sugars with production of gas: found in soil and water, also in intestinal and respiratory tracts of man, e.g. *Kl. pneumoniae*, causing a very fatal type of pneumonia.

PASTEURELLA

Small Gram-negative rods, often showing marked bipolar staining: little or no fermentation of sugars in complex media:

pathogens of animals which cause disease in man, e.g. *Past. pestis* (from rats) causing plague.

BRUCELLA

Small Gram-negative cocco-bacilli, aerobic or microaerophilic, growth often improved by adding CO_2: little or no fermentation of sugars in complex media: pathogens of animals which cause disease in man, e.g. *Br. abortus* from cattle, causing undulant fever.

HAEMOPHILUS

Small Gram-negative non-motile rods, often very pleomorphic: may be capsulated: require for growth accessory factors found in blood, yeast or plant tissues, or produced by other bacteria: parasitic in man and animals, some pathogenic, e.g. *H. influenzae.*

BORDETELLA

Small oval Gram-negative cocco-bacilli: non-motile: grow rather slowly and may require high proportion of blood in medium on first isolation: do not ferment sugars in complex media, e.g. *Bord. pertussis* (*Haemophilus pertussis*) the causative organism of whooping cough.

ACTINOBACILLUS

Small Gram-negative non-motile rods: sometimes occur in long chains or unbranched filaments: cause actinobacillosis in cattle (chronic disease resembling actinomycosis): one species, *A. actino-mycetem-comitans* may be found in association with *Actinomyces israelii* in human actinomycosis.

MORAXELLA

Short, rather fat Gram-negative rods tending to occur in pairs end to end: non-motile: aerobic: produce no acid from carbo-hydrates in complex media, but usually liquefy gelatine or inspissated serum, e.g. *Moraxella lacunatus*, the cause of angular conjunctivitis in man.

BACTEROIDES

Anaerobic Gram-negative rods, sometimes motile: found in intestinal tract of man and animals: may be pathogenic, e.g. in appendicitis.

FUSOBACTERIUM

Anaerobic Gram-negative spindle-shaped (fusiform) bacilli, non-motile: present in mouth and intestine of man and animals: e.g. *Fusobacterium fusiforme* (= *Fusiformis fusiformis*) which causes Vincent's angina in association with *Borrelia vincentii.*

SPHAEROPHORUS

Anaerobic Gram-negative pleomorphic rods, sometimes giving very long filaments: found in intestinal tract of man and animals: potential pathogens and may cause necrotic lesions with septicaemia, e.g. *Sphaerophorus necrophorus* (= *Fusiformis necrophorus: Bacteroides necrophorus*).

STREPTOBACILLUS

Gram-negative rods, often filamentous with oval swellings. Aerobic and facultatively anaerobic: non-motile. *Streptobacillus moniliformis* is a normal inhabitant of the nasopharynx in rats, gives one variety of rat bite fever in man.

STAPHYLOCOCCUS

Gram-positive cocci, arranged in clusters on solid media, and in pairs, short chains or small groups in fluid media: non-motile, grow well on ordinary media, colonies being white, yellow or golden in colour: found mainly on body surfaces, some species highly pathogenic, e.g. *Staph. pyogenes* (= *Staph. aureus*).

MICROCOCCUS

Gram-positive cocci occurring in pairs, fours or small clusters: non-motile, aerobic, grow well on all ordinary media: mainly saprophytes in air, dust, water, soil, etc., e.g. *Micrococcus luteus.*

SARCINA

Gram-positive cocci dividing in three planes producing cubical packets of eight cocci: non-pathogenic, e.g. *Sarcina ventriculi.*

AEROCOCCUS

Gram-positive cocci occurring in pairs, fours or small clusters: facultatively anaerobic: gives α-haemolysis on blood agar: common in air, non-pathogenic, e.g. *Aerococcus viridans.*

NEISSERIA

Gram-negative cocci, often arranged in pairs: aerobic: grow poorly on simple media: strict parasites, usually in nasopharynx, several species pathogenic, e.g. *Neisseria gonorrhoea* (gonococcus).

VEILLONELLA

Minute Gram-negative anaerobic cocci (average diameter 0·3 μ): difficult to maintain in artificial culture: found in mouth and alimentary tract: occasionally pathogenic.

STREPTOCOCCUS

Gram-positive cocci occurring in pairs and chains: individual cocci may be distinctly oval: may be capsulated: a few strains motile: usually aerobic and facultatively anaerobic, but some strict anaerobes: growth on ordinary media much improved by blood or serum: many species parasitic, some pathogenic, e.g. *Streptococcus pyogenes.*

LACTOBACILLUS

Gram-positive non-motile, non-sporing rods: often microaerophilic or anaerobic: ferment carbohydrates vigorously producing lactic acid with or without other organic acids: found in fermenting vegetable products as well as mouth and intestines of man and animals, e.g. *Lactobacillus acidophilus.*

CORYNEBACTERIUM

Gram-positive rods of irregular outline and uneven staining, not acid fast, non-sporing, usually not motile: many are parasites of man and animals, some pathogenic, e.g. *Corynebacterium diphtheriae*, causing diphtheria.

LISTERIA

Small, motile, non-sporing Gram-positive rods with a tendency to form long filaments: growth poor on solid media, e.g. *Listeria monocytogenes* (= *Listerella monocytogenes*) an uncommon cause of infection in man.

ERYSIPELOTHRIX

Slender Gram-positive non-motile rods, non-sporing: may form long filaments, e.g. *E. rhusiopathiae*, found in the nasopharynx of pigs, causes erysipeloid in man.

BACILLUS

Gram-positive rods, often forming long chains of bacilli: usually motile: produce round or oval heat-resistant spores, aerobic and facultatively anaerobic, grow well on ordinary media: mostly saprophytes from soil, water, dust, etc.: only species pathogenic to man is *B. anthracis*, causing anthrax in man and animals.

CLOSTRIDIUM

Gram-positive rods (often appear Gram-negative in old cultures), anaerobic or microaerophilic: produce round or oval heat-resistant spores: most species motile: primarily saprophytes from soil, but some produce powerful exotoxins and are pathogenic, e.g. *Cl. tetani*, causative organism of tetanus.

MYCOBACTERIUM

Non-motile aerobic irregular rods: Gram-positive but usually difficult to stain because of high lipid content: strongly acid-fast in Ziehl–Neelsen stain: growth on artificial media may be very slow: some species highly pathogenic to man, e.g. *Mycobacterium tuberculosis*.

NOCARDIA

Non-motile aerobic Gram-positive organisms, usually growing as branching filaments which break down to give rods resembling corynebacteria in appearance: often weakly acid fast in Ziehl–Neelsen stain: saprophytes in soil, but some types may cause disease in man, e.g. *Nocardia asteroides*.

ACTINOMYCES

Anaerobic or microaerophilic Gram-positive rods which show branching: non-motile, non-sporing, non-acid fast parasites from mouth of man and cattle: can cause disease if implanted in tissues, e.g. *Actinomyces israelii*, cause of actinomycosis in man.

STREPTOMYCES

Organisms growing predominantly in the form of long Gram-positive branching filaments, often with aerial hyphae carrying spores: many strains produce antibiotic substances which inhibit the growth of other organisms: normally soil saprophytes, occasionally associated with chronic ulcerative conditions, e.g. *Streptomyces madurae* (=*Nocardia madurae, Actinomyces madurae*).

BORRELIA

Spirochaetes with coarse shallow spirals easily stained by ordinary methods: Gram-negative, motile, grow with difficulty in artificial media: some species occur as parasites on mucous surfaces, others are pathogens spread by biting insects (lice or ticks), e.g. *Borrelia recurrentis*.

TREPONEMA

Spirochaetes, with fine regular coils, which stain poorly if at all with ordinary bacteriological stains: motile: non-pathogenic species grow in artificial media: some species are parasites on mucous surfaces, others pathogenic, e.g. *Trep. pallidum*, causing syphilis.

LEPTOSPIRA

Spirochaetes with very fine regular coils and hooked ends: actively motile: readily cultured in media containing serum: stain poorly with ordinary bacteriological stains: some species saprophytes in water, pathogenic species are carried by various rodents from which man is infected, e.g. *Lepto. icterohaemorrhagiae* from the rat, giving leptospiral jaundice in man.

Part II

SYSTEMATIC BACTERIOLOGY

Chapter XIII

Staphylococcus

Spherical Gram-positive cells arranged in grape-like clusters, or in pairs or short chains. Opaque, pigmented growth on solid media. Catalase positive. Oxidase negative. Some strains pathogenic.

Staphylococci were observed by both Koch and Pasteur. Ogston, a surgeon of Aberdeen, found them in a large proportion of abscesses and proposed the name Staphylococcus to indicate their grape-like orientation. Staphylococci are ubiquitous—they are present on the skin, anterior nares, alimentary tract, and also on clothing, in air sewage, milk, water, and generally in the environment of man and animals. Fleming found that the bath water after a normal socially clean person had bathed could contain 100,000 staphylococci per ml. Fortunately the great majority are saprophytic commensals, but the small proportion which are pathogens are very versatile in their pathogenicity.

The two main varieties may be distinguished by their pigment production on solid media. The majority of the pathogenic strains, *Staph. aureus* (*Staph. pyogenes*), produce a yellow pigment which varies from a pale yellow to a golden orange colour depending partly on the medium and the conditions of culture. The saprophytic strains, *Staph. albus*, produce a white pigment. Only occasional albus strains are found to be pathogenic. *Staph. citreus* which produces a lemon yellow pigment is a rare type and usually saprophytic.

STAPH. AUREUS.

MORPHOLOGY AND STAINING REACTION

The cells are spherical, usually 0·8–1·0 μ in diameter. On solid media they are arranged in grape-like clusters but in liquid media may appear singly, in pairs, or in very short chains. Capsules have been detected in very young cultures of some strains, but are not found in cultures 18–24 hr old. They are Gram-positive, but some Gram-negative cells may be found in old cultures.

CULTURE

Staph. aureus grows readily on the usual media within a wide range of temperature, 12°–44°C. The optimum temperature is 37°C. They grow best aerobically but are facultative anaerobes. They grow less abundantly anaerobically and then fail to produce the usual pigment. In fluid media growth is rapid; in 24 hr there is general turbidity with a variable amount of deposit which disperses on shaking. After further incubation a ring growth appears at the surface. The characteristic pigment is not produced in fluid media. On solid media a heavy growth soon forms. Isolated colonies are circular, moist, opaque, smooth and shining. Pigment is formed and the colour becomes more intense if cultures are left at room temperature. Nutrient agar, potato, and coagulated serum are favourable media for pigment production.

RESISTANCE

Staphylococci are easily killed by heat at 60°C in ½–1 hr. They are particularly sensitive to analine dyes such as gentian violet, crystal violet and malachite green, and to quaternary ammonium compounds. Very low concentrations of violet dyes incorporated into media will inhibit the growth of staphylococci and prevent any associated streptococci from being overgrown. Staphylococci are resistant to low temperatures and to drying.

BIOCHEMICAL ACTIVITY

The fermentation of sugars is variable. *Staph. aureus* usually ferments glucose, lactose, maltose, sucrose and mannite to acid without gas. Gelatin is liquefied. Indole is not produced. Catalase is produced and can be detected by adding H_2O_2 to the growth. Phosphatase is produced and is demonstrated by growing the organism on agar containing phenol-phthalein phosphate. Phenol-phthalein is liberated, and if the plate is exposed to ammonia vapour the colonies turn pink.

Pathogenic staphylococci produce an enzyme, coagulase, which will clot fibrin. It may be demonstrated by a slide test or a tube test. On a slide, a drop of suspension of the organisms in saline is added to a drop of plasma and stirred with a needle. The staphylococci are rapidly agglutinated, giving a granular appearance within

2 min. For the tube test, a colony of staphylococci is added to about 1 ml of a 1/10 dilution of plasma. The reaction will occur at room temperature but is accelerated at 37°C. A clot forms within $\frac{1}{2}$–2 hr. Control tubes of coagulase negative and coagulase positive organisms should be put up at the same time. Coagulase is antigenic, and there are several antigenic variants but all produce similar effects on plasma.

Some strains of *Staph. aureus* digest fibrin; the digestion is activated by a staphylokinase. Deoxyribonuclease and hyaluronidase are also produced by pathogenic staphylococci.

Toxins. *Staph. pyogenes* produces a number of soluble toxins which may be detected in liquid or semi-solid media under certain conditions. Alpha toxin (α-toxin) is the most important. It is dermonecrotic if injected intradermally, and is rapidly lethal for rabbits if injected intravenously. Rabbit erythrocytes, but not human erythrocytes, are haemolysed, and are used for detection and assay of the α-toxin. The toxin can be converted to a toxoid by the action of formalin, and the toxoid has sometimes been used in carefully graded doses in the treatment of chronic staphylococcal infections.

β-lysin is also inactive on human red cells but produces "hot-cold" lysis of sheep cells. Preliminary incubation at 37°C followed by a few hours in the cold will produce lysis of sheep cells. γ and δ lysins have also been described but have not been as intensively studied as the α-toxin.

Leucocidins which attack human polymorphs are produced by pathogenic strains. At least three types have been described. The separation of these from each other and from the haemolysins is difficult, and has given rise to much confusion in the literature.

Enterotoxin. Some strains produce an enterotoxin which is responsible for an appreciable number of outbreaks of food poisoning. It produces vomiting and diarrhoea after a short incubation period of about 4 hr. The enterotoxin is more heat stable than the haemolysins, and is not destroyed by pasteurization of milk. There is no satisfactory *in vitro* test for this toxin, and no reliable animal test. Kittens have been used, but the results are irregular. Man seems to be the most susceptible subject, but susceptibility is variable.

Pathogenicity. *Staph. aureus* may be found as commensals, frequently in the nares and on the skin of carriers without any sign of infection. It is, however, an important pathogen which is responsible for a great variety of pyogenic infections in man and animals. It is the usual causative agent in skin lesions such as sycosis, boils, carbuncles, impetigo contagiosa, and pemphigus neonatorum. It is a common cause of osteomyelitis, peri-renal abscess, septicaemia, and also many cases of pneumonia, otitis media and endocarditis. It is almost invariably present in wound infections. Some strains which produce an enterotoxin are responsible for one type of gastroenteritis.

DIAGNOSIS

Staphylococci are readily identified by their morphology, orientation, and staining reaction in smear preparations of pathological specimens, and by cultural reactions in ordinary media on which growth occurs readily. In heavily contaminated material, such as faeces, staphylococci may be isolated by growing in a medium containing 7–10 per cent sodium chloride. Confirmation of the pathogenic character of the organism isolated may be made by the coagulase test, phosphatase test, and fermentation of mannite.

In outbreaks of staphylococcal infection the strains isolated may be identified and compared by the phage typing technique. A broth culture of the organism is poured over an agar plate to produce a uniform inoculation which will result in a confluent growth or "lawn" after incubation. The plate is dried, and drops of 25 different phages are applied, preferably by the simple apparatus devised by Tarr. Twenty-five wire loops of which the shafts run freely in a metal plate are sterilized by flaming, then dipped into a plastic plate with 25 wells each of which contains a separate phage. The inoculated agar plate is then placed in the same position as the plastic phage reservoir, and the loops which have retained a drop of phage are lowered to the plate, which absorbs the phage. The loops are then sterilized by flaming, cooled, dipped again in the phage reservoir plate, and applied to the next culture. By this method, from 30 to 50 tests can be set up within 1 hr. The plates are incubated, and the phage pattern is clearly visible by the inhibition of growth in the spots where the staphylococcus is sensitive to the phage. The large number of possible combinations of patterns produced by 25 phages will quickly demonstrate the identity or non-identity of a number of strains isolated. For investigation of epidemics, and determining the source, which may be an unaffected carrier or a nurse or doctor with minor skin lesions, phage typing is an essential technique.

THERAPY

The sulphonamides, although active *in vitro*, have given disappointing results in the treatment of staphylococcal infections.

Vaccine therapy has been tried for many years but has never been very successful and has now been largely abandoned. When toxins were identified, toxoid was prepared from the α-toxin and has been used in cases of chronic and recurring infections to raise the immunity of the patient. Although some success was achieved in a few cases, the advantage of toxoid therapy is doubtful.

Antibiotic therapy has now largely replaced these attempts at specific therapy. When penicillin was first introduced, most strains of staphylococci were sensitive, and many dramatic cures were achieved in both acute and chronic cases. Penicillin-resistant strains were found occasionally, and their number has been steadily increasing. About 75 per cent of staphylococcal infections now arising in hospital are due to penicillin-resistant strains, although the majority of infections arising in non-hospitalized patients are probably still sensitive. Many strains of staphylococci produce an enzyme penicillinase, which destroys the penicillin before it can act on the organism. Some of the semi-synthetic penicillins, such as methicillin, are not attacked by penicillinase, and are effective against organisms which produce this enzyme. Many other antibiotics, such as streptomycin, erythromycin, and tetracycline may be used in therapy, but resistance to these and other antibiotics develops within a short time. For rational therapy, it is therefore essential to isolate the organism, and test its sensitivity to a wide range of antibiotics and to choose the most suitable. Penicillin is still preferable, as it is the least toxic of all antibiotics —except in those patients who have developed an allergy to it. Without testing the sensitivity it is impossible to predict the best method of therapy.

STAPH. ALBUS

This organism is a saprophytic commensal invariably present in the skin. Its morphology, staining reaction and cultural characteristics are similar to those of *Staph. aureus*, but it differs from the pathogenic type in that it produces a white pigment, does not produce coagulase, does not ferment mannite, and does not liquefy gelatin as rapidly as *Staph. aureus*. It is rarely pathogenic.

MICROCOCCI

Micrococci are Gram-positive cocci, sometimes forming pigmented colonies. The optimum temperature for growth is 22°–25°C. They are generally non-pathogenic to man, but may be found as secondary invaders or contaminants of skin and the naso-pharynx. One member, *Micrococcus tetragenus* was first isolated as a secondary invader in pulmonary tuberculosis. It divides in two planes at right angles forming tetrads, and forms a capsule in animal tissues. It is probably not very pathogenic for man, but it is highly pathogenic for white mice.

SARCINA

The sarcina are frequent air contaminants. They are Gram-positive cocci dividing in three planes at right angles, thus forming cubical packets. They grow readily on solid media producing large colonies with a bright yellow pigment. They are not pathogenic, but one variety, *Sarcina ventriculi* may frequently be found in the stomach.

Chapter XIV

Streptococcus

Gram-positive cocci, spherical or oval, dividing in one plane forming chains of varying lengths. Catalase negative. Growth usually slight on artificial media. Most species are aerobic or facultatively anaerobic. Few species are strict anaerobes on first isolation. Normally parasitic on man or animals. Some species highly pathogenic.

There is a very large number of streptococci with great variation in their cultural requirements, metabolic activity, and pathogenicity. Various unsatisfactory attempts at classification have been devised and rejected, but a practicable system has been found, particularly for human pathogens. The basic primary classification elaborated by Brown depends on haemolysin production. The best results were obtained by poured plates of horse blood agar containing deep colonies of streptococci. However, it is more convenient to use surface growths and this is now the usual method. The species of blood used for this purpose is important. Erythrocytes of different species vary considerably in their susceptibility to bacterial haemolysins. The classification of streptococci is based on the use of horse blood. A base layer of nutrient agar is poured in the plates and when this has solidified a thin layer of agar containing 5 per cent of horse blood is poured on top of it.

Four distinct types of haemolytic action may be seen with different types of streptococci:

Alpha haemolysis. Individual colonies are surrounded by a zone of green discoloration. This is possibly due to the formation of a green pigment from methaemoglobin. Some of the red cells in this area are seen to be disrupted. Alpha haemolytic strains are called *Strep. viridans*.

Beta haemolysis. The colonies are surrounded by a clear zone. The red cells in this area are completely destroyed, and the haemoglobin diffuses away through the agar leaving a clear transparent area.

Gamma haemolysis. There is no change in the red cells, which are left intact. Organisms of this type are called non-haemolytic streptococci.

Alpha prime haemolysis. The colonies are surrounded by a zone of haemolysis, but in a very narrow ring immediately around and beneath the colony the red cells are still intact. Alpha prime streptococci are uncommon, but should not be confused with the beta haemolytic variety.

CULTURAL CHARACTERISTICS

Many species of streptococci, particularly the more highly pathogenic types do not grow well on ordinary media. Even on media enriched with blood, serum or ascitic fluid growth is never luxuriant. The average colony size is 1–2 mm in diameter after 24 hr incubation at 37°C, and does not greatly increase after prolonged incubation. As a rule, streptococci do not exhibit marked metabolic activity. Carbohydrates are fermented to acid only. Proteolytic activity is weak or absent. Indole is not formed. Pigment is not produced.

Streptococci of a rather more robust type are members of the normal intestinal flora. *Strep. faecalis* strains grow more readily on ordinary media and are more resistant to heat and other destructive agents.

There is also a large group of anaerobic or microaerophilic streptococci which have not been studied to the same extent as the aerobic variety. They are found in the intestine, and frequently in the vagina. Out of their normal habitat they can be pathogenic, and are found in a considerable proportion of brain abscesses, and in some cases of puerperal fever.

ALPHA HAEMOLYTIC STREPTOCOCCI. *STREP. VIRIDANS*

These streptococci are normally present in the upper respiratory tract. They are usually found as secondary invaders in respiratory infections. They are present in apical dental abscesses, and may be isolated by blood culture for a short time after tooth extraction. This is an important factor in the development of subacute bacterial endocarditis. There is much evidence that the organisms in the blood stream initiate bacterial endocarditis on an already damaged valve.

BETA-HAEMOLYTIC STREPTOCOCCI

There is great variation in the pathogenic potentiality of beta-haemolytic streptococci. Some are highly pathogenic and others are probably harmless commensals. Order has been established out of this confusion by the work of Griffiths and of Lancefield, who investigated the problem independently. Griffiths isolated haemolytic streptococci from a large number of severe streptococcal infections such as scarlet fever, erysipelas, and puerperal fever, and developed an agglutination test for classification. He found 27 different types by his method, but found there was no correlation between agglutination type and clinical disease. Scarlet fever or puerperal fever could be associated with a number of different types, and the same type could be found as the causative agent of different diseases. Lancefield elaborated another test for classification. She extracted the specific carbohydrate from the streptococci and used this as antigen in a precipitin test. The antigens extracted from streptococci isolated from most acute human infections proved to be similar, and these organisms were designated Group A. Most of the streptococci typed by Griffiths fell into Lancefield's Group A. The combination of these two methods provided a most useful tool for the epidemiologist. In an epidemic of streptococcal infection it became possible to isolate the organism from a number of cases, determine the group and type, and to search for the contact or carrier who may have been responsible for initiating the infection.

Subsequent work along the same lines soon established order in other haemolytic streptococci. Mastitis in cows is usually due to a beta-haemolytic streptococcus which contaminates milk, and may cause disease in humans. These streptococci had a different carbohydrate antigen and were designated Group B. Streptococci which cause the disease "strangles" in horses fell into Group C. Organisms of this group occasionally cause disease in man.

Group D consists of the beta-haemolytic enterococci which are normally found in the human intestine, *Strep. faecalis*. These are frequently found in human infections, sometimes as secondary invaders in cystitis and pyelitis.

In all, 17 different groups have now been established and identified by their carbohydrate antigen. The most important human pathogens fall into Group A. Groups B, C, D, and G are also found as human pathogens but are less virulent than Group A. Beta-haemolytic streptococci of other groups rarely cause infections in man.

ANTIGENS OF *STREP. PYOGENES*

Lancefield extracted the carbohydrate antigen of streptococci by heating in acid. It can also be extracted by Fuller's method of heating in formamide or by Maxted's enzymic digestion with an enzyme elaborated by *Streptomyces albus*. All three methods are reliable for extracting the specific carbohydrate. This carbohydrate is not antigenic itself when separated from the bacterial cell, but acts as a haptene to combine with antibody and produce a specific precipitate.

The T antigen is a protein which is antigenic both in the intact cell and when isolated. The antibody is not protective. The T antigen is not entirely type specific, and is shared by more than one type. The agglutination test for typing depends to some extent on this antigen, so some non-specific reactions may occur.

The M antigen is an alcohol soluble protein which is type specific, and antibody to it confers type specific immunity. It is associated with virulence, and is present in freshly isolated pathogenic strains. In subculture it may gradually disappear.

Other antigenic substances are present in the streptococcal cell, but have not been fully investigated.

ENZYMES OF *STREP. PYOGENES*

Fibrinolysin. Pathogenic strains produce a fibrinolysin (streptokinase) which will liquefy clotted normal human plasma. Preparations of this enzyme have been used therapeutically to liquefy loculated pus in the pleural cavity.

Hyaluronidase. A few strains of Group A streptococci are capsulated. These capsules consist of hyaluronic acid. Non-capsulated strains may produce hyaluronidase which acts as a spreading factor and extends the area of infection.

TOXINS OF GROUP A BETA-HAEMOLYTIC STREPTOCOCCI

Haemolysins. *Strep. pyogenes* produces various toxins and enzymes. Two distinct haemolysins are present in filtrates of cultures of *Strep. pyogenes.* Most strains produce Streptolysin O and Streptolysin S, but occasional strains may fail to produce one or other of these. Streptolysin O is oxygen labile, and is therefore evident only in anaerobic cultures. Streptolysin S is stable in oxygen. To detect beta-haemolytic streptococci in pathological specimens it is therefore wise to set up two cultures and to incubate one aerobically and the other anaerobically. Both toxins are antigenic, and toxic to animals. Patients infected may develop antibody to Streptolysin O, and the assay of this antibody is sometimes useful in the diagnosis of rheumatic fever. Normal adults usually have a low titre—up to 200 units per ml. There is no titre which is diagnostic of rheumatic fever.

Leucocidin. Leucocidin is produced during the early stages of growth. Polymorphonuclear leucocytes are susceptible to it.

Erythrogenic toxin. Some strains of *Strep. pyogenes*, and particularly those isolated from cases of scarlet fever produce a toxin which is responsible for the rash in scarlet fever. This is a comparatively heat-stable toxin, and is antigenic. It is commonly known as the Dick toxin, since George and Gladys Dick established that a haemolytic streptococcus was the aetiological agent of scarlet fever. After intradermal injection of a suitably diluted dose of toxin, a local erythematous patch develops and reaches its maximum in 24 hr. A skin-test dose (STD = unit of toxin) is the amount of toxin necessary to produce an erythematous patch 1 cm in diameter about 24 hr after intracutaneous injection in a susceptible individual. This test, known as the Dick test, is sometimes used to detect individuals susceptible to scarlet fever. Caution should be observed in employing this test, as too large a dose may produce an extensive rash.

The rash of scarlet fever may be blanched over a small area by the intradermal injection of antistreptococcal serum, or serum from a convalescent case of scarlet fever. This is known as the Schultz–Charlton reaction, and has been used as a diagnostic test, but it is of limited value and is now rarely used.

PROPHYLAXIS

Infections due to *Strep. pyogenes* may be initiated and spread by a carrier who probably shows no clinical sign of infection. Nasal carriers are far more dangerous than throat carriers. In case of an outbreak in a ward or an institution it is desirable to take both nasal and throat swabs from all members of the staff as well as the population of the ward or institution. By determining the group and type of the organisms isolated the source of the infection may be discovered, and precautions taken to prevent further spread.

Individual patients who suffer from chronic or recurrent attacks of disease such as rheumatic fever may be protected by regular dosage of sulphonamides or penicillin. Some work has been done on specific immunization with the M antigen of *Strep. pyogenes*, but the possible advantages and disadvantages of this procedure are still in the experimental stage.

Patients whose heart valves have been damaged by rheumatic fever are known to be susceptible to bacterial endocarditis. As Okell has shown, the infection of the valves with *Strep. viridans* is most likely to arise in the course of bacteraemia following dental extractions. Hobson and Juel-Jensen have investigated this matter and advise complete removal of all teeth in all cases of endocarditis, under heavy antibiotic cover. The 10-year follow-up of their cases has confirmed the benefit of this operation.

DIAGNOSIS

Swabs from throat, nose, pus and various lesions should be plated in duplicate on horse blood agar and incubated aerobically and anaerobically. If there appears to be a mixed infection, they should be plated on blood agar containing 1/300,000 gentian violet to inhibit staphylococci and prevent the streptococci being overgrown by other organisms. Colonies of streptococci should be subcultured to obtain sufficient growth for Lancefield grouping and subsequent typing if required. A simple method of detecting Group A streptococci, particularly for small laboratories with inadequate facilities for complete examination is to place a bacitracin disc on the primary or subculture plate. Bacitracin inhibits the growth of the great majority of Group A strains. In outbreaks and epidemics it is desirable to type all strains isolated. In view of the nephritogenic character of certain types, e.g. Type 12, it is important to be aware of possible complications.

THERAPY

Infections with *Strep. pyogenes* usually respond well to chemotherapeutic agents. The sulphonamides give good results after oral administration, but some strains may become resistant. Penicillin is probably the best antibiotic. *Strep. pyogenes* remains sensitive or only slightly more resistant after prolonged therapy. Several other antibiotics are available for use if the patient is penicillin sensitive.

In subacute bacterial endocarditis due to *Strep. viridans*, penicillin gives good results if used in very large doses, but *Strep. viridans* is very much more resistant than *Strep. pyogenes*. A combination of penicillin and streptomycin is frequently used in these cases.

In infections due to other streptococci, sensitivity tests should be done to choose the appropriate therapy.

Chapter XV

Pneumococcus—Strep. Pneumoniae

Gram-positive cocci which tend to be arranged in lanceolate pairs in tissue or pus. In culture short chains are usually present. Growth is best on media containing blood or serum. Capsules are usually present. The organisms are soluble in bile or bile salts.

In tissues or pus the pneumococcus is usually oval or lanceolate in shape, and arranged in pairs with the broad ends opposed. In culture, particularly in fluid media, short chains are formed and the cocci in the middle of the chain become flattened or more spherical in shape. Capsules are usually present and may enclose a pair or a whole chain of cocci.

CULTURAL CHARACTERISTICS

Growth is best on media enriched with blood or serum. The optimum temperature is 37°C. The organism is a facultative anaerobe. Occasional strains require an atmosphere containing 5 to 10 per cent CO_2. On solid media colonies are round, smooth, and semi-transparent. On blood agar freshly isolated pneumococci produce a raised convex colony which after a few hours undergoes some autolysis in the centre, which becomes flat while the edge is sharply raised, giving the characteristic draughtsman appearance. The colonies are generally mucoid, and this property is greatly exaggerated in Type 3 which produces a large amount of capsular substance. On blood agar α-haemolysis is produced. On heated blood agar the medium is bleached and there is frequently a greenish tinge at the edge of the bleached area. The pneumococcus ferments a number of carbohydrates, and differs from other streptococci in fermenting inulin. It has no proteolytic activity.

Pneumococcal cultures tend to autolyse, and at this stage some of the cocci are Gram-negative. The growth is readily soluble in bile or in a dilute solution of sodium desoxycholate. The reaction of the culture should be neutral or slightly alkaline, as acid will precipitate the bile salt. The pneumococcus is also sensitive to ethyl cupreine hydrochloride (optochin). A disc saturated with optochin on solid media will inhibit the growth of pneumococci around it. These tests are useful to differentiate the pneumococcus from *Strep. viridans*.

SEROLOGY

Pneumococci may be identified by the capsular swelling reaction of Neufeld (Quellung Reaktion). The two main antigenic components of the pneumococcus are (1) a nucleoprotein, and (2) a complex carbohydrate. The protein is common to all types, but the capsular carbohydrate is specific for each type. Up to the present time at least 75 distinct types have been identified. Antiserum against each type may be prepared by using the whole culture as antigen for immunization of rabbits. The capsular substance is a haptene which reacts with its specific serum. A drop of a thin suspension of pneumococci or a drop of sputum or pus is placed on a slide and mixed with a drop of the appropriate antiserum and covered with a coverglass. A swelling of the capsule takes place almost immediately and may be seen as a greatly enlarged halo with a sharply defined margin surrounding singles, pairs, or chains of pneumococci. This technique serves not only as a good diagnostic test, but also identifies the particular type of pneumococcus. Type identification was of great importance when serotherapy was employed in pneumococcal pneumonia.

The specific carbohydrate of the capsule is closely associated with virulence. It can be detected in the blood and urine of patients with pneumonia.

Important experimental work on the pneumococcus by Fred Griffiths, 1927, has contributed to the development of modern genetics. He showed that if a mouse was injected with a killed culture of Type 2, and subsequently with a living culture of Type 1, Type 2 pneumococci could be cultured from the mouse tissue at autopsy. This transformation of type has subsequently been confirmed many times. This phenomenon was later investigated by Avery, who showed that the transforming principle consisted of DNA from the killed pneumococci. It is not known whether this transformation of types occurs in clinical infections.

PATHOGENICITY

In man the pneumococcus is frequently found as a commensal in the naso-pharynx of healthy individuals. It is also an important pathogen responsible for many diverse conditions such as lobar pneumonia, broncho-pneumonia, empyema, otitis media, meningitis, endocarditis, sinusitis and primary peritonitis of young children. The pathogenicity of the pneumococcus for laboratory animals is variable. Rats are highly resistant, while white mice and rabbits are extremely susceptible. In white mice, intraperitoneal injection of a fresh virulent strain is followed by purulent peritonitis, septicaemia and death in 24–48 hr.

PNEUMONIA

Lobar pneumonia in about 90–95 per cent of cases is caused by the pneumococcus. Although many people are carriers of the pneumococcus in the naso-pharynx, pneumococcal infection is usually due to a different type from the carrier strain. The organism is probably conveyed from healthy or convalescent carriers by droplet infection. Lobar pneumonia is rarely a primary disease; it is preceded by some infection of the upper respiratory tract such as the common cold, laryngitis or influenza. These infections may facilitate the spread of the pneumococcus from the naso-pharynx to the lung. The severity of the infection is dependent largely on the type of the infecting pneumococcus and the age and general condition of the patient. Pneumococcal empyema is often a complication of pneumonia. Otitis media and meningitis occur most frequently in young children but may affect patients of any age. Primary pneumococcal peritonitis occurs most frequently in female children.

DIAGNOSIS

Material sent for diagnosis is usually pus or sputum. The identification of the organism and type may be rapidly made by direct microscopic examination if a set of specific typing sera is available. Cultures should be made aerobically and anaerobically on blood agar. Characteristic colonies may be picked off and subcultured for serological typing, tests of inulin fermentation and sensitivity to optochin and bile salts. The typing is of less importance since serotherapy has been replaced by antibiotic and chemotherapy, but a capsule swelling test on the organism in the pathological specimen provides the most rapid means of diagnosis.

THERAPY

The pneumococcus is usually very sensitive to the sulphonamides, particularly sulphadiazine and sulphadimidine. It is highly sensitive to penicillin, which should be considered the treatment of choice. In pneumococcal meningitis penicillin should be given intrathecally in doses not exceeding 20,000 units to avoid cerebral irritation. In patients who are penicillin sensitive, many other antibiotics in common use are effective.

Chapter XVI

Neisseria

Gram-negative cocci, usually arranged in pairs with flattened surfaces adjacent. Strict parasites growing poorly on simple media. Catalase positive. Oxidase positive. Aerobic. Normal habitat of most species is the nasopharynx. Some species are pathogenic.

Neisser (1879) described the first member of this genus, the gonococcus, which he observed in the urethral pus of patients with gonorrhoea. It was subsequently cultivated on artificial media, and its aetiological relationship with gonorrhoea conclusively established. Weichselbaum (1887) isolated the meningococcus from the cerebro-spinal fluid of cerebro-spinal meningitis cases. Pfeiffer (1896) described the *Micrococcus catarrhalis*, and von Lingelsheim (1906) described *Micrococcus pharyngis*. With the exception of the gonococcus, the natural habitat of most species is the naso-pharynx of healthy or infected persons.

NEISSERIA GONORRHOEA

Morphology. The gonococcus is an oval or spherical coccus, 0·6–0·8 μ in diameter, usually arranged in pairs with the adjacent sides flat or concave so there is a small oval space between them. The long axis of the gonococcus is at right angles to the axis joining the two cocci. This arrangement is usually found in pus, when the majority of the cocci may be inside leucocytes. Under conditions of artificial cultivation the cocci tend to become spherical, and varied arrangements may be found such as pairs, small groups and single forms.

CULTURAL CHARACTERISTICS

The gonococcus is not easy to cultivate. Growth is very slight or absent on most simple media; enrichment with some animal protein such as blood, serum or ascitic fluid provides a satisfactory medium which in all cases must be freshly prepared and moist. Chocolate agar is a very satisfactory medium. The optimum temperature is 37°C, and the temperature range is small. Aerobic conditions are required, but on first isolation growth is frequently improved by the presence of 10 per cent CO_2 and water vapour.

The optimum reaction of the media is pH 7·4–7·6, but growth may occur over a wider range provided the medium and atmospheric conditions are satisfactory.

In fluid media, e.g. serum-broth, growth is slow and poor. In 24–48 hr a slight granular deposit forms with little or no general turbidity.

On solid media the appearance of colonies, on first isolation, is sometimes delayed for 48 hr. They are small, 0·6–1 mm in diameter, moist, greyish-white and semi-transparent, tending to become more opaque and somewhat granular on further incubation. In some cases the edges of the colonies may exhibit marked scalloping, while the colonial appearance may be extremely variable.

BIOCHEMICAL ACTIVITY

The gonococcus exhibits very little biochemical activity. It ferments glucose, but not maltose to acid when grown in Hiss's serum water. It produces catalase and oxidase. No exotoxin is produced, but there is some evidence of an endotoxin.

RESISTANCE

The gonococcus is readily destroyed by bactericidal agents. It is susceptible to desiccation and is killed by heat at 55°C in 5 min; silver compounds have a strong bactericidal action, and in the past were extensively used in the therapy of gonococcal infections.

Owing to the low degree of resistance and the tendency to autolysis the conservation of stock-cultures is difficult. Cultures should be prepared on moist suitable media, incubated at 37°C, and subcultured every 7–10 days, or dried *in vacuo*.

SEROLOGY

Agglutination, agglutinin-absorption and complement fixation have been employed by many workers to sub-group the gonococci. The results, however, have been far from conclusive. A large variety of sub-groups appear to be present, but the differentiation of many of these is not sharply defined. In spite of this many workers have been able by agglutination to make a primary subdivision into two main serological groups (types I and II). Most of the strains isolated from acute cases fall into type I, whereas strains from chronic cases tend to belong to type II. It is considered that the transition of type I strains to type II is a gradual process resulting in the formation of numerous sub-types, which are responsible for the frequent irregularities.

The gonococcus appears to have a complex antigenic structure. Fractionation tests suggest that a group-polysaccharide and a group nucleoprotein are present together with certain undefined type-specific factors.

PATHOGENICITY

The gonococcus gives rise to spontaneous infection only in man, gaining entrance by either the genito-urinary mucosa or the conjunctiva and producing eventually a purulent infection.

Its offensive weapons have been incompletely examined, as laboratory animals are refractory to infection; gonorrhoea has, however, been reproduced by the inoculation of the gonococcus into the human urethra. The intraperitoneal injection of either a living or a lysed culture into guinea-pigs and rabbits gives rise to a definite, but non-specific, disturbance which may end fatally. The responsible substance is probably of the nature of an "endotoxin".

GONORRHOEA

Gonorrhoea is an ancient disease, in which infection is conveyed directly by sexual intercourse. Following implantation of the gonococcus on the urethral mucosa there is a variable incubation period usually of two to three days, after which a mucoid discharge, which rapidly becomes purulent, appears. The organism penetrates between the columnar cells of the mucous membrane, and inflammatory changes occur around the mucous glands and in the submucosa. If adequate treatment is not undertaken immediately, the infection may spread, either by continuity along adjacent surfaces or through dissemination by the lymphatic system or blood stream.

Spread by continuity is a frequent event. Lesions of the posterior urethra, prostate, epididymis and testicles are not uncommon in the male. In the female the infection passes to the vagina, which in adults appears to be relatively resistant, and inflammatory changes are often produced in Bartholin's gland and the cervix of the uterus. Spread along the endometrium is not common, but when it occurs infection may pass to the ovaries and peritoneum by the Fallopian tubes, where salpingitis or pyosalpinx may be produced. The rectum may also be involved, in which event a purulent proctitis develops.

The generalized infection is more frequent in males than females. The gonococcus enters the blood stream, from which it may be cultured, and tends to localize in joints and the cardiac endothelium. Before the introduction of chemotherapy, gonococcal arthritis was a relatively common complication; it was considered by some workers to occur in about 1 per cent of all cases of gonorrhoea. Endocarditis occurs less frequently but is a more serious complication. In a few isolated cases metastatic lesions are found in the pleural cavity, peritoneum and meninges.

In the absence of satisfactory treatment, the organisms may remain viable in the tissues for many years and may not be detected by routine bacteriological examination. After the acute symptoms subside the disease takes a chronic course, in which event the urethral discharge may be scanty. These cases are, nevertheless, capable of infecting others, and any subsequent temporary lowering of the general resistance may be followed by an acute exacerbation of the infection

In children the gonococcus may give rise to a vulvo-vaginitis; in these cases the infection is usually conveyed by contaminated towels, clothes, linen, etc. The infection tends to remain localized in the vagina, which appears to be more susceptible than in adults, but occasionally it may spread to the peritoneum.

When the mother is suffering from gonorrhoea, infants may acquire, during delivery, a purulent infection of the conjunctiva, termed "ophthalmia neonatorum". This condition may also be caused by other organisms and it decreased considerably as a result of the introduction by Credé of the prophylactic instillation of some silver compound into the eye immediately after birth. Penicillin is now used for this purpose.

DIAGNOSIS

Three main methods may be employed in the bacteriological diagnosis of gonorhoea:

(1) Microscopical examination of smears of pus stained by Gram's method.
(2) Culture of swabs from lesions.
(3) Serological methods.

(1) Microscopy: Microscopical examination of stained smears is the simplest method, and frequently the only one used. In acute cases when there is no secondary infection, the Gram-negative diplococci are readily seen both intracellular and extracellular. Many of the pus cells are free of organisms, but some contain very large numbers which eventually destroy the leucocyte, and the gonococci may then be seen free until they are taken up by another leucocyte. In chronic cases in the male the organisms may be found after prostatic massage, but the picture is confused by the

presence of other organisms, and is never so clear as in acute cases.

In females, smears from the urethra, cervix and exudate from the ducts of Bartholin's glands, may prove satisfactory in the early stages, but the large number of other organisms usually present makes the identification of the gonococcus much more difficult.

(2) Culture: The best diagnostic method is the culture of the gonococcus, and its isolation in pure culture for confirmation. The collection of material, transfer to the laboratory, selection of medium, and conditions of incubation require careful attention. In acute cases in the male it is usually easy to obtain pus uncontaminated with other organisms. In the female, material should be collected from both urethra and cervix, and this is almost always contaminated with saprophytic commensals or organisms of secondary infection. As the gonococcus is very susceptible to drying, cultures should be made immediately, but as this is not always possible, swabs should be plunged immediately into Stuart's medium, which will preserve them up to 24 hr during transfer to the laboratory. Of the media normally available as a routine, chocolate agar is probably the most suitable if the specimen is not too contaminated. A better medium, particularly for cultures from females or chronic male cases where contaminants and secondary infection is common, is described by Thayer and Martin, and modified by Berger (see page 416). This medium contains antibiotics which suppress the growth of coliforms and some Gram-positive cocci without affecting the growth of the gonococcus. The detection of gonococcal colonies in mixed cultures is facilitated by the oxidase reaction. A freshly prepared solution of 1 per cent tetramethyl-p-phenelynediamine is poured on the plate and immediately poured off. Colonies of gonococci immediately develop a bright purple colour. Subcultures should be made immediately as the organism tends to die soon after the addition of the reagent. A film should be made to confirm the presence of Gram-negative diplococci, and the identification of the organism should be made on the subculture. This is sometimes necessary where legal evidence of infection is required.

NEISSERIA MENINGITIDIS, MENINGOCOCCUS

Morphology. The morphology and staining reaction of the meningococcus are practically identical with that of the gonococcus. In the cerebro-spinal fluid it is frequently found intracellularly as diplococci, but in culture may be in pairs, singly or in irregular groups. Autolysis occurs in culture and blown-up pale staining ghost forms are frequent

CULTURAL CHARACTERISTICS

The meningococcus is usually easier to grow than the gonococcus. On first isolation enriched media, such as chocolate agar, blood agar or serum agar, are necessary, but, after several subcultures, growth may occur on the ordinary media; it is, however, never profuse.

The optimum temperature is 37°C, and growth may be obtained over a range of 30°–40°C. It is a strict aerobe, there being no growth under strict anaerobic conditions. It has been found that growth is more marked in the presence of 10 per cent CO_2; this is particularly important in the isolation of the organism from cerebro-spinal fluid on a blood agar plate. The most favourable reaction of the media is pH 7·4–7·6.

On solid media, in 24–48 hr, the meningococcus frequently forms characteristic colonies which are smooth, translucent and round (1–2 mm diameter) with a regular edge and a bluish-grey tinge. They are readily picked off and emulsify without difficulty. On further incubation the colonies become opaque, slightly yellowish and may be differentiated into a central granular area with a distinct peripheral zone.

Table I

Fermentation Reactions of Certain Members of the Genus Neisseria

Organism	*Glucose*	*Maltose*	*Sucrose*
N. catarrhalis	−	−	−
N. gonorrhoea	+	−	−
N. meningitidis	+	+	−
N. pharyngis	+	+	+−

In fluid media growth occurs as a slight general turbidity with a granular deposit.

BIOCHEMICAL ACTIVITY

Glucose and maltose in Hiss's serum water are fermented to acid. Oxidase and catalase are produced. An endotoxin is present in lysed cultures, which will produce fatal toxaemia in laboratory animals after intraperitoneal injection.

RESISTANCE

The meningococcus is readily destroyed by bactericidal agents. It is killed by heat at 55°C in 5 min and 1 per cent phenol in 1 min. Methylene-blue has a specific inhibitory action in high dilutions.

Marked autolysis takes place after incubation at 37°C for several days; this effect is considered to be due to the action of an intracellular enzyme.

Stock cultures are difficult to conserve owing to the marked readiness with which the organism dies out. Survival is longest at 37°C, and on Dorset's egg medium, as on this medium growth is rather poor and autolytic changes are delayed. The use of semi-solid blood agar has also given satisfactory results; drying *in vacuo* is satisfactory.

SEROLOGY

There have been several schemes of classification of the meningococci into different groups, but none of these has met with universal acceptance. The latest system adopted by a sub-committee of the Nomenclature Committee of the International Association of Microbiologists divides the organisms into four groups designated A, B, C and D. This is not entirely satisfactory, as there are still some types which do not fit into any group. It has been shown that the antigens of the organisms of one group may be changed to that of another by growing them in the presence of the deoxyribonucleic acid extracted from the organisms of another group.

PATHOGENICITY

The common effect of the meningococcus is probably to produce a mild rhino-pharyngitis, which normally resolves without any particular therapy. The very serious diseases caused by this organism are probably complications arising in the course of an otherwise mild infection. Cerebro-spinal meningitis cases occur usually in overcrowded institutions with inadequate ventilation. Outbreaks are common in war time, when fatigue, overcrowding and other factors may lower the resistance of individuals. Epidemics are also frequent in native populations in Africa during the dry season. Before and during epidemic periods the carrier rate may be very high, from 25 to 80 per cent, but the incidence of meningitis rarely exceeds 1 or 2 per cent. The route of infection of the meninges has not been definitely established. There is generally an invasion of the blood stream, as is indicated by the petechial rash—hence the name spotted fever. Meningococci may be found in these petechial spots. It is unknown whether the meninges are infected by the organisms in the blood stream, or whether the organisms reach the meninges by the perineural lymphatics of the olfactory nerve via the cribiform plate, and then produce a bacteraemia.

Meningococcal septicaemia may occur without meningitis. Chronic meningococcal septicaemia is characterized by recurrent pyrexia, muscle and joint pains and a rash, often petechial, over the limbs and back of the body. The rash disappears in a few days but tends to reappear in the next pyrexial period. The disease runs a chronic course and may persist for many months if untreated.

The acute form of meningococcal septicaemia is a fulminating condition with extreme collapse and a high mortality rate. It is usually associated with the Waterhouse–Friderichsen phenomenon in which there is bilateral haemorrhage into the adrenals.

DIAGNOSIS

(1) **Cerebro-spinal fluid.** In cases of meningitis the fluid is turbid, containing a large number of polymorphonuclear leucocytes, and is under greatly increased pressure. The meningococci are present in variable numbers. In some cases there are large numbers of Gram-negative diplococci, both intra- and extracellular. In other cases the organisms presumably undergo autolysis and there may be great difficulty in finding typical meningococci. It is important to examine the fluid before any therapy is given, as even a very small dose of sulphonamide or antibiotic greatly increases the difficulty of making a diagnosis.

The organisms may be cultured on a suitable medium such as chocolate agar, blood agar, or Loeffler's serum, aerobically in an atmosphere containing 5–10 per cent CO_2. They may be identified by fermentation tests and agglutination test.

(2) **Blood culture.** In the pyrexial period and before treatment, blood culture is sometimes successful, but it may be necessary to repeat this several times to obtain a positive result.

(3) **Naso-pharyngeal swabs.** The meningococcus may usually be found in the naso-pharynx of cases and carriers, but the presence of the normal flora makes the isolation more difficult. Specimens are best taken with a West's swab, which is protected by a curved glass tube from the oral flora. The tube is passed to the back of the mouth, and the curved end pointed to the naso-pharynx. The swab is then pushed through and rubbed on the mucous membrane, and withdrawn into the glass tube. Cultures may be made on chocolate agar or blood agar, but are best made on Berger's medium as described for the gonococcus and incubated in an atmosphere containing 10 per cent CO_2. Colonies of meningococci may be detected by the oxidase reaction. They should be subcultured immediately and pure cultures should be confirmed by fermentation and agglutination tests.

PROPHYLAXIS

There is no specific method of immunization. Dissemination of the meningococcus should be reduced by prevention of over-crowding and provision of adequate ventilation. In the event of an epidemic of meningitis, chemoprophylaxis with sulphadimidine snuff taken in three or four doses over a period of two days has proved successful, but it is essential that the whole population at risk should be treated at the same time, as the effect is temporary, and individuals who have been treated will be exposed to recolonization or infection by untreated individuals.

THERAPY

Chemotherapy with various sulphonamides has proved very successful in meningococcal meningitis. Early, intensive and regular administration has greatly reduced the mortality rate. Recently, there is evidence that some meningococci, particularly Type B, have developed resistance to sulphonamides. The meningococcus is very sensitive to penicillin, which may be administered intramuscularly and intrathecally, and also to other antibiotics. In areas where sulphonamide resistance has developed, it will be necessary to replace sulphonamides with antibiotics.

NEISSERIA CATARRHALIS

N. catarrhalis is a Gram-negative coccus usually arranged in pairs or tetrads. On first isolation from the tissues it grows rather poorly on simple media which should be enriched with serum or blood. After several subcultures it grows readily on the ordinary media. On agar, variation of the colonial appearance may be found; one common type is definitely roughish and another is smooth, resembling that formed by the meningococcus. In broth growth occurs as a granular deposit. The optimum temperature is 37°C, but multiplication may take place at 25°C. Its biochemical activities are very limited, carbohydrates are not fermented (Table I).

N. catarrhalis is normally present in the naso-pharynx of man, but it has a low pathogenicity for man and animals. It is usually found as a secondary invader in lesions of the upper respiratory tract, and is therefore usually included in vaccines prepared to counter the secondary effects of colds and influenza. Its main claim to importance is that it must be excluded in the examination of naso-pharyngeal swabs for the meningococcus.

NEISSERIA PHARYNGIS (MICROCOCCUS PHARYNGIS FLAVUS AND SICCUS GROUP)

A number of Gram-negative cocci, some forming yellowish colonies on agar, have been isolated from the upper respiratory tract of man. These have been incompletely investigated but they have been tentatively placed into different groups, according to their cultural and fermentation reactions; some three or six groups have been described by different workers. This sub-grouping is unsatisfactory, and the various cocci are probably best considered tentatively with the other indefinite Gram-negative cocci found in this situation as one heterogeneous group, *N. pharyngis*. They are normally present in the upper respiratory tract, where they may exercise a mild pathogenic action as secondary invaders; they must be differentiated from the meningococcus, and they consequently interfere with the isolation of the latter organism from the naso-pharynx.

VEILLONELLA

There is a group of Gram-negative cocci which will grow on the ordinary media but only under anaerobic conditions; these are included in the genus *Veillonella*. They are commensals usually found in the upper respiratory and alimentary tracts; they are doubtful pathogens.

Chapter XVII

Coliform Bacteria

The lower intestinal tract of man and animals contains a vast number of organisms of various genera. Normally, probably 40 per cent of the weight of faecal excretions consists of bacteria, some of which may be recent acquisitions which have survived the acid of the stomach, but the majority are descendants of the flora of a much earlier age. The profuse normal flora does no harm, but on the contrary produces some useful components of the vitamin B complex which are absorbed. The organisms most readily grown from this vast population are generally known as the coliform bacteria—a vague term which indicates their origin in the colon but gives no indication of the enormous number of species of different genera which it includes. The majority of these coliform organisms are saprophytes, but there are numerous pathogens which are also included in the term coliform bacteria or enterobacteria.

The common characteristics of the coliform bacteria are: Gram-negative rods which grow readily on simple media. Motility is variable. They are biochemically fairly active. They metabolize a variety of carbohydrates. Some are proteolytic. Some are able to utilize simple sources of carbon or nitrogen. Endotoxins are present in some.

It is impossible to differentiate these organisms by microscopy. The differentiation of the pathogens from the saprophytes and the identification of both involves various biochemical and serological examinations. Even those organisms normally regarded as saprophytes frequently require detailed examination and classification as they may at times be important secondary invaders in other infections, and in some cases particular strains have been found to have developed pathogenic characters.

Some of the genera which are included in the term coliform bacteria are: ***Escherichia***, ***Klebsiella***, ***Citrobacter***, ***Hafnia***, ***Serratia***, ***Proteus***, ***Pseudomonas***, ***Shigella***, ***Salmonella***, ***Alkalescens***, ***Arizona***. The more important members will be described in greater detail.

In the examination of faecal specimens, cultures are made on a selective medium which restricts the growth of Gram-positive organisms and favours the growth of coliform bacteria. Lactose is incorporated in the selective medium, with an indicator, as most of the common saprophytes ferment lactose to acid or acid and gas, and most of the pathogens are non-lactose fermenters. A preliminary differentiation is thus made on isolation in the first culture, and single colonies of probable pathogens may be picked off for further examination by biochemical and serological tests. Useful differential tests are:

Urease production—splitting urea with production of ammonia.
Carbohydrate fermentation—tested on lactose, glucose, maltose, mannitol, sucrose, dulcitol and others if required.
Voges–Proskauer reaction — (VP) production of acetyl-methyl-carbinol.
Protease—liquefaction of gelatin.
Indole production from tryptophane.
Hydrogen sulphide production.
Growth or inhibition of growth in KCN medium.
Growth in medium containing citrate as the sole source of carbon.

SEROLOGY

Further differentiation of species is made by agglutination tests. Final identification in the case of the Shigella and Salmonella can only be made by serological tests. Every organism contains many antigens, and some of these antigens are common not only to other members of the same genus, but also to coliform organisms of other genera. It is essential, therefore, to use agglutinating sera from which antibody to common antigens has been absorbed in order to achieve specific reactions. It may be necessary to determine several somatic and flagellar antigenic factors in an organism to complete the identification.

ESCHERICHIA

E. coli is widely distributed in nature, being present in soil, surface water, and almost invariably in the faeces of man and animals. There are many different types which may be differentiated by biochemical and serological tests, but probably all are of faecal origin. They ferment lactose, glucose, mannitol, and sometimes sucrose to acid and gas, and usually produce indole. They have no appreciable proteolytic activity.

One sub-group of *E. coli* is of practical importance in the bacteriological examination of water (see Chapter XXXIV). Strains which ferment lactose to acid and gas at a temperature of 44°C are considered to be due to fresh faecal pollution, and to indicate the possibility that more pathogenic organisms such as salmonella could also be present. The detection of coliform organisms is technically simple, and the presence or absence of coliforms may be used as an important indication of the hygienic quality of water and food supplies.

Another group of *E. coli* is of importance as a cause of gastro-enteritis in young infants. These strains are identified serologically by their somatic antigens. *E. coli* may have flagellar "H" antigens, surface "K" antigens and "O" somatic antigens. The K antigens which appear to be related to pathogenicity may lead to difficulty in typing with O antisera. Serotypes 026, 055, 0111, 0119 and 0128 and probably several others are associated with gastro-enteritis.

With the exception of the enteropathogenic strains, *E. coli* is not pathogenic in the intestine, but elsewhere it has pathogenic properties. It is frequently a cause of, or a secondary invader, in cases of cystitis, pyelitis, pyelonephritis and peritonitis.

RESISTANCE

In vitro, *E. coli* is sensitive to heat and the usual bactericidal agents. Its profuse growth in culture can be partially inhibited by tetrathionate or selenite, and by brilliant green or malachite green in media for the isolation of other types of organisms which might be overgrown by *E. coli*.

For therapeutic purposes in *E. coli* infections, it is essential to isolate the strain and test its sensitivity with a series of antibiotics. Resistance to antibiotics usually develops rapidly.

KLEBSIELLA

Several species are included in this genus, but the taxonomic position of some of them is rather doubtful. They grow profusely on simple media and produce a large amount of capsular material which gives the colonies a typical mucoid appearance. They rapidly ferment many carbohydrates to acid and gas. Most species are pathogenic. The type species is *Klebsiella aerogenes*.

K. aerogenes is usually present in faeces, but in much smaller numbers than *E. coli*. It is also commonly found on vegetation and in soil. It is more resistant than *E. coli*, and may survive longer out of the intestine. Like *E. coli* it may be found in cases of cystitis, pyelitis and peritonitis.

K. friedländeri is very similar to *K. aerogenes*. The normal habitat of *K. friedländeri* is the naso-pharynx and the respiratory system, but may be found occasionally in the faeces. It is the cause of a very severe type of pneumonia, and sometimes causes infection of the sinuses and otitis media. It is generally more virulent than *K. aerogenes*. A very small dose of a culture injected intraperitoneally will kill a mouse in about 24 hr.

K. rhinoscleromatis produces a progressive granulomatous lesion of the nose, which occurs mainly in Mediterranean countries.

K. oezena has been isolated from cases of oezena, although its pathogenic importance in this disease is uncertain.

K. cloaca has been placed in this genus, although the validity of this classification is doubtful. It differs in many respects from the type species. It is motile, liquefies gelatin, and is not always capsulated.

CITROBACTER

The citrobacter (Ballerup-Bethesda group) are intestinal organisms which, like *E. coli* and *K. aerogenes*, may be present in urinary infections, peritonitis and pyogenic infections. They can utilize citrate as the sole source of carbon, and are not inhibited in KCN medium. They have some antigenic relationship to the salmonella, having similar somatic and Vi antigens, but differ from the salmonella in being able to grow in KCN medium.

SHIGELLA-DYSENTERY GROUP

The shigella genus is divided into four sub-groups: *Sh. dysenteriae*, *Sh. flexneri*, *Sh. boydii* and *Sh. sonnei*. Each sub-group consists of a number of serologically distinct types. They are related to the escherichia and share some somatic antigens with them. All the shigella are non-motile. They ferment some carbohydrates to acid, but not to gas with the exception of some organisms of the flexner sub-group. Only one organism, *Sh. sonnei* is a late lactose fermenter. The organisms of the *Sh. dysenteriae* sub-group ferment glucose but not mannitol. All others ferment both sugars.

SUB-GROUP A

Sh. dysenteriae Type 1 (*Sh. shiga*).

Sh. shiga was first isolated by Shiga in Japan. It is the cause of the most severe type of dysentery, which is common in the East, but very rare in Great Britain. Filtrates of autolysing cultures contain a soluble toxin. This is not an exotoxin secreted by living cells, but is found after autolysis of the cells. It was at one time considered to be a neurotoxin, as its action in experimental animals appeared to be on the central nervous system. However, it has now been shown that its action is primarily on the small blood vessels of the nervous system. The toxin can be converted to a toxoid by treatment with formalin. The toxoid is antigenic, and an antiserum can be prepared for therapeutic use. Shiga dysentery is the only type in which serotherapy is applied.

Sh. dysenteriae Type No. 2 (*Sh. schmitzii*).

Sh. schmitzii is serologically different from *Sh. shiga*, and also differs by the production of indole. It is comparatively rare in Great Britain, but is found in various European countries.

Sh. dysenteriae Types 3–10 (Large–Sachs group)

These eight types were formerly known as the Large–Sachs group. Biochemically they differ slightly from Types 1 and 2, and are serologically distinct. They are uncommon in this country.

SUB-GROUP B

SHIGELLA FLEXNERI

This group of dysentery bacilli consists of at least six types, and two degenerate variants which have lost certain somatic antigens. They generally ferment mannitol to acid, and produce indole. Serotype 6 (Newcastle and Manchester sub-types) also produce gas in certain peptone media. These organisms have some common antigenic components, but can be distinguished serologically by their type specific antigens.

SUB-GROUP B

SHIGELLA BOYDII

This group of dysentery bacilli consists of 15 different serotypes. They all ferment mannitol. Indole production is variable. The types are antigenically distinct, and are identified by agglutination tests.

SUB-GROUP D

SHIGELLA SONNEI

There is only one organism in this group. It ferments mannitol, and is a late fermenter of lactose, and sometimes sucrose. It does not produce indole. On isolation from fresh cases *Sh. sonnei* is smooth, but in convalescent cases and in laboratory strains a rough type develops. *Sh. sonnei* is the most common dysentery organism in Great Britain. Outbreaks in schools and residential institutions are fairly common.

PATHOGENICITY

Dysentery is essentially a disease of man, but is occasionally found in monkeys. Ingestion of dysentery organisms by laboratory animals has little effect. Subcutaneous, intraperitoneal or intravenous inoculation of cultures may result in death due to intoxication, but dysentery does not develop. In bacillary dysentery the organisms are introduced by the mouth, usually by means of contaminated food or water. After an incubation period of 12 hr to several days, abdominal pain and diarrhoea appear. These symptoms increase in severity; colic, tenesmus and the appearance of blood and mucus in the faeces are common features. In some cases, particularly in *Sh. shigae* infections, nervous symptoms, such as muscular twitchings and prostration, are seen. The organisms usually remain localized to the large intestine, and are rarely isolated from the blood stream and never from the urine.

In some cases the lower part of the ileum may be involved. In the initial stages there is catarrhal inflammation and congestion of the mucosa, which subsequently undergoes necrosis, sloughing and ulceration. *Sh. shigae* infections tend to be more serious than *Sh. flexneri* or *Sh. sonnei* infections; the mortality rate is greater, in some outbreaks up to 2 per cent, and the signs of toxaemia are more pronounced. The development of a chronic stage is not uncommon. On recovery, some cases remain faecal carriers.

Carriers are the usual source of infection, and the main vehicle of infection is food which becomes contaminated by handling.

DIAGNOSIS

The bacteriological diagnosis of bacillary dysentery is made by culture of the faeces or rectal swabs as early as possible during the infection. On macroscopical examination, faeces are usually fluid, containing mucus and blood. After 24–48 hr pus cells appear and may be very numerous. Deoxycholate-citrate medium is satisfactory for culture. Colonies after 18 hr incubation are about 1–1·5 mm in diameter, colourless, round, smooth and often sticky when touched with a needle. Typical colonies are subcultured on to a non-inhibiting medium and subjected to the biochemical tests of fermentation and indole production which will give an indication of the serological tests required for final identification.

PROPHYLAXIS

General hygienic measures, control of sanitation and flies, and protection of food from contamination should receive careful attention. Detection and control of carriers is important. Specific prophylactic measures such as vaccination have not been employed to any extent, except in the case of Shiga dysentery, where immunization with toxoid is possible.

THERAPY

In Shiga dysentery, specific antitoxin is of value. In other types results have been unconvincing, possibly because of the difficulty of preparing a serum with an adequate concentration of antibody to the large number of antigenic types which may be encountered.

The sulphonamides have proved very successful in the treatment of bacillary dysentery. The derivatives most widely used are sulphaguanidine and succinyl-sulphathiazole, neither of which is readily absorbed from the intestinal tract. Large doses are necessary, usually more than 50 g for each course. The clinical response is good, but in some cases the organisms still persist for prolonged periods. These clinically cured carriers constitute a hazard.

Antibiotic therapy may be useful but the sensitivity of the strain isolated should be tested to determine the appropriate antibiotic. Resistance to antibiotics and to sulphonamides is not uncommon.

SUMMER DIARRHOEA

A clinical condition closely related to dysentery is summer diarrhoea, which may occur in epidemic form during the summer and autumn months. It was at one time a widespread and serious disease of young children, but the incidence has decreased considerably in recent years as a result of improved hygienic conditions.

In spite of the great prevalence of summer diarrhoea, no specific aetiology has been determined. Many different organisms have been incriminated: *Proteus vulgaris*, *Proteus morgani*, *Ps. pyocyanea*, *Cl. welchii*, and various dysentery bacilli. The seasonal incidence is due largely to the greater possibility of bacterial multiplication in milk and various foods during hot weather. Very large doses, even of normally saprophytic organisms, may produce toxic by-products or endotoxins in sufficient amounts to irritate the gastrointestinal tract of young children.

SALMONELLA

Gram-negative rods. Usually motile. Grow readily on simple media. Most strains produce acid and gas in glucose and mannitol, but do not ferment lactose. Urease negative. V-P negative. H_2S positive. Citrate positive. KCN negative. Serotypes are closely related by somatic and flagellar antigens. Primarily intestinal parasites of man and animals. Pathogenic to many species, and produce enteric fever or food poisoning in man.

MORPHOLOGY AND CULTURE

Salmonella are similar to other coliform bacteria, and cannot be differentiated microscopically. They grow readily on simple media, and are active biochemically. Biochemical reactions provide useful preliminary evidence of identification, but final identification must be made by serological methods.

SEROLOGY

Bruce White introduced a system of antigenic analysis for the identification of the few types at that time. This scheme was greatly extended by Kauffmann and other workers and is now known as the Kauffmann–White scheme. There are now over 1,000 species of salmonella known. The convention has been adopted that these should be named according to their geographical origin. There are few important cities which are not commemorated in the salmonella list.

The antigenic components consist of the somatic O and flagellar H antigens; both of these are variable. Like other bacteria, a change from the smooth to the rough form may occur, with a loss of some somatic antigen. A motile form may lose flagella and become non-motile. Furthermore, flagella may undergo a change of phase, and then have different antigens. These are known as specific and group phases, now commonly called Phase 1 and Phase 2. For identification of a salmonella, it is necessary to determine the O antigen, and the flagellar antigens of both phases.

The O antigens are numbered in arabic numerals. Phase 1 flagellar antigens are indicated by small letters—a, b, c, etc., up to z, after which they are z_1, z_2, z_3, etc. Phase 2 antigens are usually indicated by arabic numerals, but in some cases the antigens of Phase 2 of one organism may be similar to the Phase 1 antigens of another. These inconsistencies have been revealed in the vast expanse of the salmonella population which was not suspected at the time antigenic analysis and nomenclature was introduced.

The possible changes in antigenic structure are illustrated in Fig. 1.XVII.

The normal smooth form may be O+H Phase 1 (specific)
or O+H Phase 2 (group).

If the organism loses flagella, the change is represented by

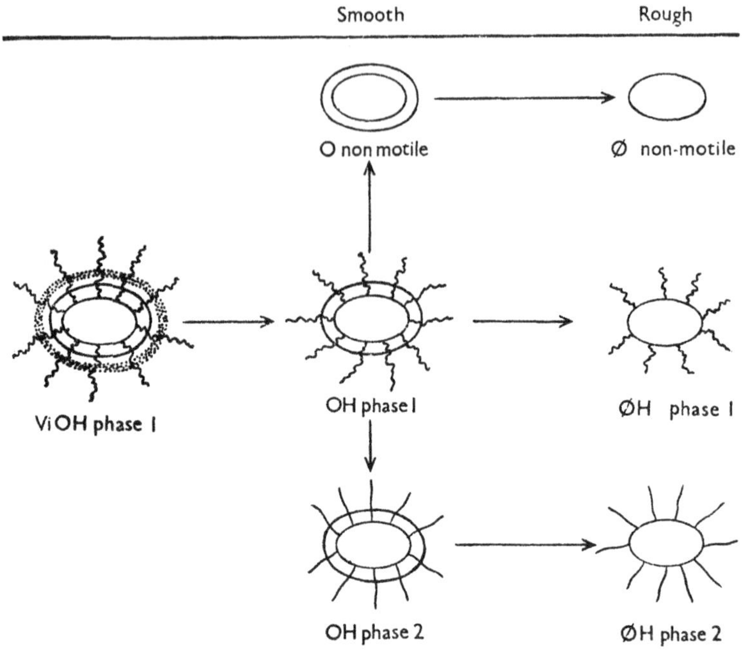

Fig. 1.XVII

O+H → O. In the rough variants, the somatic antigen is represented by Ø, and the change to rough is indicated by

O+H → Ø+H rough, motile
or O+H → Ø rough, non-motile.

Salmonella which have similar O antigenic components are grouped together, and these groups are indicated by the letters A, B, C, etc. (Table 1). For example, in Group B, O antigens 4, 12 are common. In Group C, antigens, 6, 7, and in Group D, 9, 12. In the identification of an organism suspected of being a salmonella, it should first be tested with polyvalent serum. The O antigens are then determined; reference to the list will then indicate the most probable H antigens. As both phases must be determined, the alternative phase may be induced by the Craigie tube technique (Fig. 2.XVII). The salmonella is grown in the presence of flagellar antiserum, which will immobilize the organisms in that phase. Organisms in the alternative phase will be able to move through the medium. A hollow glass tube is placed in a tube of very soft agar to which a small amount of flagellar antiserum has been added. The inoculum is placed in the central

Table I
Antigens of Salmonella—Extract of Kauffmann–White Scheme

Group	*Species*	*"O" somatic antigens*	*"H" flagellar antigens* *Phase* **1** *Specific*	*Phase* **2** *Group*
A	*S. paratyphi A*	1, 2, 12	a	—
B	*S. paratyphi B*	1, 4, 5, 12	b	1, 2
	S. stanley	4, 5, 12	d	1, 2
	S. typhi-murium	1, 4, 5, 12	i	1, 2
	S. heidelberg	1, 4, 5, 12	r	1, 2
	S. chester	4, 5, 12	e, h	e, n, x
C	*S. paratyphi C*	6, 7, Vi	c	1, 5
	S. oranienburg	6, 7	m, t	—
D	*S. typhi*	9, 12, Vi	d	—
	S. enteritidis	1, 9, 12	g, m	—

tube, and after incubation motile organisms of the alternative phase may be subcultured from the medium outside the central tube.

A few types, of which *S. typhi* is the most important, may have a surface Vi antigen when first isolated, and this may interfere

Fig. 2.XVII. Diagram to illustrate the Craigie tube technique. Inoculation is made at X and subculture taken from Y. The arrows represent the route of the specific phase *Salmonella*.

with the agglutination reactions of the O and H antigens. The Vi antigen diminishes on subculture in simple media. *S. typhi* differs also in having no Phase 2, and in its biochemical reactions. It does not produce gas in fermentation of carbohydrates.

PATHOGENICITY

The members of the salmonella group are essentially pathogenic to man and animals. In man two important clinical conditions may be produced: (a) enteric fever caused mainly by the typhoid and paratyphoid bacilli, and (b) gastroenteritis resulting from infection with organisms associated with food poisoning, e.g. *S. typhimurium* and *S. enteritidis*.

In experimental animals a condition analogous to the human infections cannot be produced. Intravenous or intraperitoneal inoculation is followed by death in 18–72 hr, and on autopsy a generalized infection is found.

ENTERIC FEVER

Enteric fever is a disease of great antiquity and world-wide distribution. In Great Britain it was prevalent to an appreciable extent until the end of the last century, when marked advances were made in the general hygienic arrangements.

Although many striking epidemiological studies had been made previously, knowledge was greatly advanced by the description of the typhoid bacillus by Eberth in 1880 and its isolation in 1884 by Gaffky. The condition subsequently attracted the attention of many bacteriologists, and it was found that *S. paratyphi* A and *S. paratyphi* B were also responsible for enteric fever (Schottmüller, 1900). Infections with the paratyphoid bacilli are, however, generally less severe and less often fatal than those with the typhoid bacillus. *S. paratyphi* B infections are much more frequent in temperate climates than infections with *S. paratyphi* A, the latter organism being mainly encountered in tropical regions.

PATHOGENESIS

Enteric fever occurs spontaneously only in man. Infections in man originate directly or indirectly from the excreta of infected humans, either from actual cases or carriers, the latter being more dangerous. The portal of entry of the causative bacteria is the mouth, and the most frequent vehicles of infection are contaminated food or drinking-water. Contamination may be due to faecal or urinary pollution of the water supply or may be conveyed by the hands or flies. Shell-fish, such as oysters, may also be responsible for the transmission of infection, as they sometimes become contaminated with sewage during growth on the river-bed. Examples of enteric fever in this country are the Bournemouth outbreak in 1936 in which 700 cases occurred and the vehicle of infection was milk, and the Croydon outbreak in 1937 for which the water supply was responsible. In both cases carriers were the original source of the infection. In the Aberdeen outbreak in 1964 there were over 400 cases arising from corned beef. The tins after autoclaving had been cooled in raw river water contaminated with sewage. Imperfectly sealed tins had sucked in the contaminated cooling water.

The course of events after entrance into the alimentary tract has not been definitely settled, but the old view that the bacilli multiplied in the intestines and then passed into the blood stream has been generally discarded. Clinical and bacteriological observations indicate that the bacteria proliferate primarily in the spleen, liver and mesenteric glands. The incubation period is long, about 14 days, and, during this time, the bacteria pass from the alimentary tract probably via the intestinal lymphatics to these tissues, where multiplication takes place. At the end of this period the bacteria pass into the blood stream and produce a bacteraemia and the associated symptoms; cases usually present as a P.U.O. (pyrexia of unknown origin), very rarely purulent meningitis is the initial manifestation. After this, bacteria appear in the intestines either by way of the gall-bladder or from Peyer's patches. The bacteria invade and multiply in Peyer's patches, which break down and slough; this process may be responsible for severe haemorrhage or perforation of the gut, both of which are serious complications. The organisms are often present in the gall-bladder, which appears to be infected from the blood stream. Cholecystitis not infrequently develops, and gall-stone formation is a relatively common sequel; this favours the persistence of the carrier state. It is interesting to note that living typhoid bacilli have been isolated from the centre of gall-stones. The bacteria may also be excreted by the kidneys into the urine. When this excretion persists and the individual becomes a urinary carrier, there is usually an associated lesion in the urinary tract, frequently in the kidney substance or renal pelvis.

CARRIERS

The faecal and urinary excretion of the causative organisms frequently persists throughout convalescence, during which it gradually becomes less marked until it finally ceases. These cases are transitory or temporary carriers. In a few instances, 2–4 per cent, the excretion, faecal and/or urinary, persists, giving rise to the chronic or permanent carrier. These cases seldom clear spontaneously. The discharge may be intermittent, but nevertheless the uncontrolled carrier is always a potential source of danger to the community. These unfortunate individuals are almost invariably responsible for the endemic outbreaks occurring in this country. Women are more frequently faecal carriers than males; this is probably due to the greater incidence of cholecystitis and gall-stones in females than in males. Urinary carriers are less common than faecal carriers, but they constitute a more serious danger to the community because of the greater facilities for disseminating the organisms.

DIAGNOSIS

The pathogenesis of the condition indicates that different specimens should be examined according to the duration of the infection. Materials giving valuable results are (1) blood, (2) faeces, (3) urine and (4) serum.

(1) **Blood.** The most useful results in the early stages are obtained from blood culture. In the first week the bacterium can be isolated in 90–95 per cent of cases, in the second week the percentage of positives is less and after the fourth week the blood is rarely positive. Five to 10 ml of blood is collected and allowed to clot. The serum is removed aseptically and used for agglutination tests. The clot is broken up aseptically with a sterile rod, or by addition of streptokinase. Add to the clot 15 ml of 0·5 per cent sodium taurocholate broth, or 15 ml of sterile bile. Incubate at 37°C and plate out daily on MacConkey agar.

(2) **Faeces.** Cultures should be made on to selective media, such as deoxycholate-citrate, Wilson and Blair, and into tetrathionate or selenite broth; if the faeces are solid, they must first be suspended in saline or peptone water. After incubation at 37°C for 24 hr, the cultures are examined. In the case of the solid medium, discrete non-lactose-fermenting colonies are picked off and subcultured in peptone water; biochemical and serological tests are ultimately employed for identification. Subcultures from the fluid media are made on to MacConkey's medium, from which non-lactose-fermenters are picked off and examined as above.

(3) **Urine.** Urine should be collected with reasonable care and examined in a similar manner to the faeces. Typhoid bacilli are frequently isolated in the second, third or fourth week and may be present in large numbers. The urine may be centrifuged to concentrate the organisms in the deposit, which is used for the examination.

(4) **Serum.** The application of the agglutination test in the

diagnosis of enteric fever has evolved considerably since the method was originally introduced by Widal. The use of a simple suspension of *S. typhi* yields information of an inconclusive nature and, as previously indicated, other bacterial suspensions are necessary to ensure a satisfactory result. The routine test at the present time contains at least six suspensions, which include the H and O factors of the three members usually responsible for enteric fever, and a Phase 2 H suspension in order to pick out any other member likely to be involved. These suspensions are standardized and tested by the Public Health Laboratory Service, and are obtainable from commercial sources (Burroughs Wellcome). Their general use tends to obviate irregularities due to unsatisfactory suspensions.

Agglutinins appear towards the end of the first week of the disease and the titre increases until the fourth week, after which there is a fall, marked at first, but slow later. Agglutinins may be demonstrated some time after recovery. In infections with the typhoid bacillus, the O agglutinin response may be more marked than the H. The titre varies considerably, but one over 1/100 with either the H or O suspension is practically diagnostic in an uninoculated individual. Agglutinins may be present in the serum of normal individuals, but these seldom give a titre above 1/25 and the O agglutinins appear to be more common than H agglutinins.

In the case of *S. paratyphi* B infections, H agglutinins are more marked and high titres are usually recorded. The O response is not infrequently poor.

Cross-agglutination with the O suspensions of *S. typhi* and *S. paratyphi* B often takes place due to the common possession of somatic antigen 12.

Agglutinins are also produced by prophylactic inoculations of the T.A.B. vaccines but this response can usually be differentiated from that manifested in the actual infection. Following inoculations of the T.A.B. vaccines, H agglutinins are elicited more readily than O agglutinins, which can frequently not be detected unless the examination is made within a short time after the injections. The titres of the H agglutinins may, however, be high, particularly if the inoculations have been recent, titres up to 1/2,560 have been observed; H agglutinins for the typhoid and paratyphoid bacilli are generally present to a similar titre. This is not usual in infection.

It is important to note that virulent strains of *S. typhi* may be inagglutinable on account of the antigenic component, termed "Vi" (p. 211). The Vi antigen is labile and further agglutination tests should be carried out with bacterial suspensions heated at

100°C. The Vi antigen evokes a specific agglutinin; this appears late in the disease and is therefore of no value in early diagnosis. Its presence is useful in the detection of carriers (*vide infra*).

PHAGE TYPING

Bacteriophages acting on the Vi antigen of *S. typhi* were originally demonstrated by Craigie and his colleagues. Four distinct phages have been identified; one, type II, exhibits a high degree of strain selectivity and can be "conditioned" to particular strains by appropriate changes of the substrate. These conditioned phages are stable and are used to type strains of *S. typhi*; at least 33 different types have been identified. Phage typing requires considerable experience and it is used mainly for epidemiological purposes, e.g. for tracing the source of infection and relationship between cases. Phage typing is now carried out on other salmonellas, e.g. *S. paratyphi* B and *S. typhi-murium*.

PROPHYLAXIS

Prophylaxis has played an important part in the elimination of enteric fever from civilized communities. General and specific measures have been adopted.

The general measures adopted in the prophylaxis of enteric fever have been numerous. Improvement in sanitary arrangements, particularly with regard to sewage disposal, has been an important step, while the control of possible vehicles of infection, such as drinking-water, milk, ice-cream, shell-fish and flies, has been another valuable contribution.

Another important factor is the control of carriers, though unfortunately, owing to the difficulty of detection, their complete elimination is impossible. When endemic outbreaks occur carriers are almost invariably responsible, and rigorous attempts should be made to locate them; cultural and serological examinations have given the most satisfactory results in identifying carriers. It has been stated that some 2–4 per cent of all cases become permanent carriers, either faecal or urinary, and that female carriers greatly outnumber males. As women are more frequently concerned with the preparation of food, the danger is apparent. As an illustration, brief mention may be made of the celebrated "Typhoid Mary" case. "Typhoid Mary" was a cook in America who, in 1901, developed enteric fever. She subsequently was employed in different houses and in many was responsible for several cases of enteric fever; in five years she was known to have caused 26 cases. She was then placed under the control of the Health

Department of New York, but in 1914 she escaped observation. In 1915 an outbreak involving 25 cases occurred in a hospital; it was discovered that the cook had disappeared and on location she proved to be "Typhoid Mary". She was then kept under stricter control and has subsequently died. It was thought that she was also responsible for other outbreaks, including one involving over 1,300 individuals.

The detection of carriers would doubtless be facilitated by the adequate bacteriological control of the cases at the time of the infection. Cases should be kept in quarantine until at least three successive bacteriological examinations of the urine and faeces, carried out at weekly or fortnightly intervals, have given negative results. It is claimed that practically all carriers of *S. typhi* possess Vi agglutinins in a titre of 1 in 10 or over. Agglutination tests must be carried out at 37°C, in $3 \times \frac{1}{2}$ in test-tubes, using a specially prepared bacterial suspension, i.e. one containing Vi antigen, and adequate controls. Any case giving a positive result should be subjected to rigorous examination of the faeces and urine.

The control and treatment of the chronic carrier is unfortunately an extremely difficult problem. Cholecystectomy in the case of the faecal carrier and the operative treatment of gross lesions in the case of urinary carriers is sometimes successful. Other measures, such as drugs and vaccines, have proved of little value. In all cases personal instruction of the potential danger and the precautionary measures to be adopted must be given by the medical officer. *Specific immunization* by means of vaccines has yielded very satisfactory results. Since the introduction by Wright of the T.A.B. vaccine the incidence of enteric fever in exposed individuals has decreased considerably. In ordinary communities the necessity for such a prophylactic procedure does not arise, but where the risk of exposure is increased, such as in military operations in foreign countries or by visits to tropical regions, vaccines have proved particularly useful.

Various methods have been used for the preparation of vaccines. Smooth virulent cultures must be used. The organisms may be killed by heat, phenol, alcohol or acetone. The Vi antigen is probably best preserved in alcohol. There is no satisfactory laboratory test for the efficacy of a vaccine. Extensive comparative field trials have been carried out in areas where typhoid is endemic, but it is difficult to establish the superiority of any particular vaccine. Some degree of protection is produced by all of them. The usual vaccine contains 1,000 million *S. typhi* and 500 million each of *S. paratyphi* A and *S. paratyphi* B. The initial dose is usually 0·5 ml given subcutaneously, followed by a second dose of 1 ml in 3–4 weeks.

THERAPY

Specific therapeutic measures have been generally unsatisfactory. The use of "Vi" antiserum has been disappointing.

Most chemotherapeutic agents have been disappointing but chloramphenicol has proved effective and it is the drug of choice; the daily dosage recommended is 3–6 g during the acute stages, which tend to settle within a few days, followed by 1–3 g for a further 10–14 days. In view of the danger of blood dyscrasias after chloramphenicol, various other antibiotics have been tried, but are probably not as effective as chloramphenicol.

SALMONELLA FOOD-POISONING

Members of the salmonella group are the main bacterial agents causing food poisoning. *S. typhi-murium*, *S. enteritidis*, *S. thompson*, *S. heidelberg*, *S. oranienburg*, *S. brandenburg* are probably the most frequent, but many other types have been isolated in various outbreaks.

The clinical picture of salmonella food-poisoning is that of an acute gastroenteritis. The time of onset varies, but is usually 12–36 hr after ingestion of the contaminated food, and is characterized by nausea, abdominal discomfort, vomiting, headache, and later diarrhoea. The symptoms are frequently associated with a moderate degree of pyrexia. In most cases these symptoms subside in a few days, and the patients are completely recovered in 6–8 days. The morbidity rate is high, but the mortality is low, approximately 1 per cent. In fatal cases the infection is generalized and the causative organism can be isolated from the intestinal tract, blood, liver, spleen and lymph nodes.

The food involved may be meat, milk, eggs or fish. Meat dishes frequently involved are pies, pastes, sausages and brawns which have been insufficiently heated during preparation, or have been subjected to much handling. Contamination of these foods results either from disease of the animal, or from an outside source during preparation of the food. *S. typhi-murium* and *S. enteritidis* are important pathogens of cattle, and tissues of an animal with a generalized infection are contaminated. Meat may also be contaminated by faeces at the time of slaughter.

Eggs from infected ducks have been responsible for several outbreaks. Many types of salmonella have been isolated from dried egg powders, egg albumen and dried coconut, and these are important sources of infection.

Salmonella infections are very common in rodents. Rats and mice have been incriminated in a number of outbreaks as the original source of infection. Infected mice in food stores and kitchens may contaminate food and spread infection.

Members of the salmonella group, particularly *S. enteritidis* have been employed in various preparations as Liverpool virus and Rattin for exterminating rats. The idea is to introduce the infecting agent into the rat population and produce epidemics. Their use has not been very successful. Acute and fatal attacks do not invariably result, and when fatal, the bodies may remain hidden and putrefy with unpleasant results. There is also a danger that the infection may spread to the human population.

The types of food usually involved in outbreaks of food poisoning are excellent bacteriological media, so even a small inoculum may very rapidly multiply. As salmonella have very little proteolytic action there may be no very obvious evidence of contamination. There is no alteration in the smell, taste or appearance of the food.

DIAGNOSIS

Whenever the clinical signs indicate food poisoning, examinations should be carried out on the following materials:

(1) Vomit
(2) Faeces
(3) Suspected food
(4) Blood for culture and for agglutinins
(5) Blood and faeces of suspected carriers, particularly food handlers.

Cultures should be made on selective media, following the same procedure as in the examination of faeces in cases of enteric fever. Food should be mixed with five times its volume of selenite broth, incubated at 37°C and plated daily on deoxycholate citrate agar. Blood cultures, except in fatal cases are usually negative. Agglutinins are not present in the early stages, but may appear after 10–14 days, and provide a useful diagnostic indication. Suspected carriers among the food handlers should be subjected to serological and bacteriological examination. If rodents are present in the food stores, they should be trapped and examined for salmonella infection

PROPHYLAXIS

The preventive measures likely to reduce the incidence of food poisoning are:

(1) Veterinary examination of animals before slaughter, and adequate meat inspection at the abattoir.
(2) Extermination of rodents in places where food is stored or prepared.
(3) Adequate cooking of food.
(4) Conservation of food at low temperatures.
(5) Control of salmonella carriers.
(6) Inspection of food handlers.

Chapter XVIII

Proteus: Providence: Pseudomonas

PROTEUS

The members of this group are Gram-negative, motile rods which grow readily on simple media and tend to spread or swarm over the whole surface of solid media in a very thin film. They are non-lactose fermenters, and are frequently present in faeces. They are pathogenic, and are commonly found in cases of cystitis and in infected wounds.

MORPHOLOGY

Proteus organisms are very pleomorphic. Long and short rods, filaments and curved forms are frequently seen in the same microscopic field. They are actively motile by lateral flagella.

CULTURAL CHARACTERISTICS

Growth occurs readily on ordinary media over a wide range of temperature. On solid media, such as nutrient agar, a phenomenon of swarming is frequently found. A thin film of growth spreads rapidly over the surface, and may be almost invisible. This property is troublesome when attempting to isolate other organisms from material in which proteus is also present. Swarming may be prevented by the addition of chloral hydrate (1/1,000) to the medium. Bile or bile salts, as in MacConkey agar or deoxycholate citrate agar will also inhibit swarming.

Table I

Biochemical Reactions of the Genus Proteus

Test	*P. vulgaris*	*P. mirabilis*	*P. morgani*	*P. rettgeri.*
Urease	+	+	+	+
Gelatin	+	+	–	–
Indole	+	–	+	+
Maltose	+	–	–	–
Mannitol	–	–	–	+
Citrate	–	–	–	+
H_2S	+	+	+	–

RESISTANCE

Proteus organisms are relatively susceptible to heat and other destructive agents.

BIOCHEMICAL ACTIVITY

The members of this genus rapidly decompose urea to NH_3. By this property it is possible to distinguish proteus strains from other non-lactose fermenters within a very short time. A colony picked off on to Christensen's medium will turn the indicator pink within an hour or two. Gelatin is usually liquefied, and H_2S is produced. All strains ferment glucose, and many ferment other sugars. Table I outlines a scheme of classification on biochemical reactions.

SEROLOGY

Proteus organisms possess both O and H antigens. Over 100 different serotypes may be differentiated by their antigenic structure, but little use is made of these reactions in diagnostic bacteriology. One strain, OX19 is used in the Weil–Felix reaction, as it is agglutinated by the serum of typhus patients. The reason for this appears to be the fortuitous possession of common antigens with the *Rickettsia* of typhus. The strain OXK serves a similar purpose in the diagnosis of scrub typhus. These proteus strains are not concerned in the aetiology of typhus fever.

PATHOGENICITY

Members of the proteus group are frequently present in faeces, soil and putrefying material. They are pathogenic to man and are often responsible for urinary tract infections. They are important as secondary invaders of wounds, and are occasionally found in chronic nasal and aural infections.

THERAPY

The therapy of proteus infections is difficult. Sensitivity to antibiotics is variable, and resistance develops rapidly. It is important to carry out sensitivity tests to determine the most suitable therapy, and to repeat the tests on organisms re-isolated at frequent intervals during treatment to detect the development of resistance.

PROVIDENCE GROUP

The providence group are probably related to both the proteus and shigella groups. They may be isolated from urine and faeces.

They are indole positive, citrate positive, and do not liquefy gelatin. Their pathogenicity is doubtful.

PSEUDOMONAS

Gram-negative rods, motile by polar flagella. Strict aerobes. Catalase positive. Oxidase positive. Produce water soluble pigments. Common in soil and water. Some are plant pathogens. One species is pathogenic for man—*Pseudomonas aeruginosa*, commonly known as *Ps. pyocyanea*.

Ps. aeruginosa was isolated by Gessard in 1882 from blue pus. These organisms are slender Gram-negative rods, actively motile by polar flagella.

CULTURAL CHARACTERISTICS

Grow well on simple media, forming water-soluble pigments which diffuse throughout the medium in well-aerated cultures. Most cultures have a characteristic sweet sickly odour.

BIOCHEMICAL CHARACTERISTICS

Ps. aeruginosa produces slight acid from glucose, and a few other carbohydrates. Gelatin is liquefied. Citrate is utilized as the sole source of carbon. Catalase and oxidase tests are positive. Growth is not inhibited in KCN media. *Ps. aeruginosa* produces a bluish green pigment which consists of two different pigments. Pyocyanin, a blue pigment is soluble in water, and may be extracted from cultures by chloroform. Fluorescein is greenish yellow, soluble in water but insoluble in chloroform.

PATHOGENICITY

Ps. aeruginosa is widely distributed in nature, and is frequently found in faeces. It is a usual contaminant of burns and wounds. It occurs in chronic infections of the urinary tract, nose and middle ear. It sometimes produces a generalized infection, which is usually fatal.

THERAPY

Ps. aeruginosa is usually resistant to most antibiotics, or develops resistance rapidly. Sensitivity tests should be made with a wide variety of possible agents in every case.

Chapter XIX

Pasteurella

PASTEURELLA

Gram-negative ovoid rods showing bipolar staining. Slight biochemical activity. Pathogenic to man and animals. The type species is *Pasteurella septica.*

Pasteurella septica is widely distributed in domestic and wild animals and birds. These organisms were previously named according to the species from which they were isolated: *Past. oviseptica* from sheep, *Past. muriseptica* from mice, *Past. aviseptica* from birds, etc. As there is no precise method of differentiation of these types they are now considered as members of one species, *Past. septica.*

Past. septica produces a serious disease, haemorrhagic septicaemia, in various species of animals and birds. The normal habitat of this organism is probably the upper respiratory tract, from which it can frequently be isolated in comparatively healthy animals. In man, it is frequently found as the cause of infection following a bite by cats, dogs, rabbits or other animals. The site of infection is inflamed, and there is usually a lymphangitis and lymphadenopathy. The organism may also be isolated, amongst others, from sputum in many cases of bronchiectasis. The importance of *Past. septica* in such chronic mixed infections is uncertain.

DIAGNOSIS

Past. septica may be cultured from the local lesion in the case of animal bites, or from sputum in some cases of bronchiectasis. Colonies on blood agar or nutrient agar are small, greyish, smooth and translucent. They may be missed in a mixed flora unless the cultures are carefully examined. The organism does not grow on MacConkey agar. It produces acid in glucose, is indole positive and methyl red negative. It will usually produce a fatal infection in white mice within 48 hr.

THERAPY

Some strains are sensitive to penicillin, and most are sensitive to streptomycin.

PASTEURELLA PSEUDOTUBERCULOSIS

Past. pseudotuberculosis is pleomorphic, Gram-negative, and shows bipolar staining. It differs from the type species in that it will grow on MacConkey agar, and is motile in broth cultures grown at 22°C. It is indole negative and methyl red positive.

Past. pseudotuberculosis is primarily a pathogen of rodents, in which it produces caseous lesions macroscopically resembling tuberculosis. Guinea-pigs are particularly susceptible, and epidemics in laboratory animals may occur.

In man, *Past. pseudotuberculosis* has been found as a cause of mesenteric lymphadenitis, producing symptoms suggestive of appendicitis. A considerable number of cases has been reported by Knapp in Tübingen, and other cases by Mair in this country. The condition is probably more common than is generally recognized.

PASTEURELLA TULARENSIS

Past. tularensis is a very small, pleomorphic, non-motile Gram-negative rod showing bipolar staining. The taxonomic position of this organism is doubtful. It has been placed by various authorities amongst the brucella, and by others in the genus pasteurella. It does not grow on the usual media without the addition of fresh tissue or cystine. It will grow on coagulated egg yolk. It produces H_2S.

Tularaemia is primarily a disease of rodents, particularly rabbits and ground squirrels in which it produces a fatal septicaemia. The infection may be spread through the respiratory tract, or through the agency of blood sucking insects.

Man is very susceptible, and many cases occur through handling animals in areas where the infection is endemic amongst rodents. Laboratory infections are not uncommon. Tetracycline and chloramphenicol have been used successfully in therapy.

PASTEURELLA PESTIS—THE PLAGUE BACILLUS

MORPHOLOGY

The plague bacillus is usually a small oval rod, $1{\cdot}5 \times 0{\cdot}6\ \mu$, with rounded ends and convex sides; swollen involution forms are frequently found, particularly in old cultures and chronic lesions. Pleomorphism is particularly marked when growth is obtained on agar containing 3–4 per cent NaCl; this constitutes an important diagnostic criterion. It is non-motile, having no flagella, and does not form spores. When freshly isolated from animal tissues a viscid capsule can be demonstrated.

Past. pestis stains readily with the ordinary aniline dyes and is

Gram-negative. The stained forms may show a definite bipolar reaction, this is seen particularly when the organisms are in the tissues; this is not due to the presence of granules, but to the fact that the intervening protoplasm is usually less intensely stained than the polar areas. Methylene-blue is the most suitable stain for demonstrating this uneven staining.

CULTURAL CHARACTERISTICS

Growth is obtained on the ordinary media, but it is slow and not very profuse; the optimum temperature is 30°–35°C, and the optimum reaction of the media is acid, pH 6·2–7·0; for primary isolation from the tissues, enriched media, such as blood agar or serum agar, give the most satisfactory results. The plague bacillus is a facultative anaerobe, but growth under strictly anaerobic conditions is poor.

In fluid media, such as broth, after 24 hours' incubation there is a slight general turbidity with a definite floccular deposit, while later a surface pellicle may form. When sterile oil is floated on the surface of the medium, growth occurs downwards from the under surface of the oil in the form of stalactites.

On solid media, such as agar, small, transparent, colourless granular, viscid colonies are formed in 24 hr; later these become opaque, greyish, irregular, and tend to be differentiated into peripheral and central zones.

RESISTANCE

The plague bacillus is readily killed by destructive agents; it is destroyed by heat at 55°C in 5 min and by 0·5 per cent phenol in 15 min. Stock cultures may be conserved at room-temperature or in the refrigerator for several weeks.

BIOCHEMICAL ACTIVITY

The saccharolytic and proteolytic activities of the plague bacillus are not very marked. Acid, but not gas, is produced in glucose, maltose, rhamnose, glycerol and mannite after incubation for several days. As growth is poor in peptone water sugar solutions, Hiss's serum-water is a more suitable medium. Litmus-milk is not altered. Gelatin is not liquefied, and indole is not produced. Broth is rendered alkaline.

SEROLOGY

All strains of *Past. pestis* are serologically homogeneous and are agglutinated by the serum prepared against one strain. The serology of the group as a whole has not been completely investigated; cross agglutination is marked with *Past. pseudotuberculosis*, but it is not definite with the haemorrhagic septicaemia group.

PATHOGENICITY

Past. pestis is primarily a pathogen in rodents, particularly wild rats, from which infection is conveyed to man through the agency of fleas, giving rise to a condition known as plague.

Experimental infection may readily be produced in a number of laboratory animals, e.g. rats, mice, guinea-pigs, rabbits and other rodents. In all, after subcutaneous inoculation a necrotic local lesion and enlarged regional glands are first produced with subsequent septicaemia and death. The guinea-pig may also be infected by nasal instillation of the organisms or by rubbing them on the shaved skin of the abdomen.

PLAGUE

Plague has been one of the most devastating diseases in the history of the human race. Reports of extensive and widespread epidemics have appeared in the literature of all ages. One celebrated epidemic, the "Black Death", swept this country and Europe in the fourteenth century and was responsible for the death of over 25 million people. At the present time the disease is rare in European communities, but large outbreaks are still reported in the East, particularly in India.

The isolation and incrimination of the plague bacillus were carried out independently by Kitasato and Yersin in 1894 during an investigation of an epidemic in Hong Kong. Subsequently several workers postulated that plague was essentially a disease of rats and that the spread to man occurred through the agency of fleas. This theory, however, was only conclusively proved by the work of the English Plague Commission at Bombay in the early part of the present century. Other rodents may also be responsible for outbreaks; epidemics in California have originated from the ground-squirrel, while in Tibet the marmot has been the primary source of infection in several outbreaks.

Epidemics have usually originated from four main endemic foci, two of which are located around the Himalaya mountains, one in Arabia and the other in Africa. In these regions plague is present in the rat population either in endemic form or as an epizootic. The rats involved are *Rattus rattus*, the small black house rat, and *R. norvegicus*, the large grey sewer rat. During an epizootic in these animals there is a plentiful supply of infected material, and it has been observed that human epidemics usually occur a short time after an epizootic in the rat.

The plague bacillus is conveyed from rat to rat and from rat to man almost exclusively by the rat fleas, particularly *Xenopsylla cheopis* and *Ceratophyllus fasciatus*. These fleas suck the blood of the rat and when the rat dies they search for another suitable host, rat if available, or, if not, man. Plague in its acute form is a septicaemic disease and the blood sucked up by the fleas consequently contains many bacilli, which multiply in the proventriculus and impede access of food to the stomach. The hungry flea settling on a fresh host sucks blood, but this further distends the oesophagus and regurgitation of the contaminated contents into the tissues of the host takes place. In other cases the plague bacilli may be implanted on the skin with the faeces or blood of the flea; they usually gain entrance to the tissues as a result of subsequent scratching of the bite. Fleas once contaminated may remain so for some time, usually 7–15 days. Survival of the bacillus is favoured by a low temperature, about 50°F, and an almost saturated atmosphere.

In man two types of infection occur, (1) bubonic plague and (2) pneumonic plague. Of these the former is the more common condition. In bubonic plague, the bacilli are introduced by the flea into the tissues. After an incubation period of 5–7 days, the cardinal symptoms are painful enlargement of the regional glands, the so-called "buboes", depression and prostration, with pyrexia. The local lesion may be extremely slight or negligible, but enlargement of the regional glands is invariably encountered. As people are usually bitten on the legs, the inguinal glands are frequently involved. In unfavourable cases septicaemia and death supervene; this sequence of events is found in 60–90 per cent of cases. Bubonic plague is spread only by the flea, direct infection does not occur. While the usual sequence is rat–flea–man, the cycle man–flea–man may also be encountered.

Pneumonic plague does not depend on the presence of fleas for its transmission; it spreads from man to man by droplet infection, and is usually found in the cold seasons. The primary focus in man appears to be those cases of bubonic plague in which pneumonic symptoms develop. The incubation period is short, 1–3 days, and the clinical signs are those of a fulminating broncho-pneumonia. Septicaemia and death almost invariably result, the mortality rate being approximately 100 per cent.

DIAGNOSIS

Bacteriological methods employed in the diagnosis of plague in man are the detection and isolation of the organisms from the various lesions, particularly the local vesicle, buboes, blood stream and the sputum in the pneumonic form. Serological tests are sometimes used; agglutinins may be present in the serum to a titre of 1/10–1/100; a negative test is, however, valueless.

The diagnosis of plague in the rat is not infrequently a procedure of great importance. In this animal the disease may be either acute or chronic. In the acute form the usual lesions are enlarged lymph glands or buboes, mottling of the liver, pleural effusion, congested spleen and subcutaneous congestion and haemorrhages. The plague bacilli are most readily isolated from the buboes, spleen and blood. In chronic cases, buboes and abscesses in the spleen and sometimes the liver are common. Except in the presence of epizootics the isolation of the organism is required in order to establish the diagnosis.

It is important that the presence of *Past. pseudotuberculosis* should be excluded, as this organism also gives rise to infection in rats. Important differential tests are the motility of this species at 22°C and its lack of pathogenicity for white rats. Criteria useful in establishing the identity of *Past. pestis* are morphology, cultural characteristics, agglutination and pathogenicity for white rats.

PROPHYLAXIS

Measures adopted in the prevention of plague are general and specific. The general measures employed in endemic foci include the suppression of rats, improvement in sanitary arrangements, removal of overcrowding and attention to personal hygiene, particularly to the avoidance of fleas. In the case of pneumonic plague, rigorous measures to counter droplet infection must be introduced.

In countries where the disease is absent or extremely rare, the entrance of infected rats must be prevented and regular bacteriological examinations of the local rat population performed. This is one of the important duties of the Port Medical Officer. Quarantine regulations and the control of incoming ships are strictly enforced. Great precautions are taken with ships arriving from ports where plague is endemic; medical inspection on arrival is carried out, shields are placed on the hawsers and dead and trapped rats are submitted for a bacteriological examination. If a rat is reported infected, fumigation of the ship is at once carried out in order to kill the rats. It is largely as a result of the precautions taken by the port authorities that in this country plague is now a rarity.

Specific prophylaxis by means of vaccines has been extensively employed in Eastern countries. The usual method of preparation is to heat at 60°C for 1 hr a 2–6 weeks' culture of *Past. pestis* grown at 25°–30°C in digest broth, 0·5 per cent phenol being added as a preservative. Unfortunately available statistics are incomplete, but the indications are that, by the use of vaccines, the incidence and mortality rate of the disease have both been reduced. Nevertheless, an entirely satisfactory vaccine has not yet been discovered. Schütze found that cultures incubated at 37°C gave more protection to rats than those incubated at 26°C. At 37°C a well-marked gelatinous envelope formed, whereas at 26°C its development was poor. The presence of this envelope was considered to be important in the preparation of an efficient vaccine, as it contained an antigen quite distinct from that contained in the body of the organism. The envelope antigen was destroyed by heat at 100°C. The function of this envelope in the immunization of man has, however, not yet been determined.

THERAPY

Specific therapy by means of sera prepared by various methods, e.g. the injection of bacterial extracts and filtrates, has only been employed on a small scale, but without any striking success.

Penicillin is without effect on infections with *Past. pestis* but promising results have been given by the more active sulphonamides and streptomycin; heavy and prolonged dosage is often necessary.

Chapter XX

Brucella

Small non-motile, non-sporing, Gram-negative short rods. Aerobic, but some types require additional CO_2. Slight biochemical activity. Growth slight on ordinary media. Strict parasites in animals and man, producing characteristic infections. The type species is *Brucella abortus.*

HISTORY

Brucella melitensis was first isolated by Bruce in 1887 from the spleen of a fatal case of Malta fever. Several years later Zammit found the reservoir of infection in goats, and that goat's milk was the source of infection in man. When the Naval Authorities prohibited the use of raw goat's milk, the incidence of Malta fever in naval personnel fell to an insignificant level, although the high incidence continued in the civil population.

Brucella abortus was isolated by Bang in 1897 from cases of infectious abortion in cattle. Many years later this organism was recognized as a cause of undulant fever in man. Evans, in 1918, demonstrated a close morphological, cultural and antigenic similarity between these two organisms. Another similar organism, *Brucella suis*, was isolated from pigs in 1914 by Traum. It also may cause undulant fever in man. The geographical distribution of *Br. suis* is more limited than that of *Br. abortus.*

MORPHOLOGY

Members of the brucella group are minute rods, $0{\cdot}6\text{–}1{\cdot}5 \times 0{\cdot}6\ \mu$, appearing as cocco-bacilli and without any characteristic arrangement.

CULTURAL CHARACTERISTICS

Growth occurs on simple media, but it develops slowly and is never profuse. Growth is improved by the addition of blood, serum or liver extract to the medium. Erythritol stimulates growth but other carbohydrates have little effect. The organisms are aerobic. *Br. abortus* differs from other members in that on first isolation an atmosphere containing 5–10 per cent CO_2 is essential. After several subcultures on artificial media growth occurs under usual atmospheric conditions. On solid media colonies are small, greyish and semi-opaque. On potato a yellowish-brown pigment is produced.

RESISTANCE

The Brucella are readily destroyed by most bactericidal agents and by heat. In milk they are killed by the usual temperature of pasteurization. The variation in the bacteriostatic action of dyes affords a useful method of differentiating the three types. The dyes may be incorporated in the medium, but a simpler method is to impregnate strips of filter paper with the dye, place them on the surface of the medium, and inoculate in streaks across the strip. Inhibition of growth is then evident after incubation (Table I).

BIOCHEMICAL ACTIVITY

Members of the Brucella genus are not very active biochemically. Carbohydrates are not fermented, indole is not produced, and gelatin is not liquefied. *Br. abortus* and *Br. suis* produce H_2S.

SEROLOGY

Freshly isolated strains maintained in the smooth phase may be differentiated by agglutination tests with monospecific sera. After cultivation for a variable period on artificial media strains may become rough, and will not be agglutinated by antisera to smooth types. In the transition stage from smooth to rough they may be agglutinated by both anti-smooth and anti-rough sera. Rough strains are thermo-agglutinable. *Br. melitensis*, *Br. abortus* and *Br. suis* all have the same two main antigens, A and M, but in different proportions. Monospecific sera for identification purposes may be prepared by careful agglutinin absorption.

PATHOGENICITY

The Brucella produce spontaneous disease, usually affecting the genital tract in goats, cattle, pigs and sheep. Infection is conveyed to man, usually by consumption of infected milk or milk products, but also by direct contact with infected tissue. Brucellosis is an occupational disease of veterinarians. In man, all types produce undulant fever. The infectivity of melitensis strains for man is high. Abortus strains have a lower degree of infectivity, and only rarely produce the acute type of disease which is common with melitensis strains. Sub-clinical and mild attacks of brucellosis are common with abortus strains, and are frequently not investigated and diagnosed. In man, melitensis and suis strains are more virulent than abortus strains.

UNDULANT FEVER

The incubation period is variable, but probably about 8–14 days after the ingestion of infected milk. The symptomatology of the the disease is extremely variable: headache, lassitude, malaise, joint pains, backache, profuse sweating and intermittent pyrexia are usually present. There may be nausea, vomiting, diarrhoea, mental depression, cough and visual disturbances. Intermittent fever persists over a long and indefinite period. The clinical picture may resemble that seen in enteric fever, malaria, tuberculosis, Hodgkin's disease, glandular fever and subacute bacterial endocarditis.

DIAGNOSIS

The most satisfactory method of diagnosis is the isolation of the infecting organism. Failing this, reliance must be placed on agglutination tests.

ISOLATION OF BRUCELLA

The organism may be isolated by blood culture during febrile attacks, or by intraperitoneal inoculation of the blood in a guinea-pig and subsequent culture of the spleen. The blood, taken aseptically, should be cultured in two flasks of liver infusion broth. One should be incubated aerobically, and the other in an atmosphere containing 10 per cent CO_2 for several weeks. Subcultures should be made at three- or four-day intervals on liver infusion agar.

AGGLUTINATION

Agglutinable suspensions must be prepared from organisms in the smooth phase. Rough cultures may give unreliable results. Agglutinins up to a titre of 1 in 3,000 or higher may sometimes be demonstrated in the serum of patients. The prozone inhibition phenomenon is common. Lower serum dilutions may show no reaction, while higher dilutions show strong agglutination. Serum dilutions up to at least 1/3,000 should always be tested. Low titres, less than 1/100, particularly in individuals who come in contact with cattle, are of little diagnostic importance. A rising titre on repeated examination suggests infection.

BRUCELLIN TEST

An intradermal test with brucellin, which is a sterile filtrate of a 4–6 weeks' old culture of brucella may produce a local allergic reaction, analogous to the tuberculin reaction, in 24–48 hr. Blood for agglutination tests should be taken before brucellin is given, as false reactions may otherwise be obtained.

PROPHYLAXIS

The most valuable measure in the prevention of undulant fever is the control of the milk supply. Pasteurization of milk will destroy the Brucella.

The elimination of abortus infection from cattle presents a difficult problem. The infection is very prevalent in Great Britain, and infected cattle excrete *Br. abortus* in the milk. Elimination of infection by slaughter as in the case of foot-and-mouth disease or tuberculosis is not practicable.

Protection of calves by vaccination with dead vaccines has not been highly successful. Better immunity is obtained by the use of an attenuated living strain of *Br. abortus*, known as S19. Calves vaccinated with this strain at the age of six months develop a considerable degree of immunity. The agglutinins they develop will largely disappear by the time they begin to produce milk, and the diagnosis of infection in cattle by the agglutination test will not be invalidated.

THERAPY

The brucella are sensitive to streptomycin, tetracycline, chloramphenicol and ampicillin, but therapy has been disappointing. In the tissues, the organisms are intracellular, and thus protected from agents to which they are sensitive *in vitro*. Combined therapy with sulphadiazine and antibiotics is more effective, but prolonged treatment is usually necessary.

Table I

Differentiation of the species of the *Brucella* Genus

Strain	CO_2 requirement	H_2S production	Growth in presence of		
			Thionin	Basic fuchsin	Methyl violet
Br. melitensis	−	−	+	+	+
Br. abortus	+	+	−	+	+
Br. suis	−	+	−	−	−

Chapter XXI

Haemophilus

Small, Gram-negative pleomorphic rods. Most species dependent on growth factors present in blood or plant tissues. Aerobic. Strict parasites. Usually pathogenic. Type species: *Haemophilus influenzae*.

Haemophilus influenzae was first isolated in 1892 by Pfeiffer who considered it to be the cause of influenza, which is now known to be caused by a virus. *H. influenzae* is normally present in the upper respiratory tract, and may produce various types of infection.

MORPHOLOGY

The usual form is a minute rod $1–1{\cdot}5 \times 0{\cdot}25\ \mu$, but long thread-like forms may occur in culture, and are frequently predominant in cases of meningitis. The morphology of a given strain is by no means constant. *H. influenzae* is non-motile and non-sporing. Many pathogenic strains are capsulated when first isolated.

CULTURAL CHARACTERISTICS

H. influenzae does not grow on simple media. Two accessory growth factors are essential. One of these, designated the V-factor is a co-enzyme. It is present in blood and various vegetable tissues, especially yeast and potato, and is also produced by many other organisms. It is heat labile, being destroyed by autoclaving at 120°C for 30 min. The other factor, X, is the iron containing haemoglobin derivative haematin. It is active in high dilution and is heat stable. It is required for synthesis of peroxidase and catalase.

Although both factors are present in small amounts, blood agar is not a satisfactory medium. The haemoglobin may be converted to haematin, by heat as in chocolate agar, or by peptic digestion as in Fildes' medium, and growth then occurs readily but is never profuse. On solid media, minute, smooth, transparent circular colonies are formed in 24 hr. Atypical strains such as those showing thread-like forms yield colonies that are more opaque and have a bluish iridescence. Colony size increases on prolonged incubation, and colonial appearance tends to be irregular. When

H. influenzae is grown on blood agar in a mixed culture with staphylococci, the growth of colonies of *H. influenzae* adjacent to colonies of staphylococci is greatly enhanced. This phenomenon of satellitism is due to the synthesis of the V-factor by the staphylococci. The V-factor diffuses into the surrounding medium.

RESISTANCE

H. influenzae is readily destroyed by bactericidal agents and by heat.

BIOCHEMICAL ACTIVITY

H. influenzae ferments glucose, maltose and laevulose to acid without gas. Proteolytic activity is very slight but many strains produce indole. All strains reduce nitrates.

SEROLOGY

The capsule of *H. influenzae* contains a specific soluble substance (S.S.S.) and the organisms can be typed by precipitin, agglutination or capsule swelling reactions. Six distinct types, a to f, have been identified. Type b appears to be the most virulent strain and is responsible for most human infections.

PATHOGENICITY

The influenza bacillus produces spontaneous disease only in man. It is, however, pathogenic to some laboratory animals. Large doses cause death by toxaemia. Nasal instillation under ether anaesthesia may produce fatal bronchopneumonia in mice and rabbits.

In man, *H. influenzae* is frequently found in the normal flora of the upper respiratory tract, but when local resistance is reduced it may become actively pathogenic. Many cases of bronchopneumonia following measles, influenza and whooping cough are caused by this organism. It may also be responsible for a number of other conditions such as antrum infections, otitis media, empyema, endocarditis, arthritis and conjunctivitis. Meningitis due to *H. influenzae* occurs chiefly in young children.

THERAPY

Streptomycin, tetracycline and chloramphenicol have been used successfully in treatment of *H. influenzae* infections. It is best to test the sensitivity of any strain isolated to determine the optimum therapy.

H. haemolyticus is a variant of *H. influenzae* which produces a haemolysin.

H. aegyptius (Koch–Weeks bacillus) is very similar to *H. influenzae*. It has been isolated from cases of catarrhal conjunctivitis.

H. parainfluenzae is similar to the rough variant of *H. influenzae*. It requires the V-factor but not the X-factor for growth. Indole is usually not produced.

H. ducreyi is the organism causing a venereal disease known as soft sore or soft chancre. It requires the X but not the V-factor for growth.

H. vaginalis is in a somewhat doubtful position taxonomically. The requirements for X and V-factors are ill defined. It is a human pathogen associated with some cases of vaginitis and urethritis.

BORDETELLA

Small Gram-negative rods. Non-sporing. Some species are motile. Carbohydrates are not fermented. Enriched media required for primary isolation. Normal habitat is the respiratory tract. Pathogenic for man or animals. Strict parasites. Type species: *Bordetella pertussis.*

Bordetella pertussis was isolated from cases of whooping cough in 1906 by Bordet and Gengou.

MORPHOLOGY

The morphology of *Bord. pertussis* is similar to that of *H. influenzae,* but pleomorphism is less marked and short oval rods predominate in freshly isolated cultures.

CULTURAL CHARACTERISTICS

Bordet and Gengou isolated the organism on an agar medium containing a high proportion of blood with potato extract and glycerine. This medium is still used for primary isolation. Colonies after 48–72 hr are small, smooth, glistening with a pearly sheen. It has been shown that the purpose of the blood and potato constituents is to absorb or neutralize toxic substances, probably fatty acids, which would otherwise inhibit growth. Simpler media containing starch or activated carbon may serve the same purpose.

On continued subculture, *Bord. pertussis* will eventually grow on serum agar or even plain agar, but this change in growth requirements is associated with a change in antigenic structure and a smooth to rough change.

RESISTANCE

Bord. pertussis is readily destroyed by bactericidal agents and by heat.

BIOCHEMICAL ACTIVITY

Carbohydrates are not fermented. Indole is not produced. Gelatin is not liquefied. Some strains produce a haemolysin.

SEROLOGY

Freshly isolated strains of *Bord. pertussis* are antigenically homogeneous. They are agglutinable by specific sera, and by the sera of convalescent patients. During growth on artificial media the antigenic structure changes. Four phases, I, II, III and IV have been described by Leslie and Gardner. Phases I and II are smooth and toxic, while phases III and IV are rough and non-toxic. In the rough phases the organisms are not agglutinable by convalescent sera.

PATHOGENICITY

Bord. pertussis is the aetiological agent of whooping-cough, a common and serious disease of childhood. It is transmitted by direct contact or droplet infection. Epidemics frequently occur in schools and residential institutions. The incubation period is variable, from a few days to a fortnight, and the onset is characterized by cough, anorexia and general malaise. The cough is at first dry and short, but later becomes paroxysmal and may be followed by vomiting. During a paroxysm there is usually a series of rapid dry coughs, followed by a violent inspiratory effort which produces the typical whoop. This stage may last for several weeks, and is followed by recovery or the onset of complications, the most important of which are bronchopneumonia, bronchitis and convulsions.

The lesions are found in the mucosa of the upper respiratory tract, which is the seat of hyperaemia and catarrhal inflammation. It is considered that the cilia of the trachea become clogged and destroyed by mucus and the organisms, and these collect instead of passing upwards and are only expelled by the paroxysm of coughing. There may be absorption into the blood stream of the endotoxins which may affect the central nervous system.

DIAGNOSIS

Bord. pertussis is usually present in the sputum during the first week of the illness, but then becomes less frequent and is rarely found after the third week. For isolation of the organism, Bordet–Gengou medium is usually satisfactory. A plate of the medium is held about 1 ft from the mouth during a paroxysm of coughing, and then incubated at 37°C for 48 hr. Post-nasal, or preferably per-nasal swabs, are better, as there is less chance of contamination with extraneous organisms. A small concentration of penicillin in the medium is useful to prevent the growth of Gram-positive contaminants.

PROPHYLAXIS

Specific immunization may be carried out by vaccines prepared from freshly isolated cultures in Phase I. Rough strains do not produce immunity. Toxic reactions may occur with some vaccines.

Chapter XXII

Corynebacterium: Diphtheria

CORYNEBACTERIUM

Gram-positive rods with tendency to club formation. Characteristic orientation in culture or in tissues. Irregularly staining segments or granules. Non-motile, non-sporing. Some species pathogenic for man or animals. Pathogenic species produce an exotoxin. Type species: *Corynebacterium diphtheriae.*

In 1883 Klebs reported the presence of the diphtheria bacillus in the diphtheritic false membrane. In 1884 Loeffler published his classical paper reporting the isolation of the diphtheria bacillus from the throats of diphtheria patients and the production of lesions in experimental animals. In 1888 Hofmann isolated a somewhat similar organism from the throats of normal individuals. This organism, *C. hofmanni* is non-pathogenic, and is the most frequent of the so-called diphtheroids occurring in man.

Several pathogenic corynebacteria have been isolated from various types of animal, and these are occasionally found in atypical lesions in man. The Preisz–Nocard bacillus, *C. ovis*, causes lymphadenitis in sheep, and may cause infection in man. It can be differentiated by cultural and toxicity tests from the true diphtheria bacilli. *C. pyogenes* causes suppurative lesions in various species of animals and may also attack man.

Numerous different types of corynebacteria may be found in the skin and mucosa of normal individuals. These diphtheroids may be differentiated from pathogenic species by their cultural and biochemical reactions and by animal toxicity tests.

C. DIPHTHERIAE OR THE DIPHTHERIA BACILLUS

As the diphtheria bacillus was first described by Klebs and isolated by Loeffler, it is sometimes referred to as the Klebs–Loeffler bacillus (K.L.B.).

Morphology and staining reactions. The morphological appearance of the diphtheria bacillus presents several characteristic features. As a rule it is a straight or slightly curved rod, $3\text{–}5 \times 0{\cdot}5\ \mu$, but pleomorphism is frequently encountered. Short forms and cells swollen at one end, to represent clubs, in the centre or at both ends are frequently found.

The arrangement is also distinctive. Single forms and groups occur; in the groups the organisms tend to be arranged at angles to each other resembling the letters L or V; combinations of these are likened to Chinese or cuneiform writing. This peculiar arrangement is probably due to the method of division and the incomplete separation at the moment of division.

The staining reaction is uneven; in the same organism deep and light staining areas are seen. Also, granules, first described by Babes (1886) and Ernst (1888), are frequently present, particularly in mitis strains; these are usually termed metachromatic or Babes–Ernst granules. They may be situated at the poles of the cell, when they are called "bipolar", or they may be scattered irregularly throughout the protoplasm. A single cell may contain anything up to half a dozen, but the usual number is two or three. In order to accentuate the granules and simplify identification, special staining methods are employed; the most useful of these are Loeffler's methylene-blue and Neisser's differential stain.

The diphtheria bacillus is Gram-positive, but it is decolorized, more readily than most Gram-positive organisms; the granules, however, retain the stain more tenaciously than the remaining protoplasm. *C. diphtheriae* is non-motile and does not form spores or a demonstrable capsule.

CULTURAL CHARACTERISTICS

Growth of *C. diphtheriae* occurs on the ordinary media, but it is more profuse on enriched media. One of the most satisfactory media in use is Loeffler's serum. The optimum temperature is 37°C, range 14°–40°C, and the optimum reaction is pH 7·2. The diphtheria bacillus is a facultative anaerobe, but only scanty growth is obtained in the absence of free oxygen.

In fluid media, such as broth, a moderate turbidity develops and frequently a well-marked pellicle forms. The appearance, however, varies with the type of the organism.

On solid media, such as Loeffler's serum, growth is profuse; after 24 hours' incubation the colonies are small, 1 mm in diameter, smooth, circular, opaque, greyish-white or creamy with a granular structure and easily emulsifiable. On further incubation the colonies become more irregular, and there is a definite differentiation into raised central and flatter peripheral portions.

Growth is not arrested by dilutions of potassium tellurite which inhibits many other organisms and, in the presence of this salt, the colonies acquire a greyish-black appearance. This substance has consequently been added to various enriched media and useful selective media have been obtained. It is important to note that the characteristic morphological appearance may not be found when growth occurs on some tellurite media.

McLeod and his co-workers (1931) employed a special medium containing fresh rabbit blood, special broth, potassium tellurite (0·04 per cent) and agar, which they termed "chocolate tellurite agar". They found, on examination of swabs from cases of diphtheria, that the diphtheria bacilli presented different colonial appearances and they were able, on this criterion, to classify the majority into two separate and distinct sub-groups, which were termed gravis and mitis; later a third group, intermedius, was recognized. A few atypical strains were also found. These observations have been confirmed by a number of independent workers and many varieties of tellurite media are now in use (see p. 411). The following characteristics of the various types have been determined; these tend to remain constant, but irregularities and variations have occasionally been recorded.

Gravis strains form grey or greyish-black, granular, striated colonies presenting a "daisy-head" appearance; ferment dextrin, starch and glycogen; produce a surface growth and granular deposit in broth; usually non-haemolytic; morphology, as determined by observation of an 18-hr culture on Loeffler serum, lightly stained protoplasm with one or two dark areas, granules are seldom found; usually isolated from severe cases of diphtheria.

Intermedius strains give small, discrete, flat, black, granular colonies with a central papilla and slightly irregular edge; there is no fermentation of starch and glycogen, while with dextrin the results are variable; produce in broth a fine granular deposit with a clear supernatant fluid and no surface growth; non-haemolytic; morphology (Loeffler's serum), predominance of barred forms; granules usually scanty but may be well-marked; frequently associated with a severe clinical condition.

Mitis strains form convex, smooth, black shining, entire colonies; there is no fermentation of starch and glycogen, while dextrin gives irregular results; in broth there is a diffuse turbidity with a non-granular deposit and rarely a soft surface growth; usually haemolytic; morphology (Loeffler's serum), classical irregular forms with granules; usually found in the milder types of diphtheria.

The clinical significance of these types is considered later.

RESISTANCE

The diphtheria bacillus is readily destroyed by most bactericidal agents, such as heat at 56°C which kills the organisms in 20 min. Drying and low temperatures are withstood for long periods.

BIOCHEMICAL ACTIVITY

C. diphtheriae exhibits little proteolytic activity. Growth takes place on gelatin without liquefaction of the medium; indole is not produced. Several carbohydrates are fermented with the production of acid only. In these tests Hiss's serum-water medium is employed and important sugars are glucose, maltose and saccharose; the two former are fermented, but there is never any fermantation of saccharose. This is a point of differential importance, as many, but not all, diphtheroids ferment saccharose. In typing the organisms, starch, dextrin and glycogen are of great value. Litmus-milk is not changed.

SEROLOGY

The most important antibody evoked by the diphtheria bacillus is the antitoxin. All strains produce the same toxin, which is neutralized by the antitoxin prepared against the toxin of one strain. The toxin is, however, distinct from that produced by the Preisz–Nocard bacillus, *C. ovis.*

Attempts have been made to sub-group strains of the diphtheria bacillus by means of the agglutination test; the results have, however, been irregular. Investigations indicate that the three sub-groups, gravis, mitis and intermedius, are not antigenically related, and that each group contains a number of strains which are serologically distinct.

TOXIN PRODUCTION

The pathogenicity of the diphtheria bacillus depends largely on its toxin production. It is important to note that there exist strains possessing all the properties of the classical diphtheria bacillus, except that they do not produce an active toxin; these non-toxigenic strains are usually considered to be avirulent. It has been shown that avirulent strains may be converted into virulent strains by the action of a bacteriophage. It is important to note that gravis and intermedius strains are nearly always virulent, whereas mitis strains are frequently avirulent.

Loeffler in his original investigations came to the conclusion, from the results of his animal inoculation tests, that the lesions were due to a bacterium acting locally and to some toxin produced locally and passing into the general system. Although definite lesions occurred in various regions, the diphtheria bacilli were isolated only at the site of inoculation. A few years later Roux and Yersin (1888–90) found that the injection of the sterile filtrate, after passage of the culture through a bacterial filter, into the guinea-pig produced the same lesions as the inoculation of the whole culture. In this way the production of an "exotoxin" by the organism was established. An antitoxin was later prepared and this neutralized the toxin. Confirmation of these results was soon forthcoming, and the introduction of antitoxin therapy was the natural sequence.

Although toxin production is a characteristic property of virulent diphtheria bacilli, the amount formed by various strains varies to a marked degree. The capacity to produce toxin *in vitro* does not appear to be definitely related to virulence, as excellent toxin-producing strains have been isolated from mild cases of diphtheria, e.g. the Park 8 strain, a celebrated toxin-producer employed in many laboratories for the preparation of antitoxin, was originally isolated from a mild case.

The production of toxin also depends to a large extent on the cultural conditions, particularly the composition of the medium. The presence of peptone is essential, but the different makes of peptone vary greatly in their toxin-inducing value; those derived from the digestion of fibrin, such as Witte's or Difco, or from autodigestion of the meat, should be selected. Acid is detrimental, the reaction should therefore be definitely alkaline, pH 7·6–8·2. Hartley's broth (q.v.) is a very satisfactory medium and is extensively employed in the production of toxin. The medium should not be overheated and, in sterilization, filtration with, or without, autoclaving at a relatively low temperature for a short time, is the common procedure.

Abundance of free oxygen is essential and, to obtain this, the medium is arranged in shallow layers with a large surface area in amounts of 2–5 litres in special flasks or cylinders, which are loosely plugged to permit free diffusion of air. The inoculum is the pellicle of a growth obtained in a "starter-bottle", which contains about 100 ml of the same medium. The diphtheria bacillus is acclimatized by passing through two or three starter-bottles before it is used for the preparation of toxin. This ensures good pellicle formation, which is necessary to obtain the maximum toxin production; this of course does not apply to the intermedius strains which do not form a pellicle. After inoculation or seeding, the flasks are incubated at 37°C for 7–11 days. At the end of this time the broth is filtered and the filtrate contains the crude toxin, i.e. toxin plus various products of digestion and disintegration of the components of the medium.

When freshly collected the toxin is relatively unstable; it is destroyed or modified readily by heat at 60°C, acids, alkalies, light and oxidation. By storage in the dark at a low temperature, 0°–4°C, it becomes more stable. This process, which is termed "ripening", is adopted when the toxin is required for testing purposes.

Toxin has been concentrated by various methods, such as salting out with ammonium sulphate or precipitation with acetic acid. It may be also obtained in a dried form and is then relatively stable. By using special media the toxin has been obtained in a pure state; it is a protein, and 0·0001 mg will kill a guinea-pig.

The potency of the toxin is standardized in the guinea-pig. Varying dilutions are injected subcutaneously, and the smallest amount killing a 250 g animal in four days is termed the minimal lethal dose, or M.L.D. Another method is to determine the minimal reacting dose, or M.R.D., which is the smallest amount producing a definite reaction on intradermal inoculation into a guinea-pig. A good toxin is one having a M.R.D. of 0·002 ml.

TOXOID OR ANATOXINE

The toxicity of a filtrate may be reduced by the addition of 0·2–0·4 per cent formalin and subsequent incubation at 37°C for 4–6 weeks. With the reduction in toxicity there is, however, no loss of antigenicity or combining power. The modified toxin, which is termed toxoid or anatoxine, is consequently a valuable immunizing agent. In its preparation the loss of toxicity is controlled by the subcutaneous injection of guinea-pigs, in which 5 ml of the toxoid should not give a reaction. The toxicity decreases rapidly at first, during incubation at 37°C, but it may take several weeks for its complete removal.

The toxoid is used for the immunization of horses in the preparation of antitoxin, and for human prophylaxis.

PATHOGENICITY

The pathogenicity of the diphtheria bacillus is largely dependent on its capacity to produce the powerful exotoxin. In the case of certain strains, however, there appears to be little doubt that other factors may be concerned. The same antitoxin neutralizes the toxins produced by mitis, intermedius and gravis strains, and yet serum-therapy in gravis and intermedius infections in man may have little effect on the course of the condition. The nature of the other pathogenic factors has not been determined, but they may be distinct from the toxin and intimately associated with the bacterial protoplasm.

Laboratory animals vary considerably in their reaction to the diphtheria bacillus according to the susceptibility of their tissues to the toxin. Rats and mice possess a high degree of resistance; this appears to be an example of natural species immunity, as the specific antitoxin is not present in the serum. Dogs and cats are moderately susceptible, while rabbits and guinea-pigs are highly susceptible. The guinea-pig, owing to its convenient size, is the laboratory animal chiefly employed in the study of this organism.

On subcutaneous or intramuscular inoculation of the guinea-pig with a small amount (0·5–1 ml) of a virulent culture, death follows in 24–27 hr. On autopsy an oedematous, slightly haemorrhagic, local lesion is found with enlarged regional glands, congestion of the lungs and spleen, pleural exudate and marked congestion of the suprarenals. The congestion of the suprarenals is pathognomonic, being present in at least 80 per cent of animals.

On intradermal inoculation a well-marked local lesion appears in 24 hr, becoming, in 48 hr, an erythematous swelling with a black necrotic centre.

In man the diphtheria bacillus is responsible for the spontaneous disease, diphtheria.

DIPHTHERIA

Diphtheria was a common and widespread disease, which has been the subject of numerous and extensive epidemiological surveys. It is interesting to note that its main epidemiological features were recognized and that general prophylactic measures had been introduced long before the isolation of the causative organism had been accomplished.

Diphtheria is particularly a disease of children; some 85 per cent of cases in this country occur in children under 15 years. The incidence is highest in the age-group 5–10 years, i.e. the age-group when children usually commence school, but after the tenth year there is a sharp decline. It is a disease of schools and institutions, where groups of susceptibles may be herded together; in such centres extensive outbreaks may occur in the absence of specific immunization. It is important to note that, in recent years, the epidemiology of diphtheria has changed considerably as a result of intensive immunization of infants and children. In many regions the disease has practically disappeared.

Infection is conveyed from one individual to another either by direct contact or indirectly by droplet infection, and as a rule the lesions develop primarily in the naso-pharynx. The diphtheria bacillus produces a more or less characteristic reaction, which results in the formation of the so-called "pseudo-membrane". This is mainly a fibrinous exudate with disintegrating epithelial cells, leucocytes, red blood corpuscles and bacteria. It appears as greyish-white or yellowish patches usually on the tonsils, uvula, palate, larynx or trachea, and is frequently difficult to detach. The neighbouring mucous membrane is extremely congested and the regional lymph glands are usually swollen and painful.

General symptoms are also present due to the absorption into the circulation of the toxin produced by the organisms at the local lesion. The organs most frequently involved are the heart and the nervous system. Cardiac failure is not an uncommon complication of diphtheria, but the exact site of damage has not been determined. The lesion of the nervous system is a peripheral neuritis giving rise to paralysis involving mainly the palate, the limbs, and accommodation of the eyes.

The suprarenals, in fatal cases, rarely show the gross macroscopical changes seen in the guinea-pig, but microscopical changes, such as infiltration and minute haemorrhages, may occur.

Diphtheria is essentially a toxaemia, the organisms have only rarely been isolated from the blood-stream during life. They tend to remain at the primary focus of infection and there produce the toxin which is responsible for many of the clinical features. The mortality rate is subject to much variation, but there is no doubt that it has been reduced considerably by the introduction of antitoxin therapy; in the pre-antitoxin period the average mortality was about 30 per cent, whereas it is now approximately 10 per cent. Death may result either from respiratory obstruction by the membrane, as seen usually in the laryngeal form, or from toxaemia. Recovery confers a high degree of immunity and is associated with the presence of antitoxin in the serum; second attacks are extremely rare.

The severity of the condition is definitely related to the character of the infecting organism; gravis and intermedius strains are more frequently associated with severe forms of the disease than mitis strains.

Diphtheria is widely distributed in the civilized world, but infection with the diphtheria bacillus does not always produce the classical features of the disease. Lesions may be produced in the nose, as a fibrinous rhinitis, in the ear, the conjunctiva, or in the skin. In the naso-pharynx a membrane is not always produced and atypical forms may occur. Epidemiological and statistical studies indicate that sub-clinical infection is a frequent event; many adults possess a high degree of resistance, and yet they give no history of previous attacks of clinical diphtheria or immunization.

Carriers may result from all types of infection; gravis forms tend to be the most persistent while intermedius infections clear up most quickly. The diphtheria bacilli are present in the throat and/or nose during convalescence, and in some they may persist after the main lesions have resolved. Healthy individuals possessing increased resistance are frequent carriers of virulent diphtheria bacilli. Endemic foci are largely maintained by carriers, and there can therefore be no marked reduction in the incidence of the disease in any population group until prophylactic immunization is general. It is considered that some 30 per cent of pre-school infants and 50 per cent of school children must be immunized before any appreciable decrease in the incidence of diphtheria is observed.

DIAGNOSIS

It is important to note that the diagnosis of diphtheria is essentially the duty of the clinician, and on no account should it be considered the task of the bacteriologist. Whenever the symptoms are suggestive, the case should be treated as one of diphtheria; the bacteriological examination is employed to confirm the diagnosis. The isolation of the diphtheria bacillus does not necessarily indicate the case to be one of diphtheria and the failure to detect it does not exclude the disease. With these reservations there is little doubt that the bacteriological examination is of great value to both the clinician and public health official in dealing with cases and carriers.

Material for the bacteriological examination is collected on a sterile swab, which should be rubbed along the membrane and not pushed blindly into the mouth. The methods of examination are by direct microscopy and by culture and subsequent microscopy.

The direct slide method is sometimes useful in frank cases, but it is not recommended for routine use; the swab is rubbed first on a sterile slide, which is stained and examined, and then on to a Loeffler's serum slope. If positive an early report can be made but an immediate negative result is valueless. Diagnosis by microscopy should invariably be confirmed by cultural methods.

The usual method is to prepare cultures, and after incubation at 37°C for 18–24 hr to carry out the microscopical examination. The standard medium is Loeffler's serum prepared as a slope, but now tellurite media are required and many different types are available, all giving satisfactory results. Each swab is rubbed on both a Loeffler's serum slope and the tellurite medium; after incubation overnight smears are prepared from the Loeffler serum culture, while the growth, if any, on the tellurite medium is examined through the hand lens. This requires considerable experience, as other organisms tend to produce blackish colonies, but with practice the typical colonial appearance of the various types of diphtheria bacilli can be detected. The results of the two examinations are considered together; in the great majority of cases there is complete agreement, but in some the microscopical examination may be negative while the tellurite reading is positive. In such cases subcultures are prepared from the tellurite on to Loeffler's serum, which is examined after incubation overnight at 37°C. Microscopy of the growth on the tellurite medium may be unsatisfactory. In some cases it is thus impossible to complete the examination in 24 hr.

Great care must be taken to differentiate the diphtheria bacillus from the diphtheroids, particularly *C. hofmanni*, which are frequently present in the naso-pharynx. The main diagnostic features of *C. diphtheriae* are the pleomorphism, irregular staining with the presence of granules, and the characteristic arrangement. It is, however, important to note that granules are frequently absent from gravis strains. In all cases, cultural, biochemical and toxicity tests should be made.

C. hofmanni is short and relatively stout, does not usually possess granules, tends to be regular (pleomorphism is uncommon) and to be arranged in parallel rows; its protoplasm stains unevenly, presenting a "barred" appearance, due to the weakly stained central protoplasm. It also grows rather more readily on the simple media, such as agar, than the diphtheria bacillus.

In the search for carriers it is necessary to obtain swabs from both the throat and nose; for this purpose the use of plates containing a tellurite medium is essential. After incubation overnight a mixed growth is almost invariably obtained, and suspicious blackish colonies are picked off on to Loeffler's serum slopes, which are incubated at 37°C for a further 24 hr. The morphological appearance of the organism is examined, and those resembling the diphtheria bacillus are kept for further examination. There is no morphological difference between virulent and avirulent strains; it is generally accepted that gravis and intermedius strains are almost invariably virulent but mitis strains may be avirulent. It may therefore be necessary, particularly in dealing with carriers, to determine whether the isolated organism is virulent or avirulent. This is carried out by means of the virulence test.

In performing a virulence test pure cultures must be obtained;

the action of these on glucose and saccharose in Hiss's serum-water is noted; if saccharose is fermented the strains are discarded, as this sugar is never fermented by the diphtheria bacillus. The virulence of the non-saccharose-fermenters is next examined in the guinea-pig by either subcutaneous or intradermal inoculation. The growth on Loeffler's serum is suspended in broth to give a slight opacity. In the subcutaneous method 0·5 ml of this suspension is injected subcutaneously into two guinea-pigs, one of which has previously been given antitoxin. The control animal should remain alive, whereas if the organism is virulent the test animal should die in 24–96 hr, and on autopsy present the characteristic appearances. When dealing with pure cultures control animals are not usually required.

In the intradermal method the injection is made into the skin of the shaved abdomens of two animals; one, the control, is given 500 units of antitoxin intraperitoneally the previous night, and the other, the test animal, may receive 125 units of antitoxin 4–6 hr after the injection of the bacterial suspension. Definite lesions should develop in 24–28 hr on the test animal, but not on the control, which is protected by the prior injection of the antitoxin. The small dose of antitoxin is given to the test animal when several suspensions are being tested; it does not interfere with the local lesions, but prevents the animal dying from toxaemia. The intradermal method is more economical than the subcutaneous test; six or eight suspensions may be tested on one animal, whereas only one suspension can be used in the case of the subcutaneous method, which is, however, more reliable.

A simpler, *in vitro* method of determining toxin production by *C. diphtheriae* is that of Elek. A filter-paper strip soaked in diphtheria antitoxin is incorporated in an agar plate, and the strain under test is streaked at right angles across the medium. During incubation the toxin diffuses outwards from the area of growth and a specific toxin–antitoxin precipitate is produced at the points of optimal concentration of the two reactants. The positive plate shows four lines of precipitate radiating from the point of intersection of the filter-paper strip and the streak of growth.

PROPHYLAXIS

Specific prophylaxis in combating such a widespread infection as diphtheria is only of practical value provided that it is possible to differentiate susceptible individuals from those possessing resistance to infection. This was made possible in the case of diphtheria by the introduction of an intradermal test by Schick, which is now universally employed and is known as the Schick test.

In the *Schick test* a small amount of toxin, corresponding to 1/50 M.L.D., and contained in 0·2 ml, is injected intradermally into the flexor surface of the forearm. In order to have a measure of combining power as well as toxicity, the Schick test dose should also be neutralized when added to 1/750th, but not to 1/1,250th, part of an international unit of antitoxin. Official standards for the preparation of the Schick toxin are governed by the Therapeutic Substances Regulations. As a control an equal amount of the same toxin, heated at 70°C for 30 min, is injected into the skin of the right forearm. It is advisable to read the results after 24 hr, four days and seven days as four types of reactions may be observed:

(1) Negative, in which signs of reaction do not appear at either situation.

(2) Positive, in which the control arm shows nothing while the test arm manifests a definite local erythema and oedema in 24–48 hr, this is usually well marked by the fourth day, then it gradually fades and presents a brownish appearance with perhaps some slight desquamation.

(3) Pseudo-reaction, in which at both sites there is a definite reaction in 24 hr; this subsides more rapidly than the true reaction, and by the fourth day there is little to be seen. This is actually a negative result, the lesion being of a non-specific nature.

(4) Combined, in which reactions (2) and (3) are both present. When read at the end of 24 hr little difference is detected between the test and control arms, but this gradually becomes more apparent, and by the fourth day it is striking; a well-defined positive result is then apparent on the test arm.

Reactions (1) and (3) are thus negative and indicate resistance to infection. It has been found that a negative result indicates the presence of approximately 1/100–1/40 unit of antitoxin in the serum. It has also been established that the presence of such an amount of antitoxin in the serum is a decided but not an absolute preventive against an attack of diphtheria.

Reactions (2) and (4) are both positive and indicate susceptibility to infection. The test dose for this reaction is selected as it is the largest amount of toxin usable without producing a reaction in individuals who have sufficient antitoxin in their serum to give a relative immunity. In view of the occasional occurrence of gravis infections in Schick-negative individuals, the test-dose may have to be increased. It has been suggested that a Schick toxin with at least three times the toxicity of the official standard should be used.

In the absence of either artificial immunization or an attack of the disease, most young children are Schick-positive. The number of positive reactors tends to decrease with advancing years due to immunity acquired by subclinical infection. At one time most adults were considered to be immune, but in surveys recently carried out on military communities and university students, it has been found that some 30–60 per cent were Schick-positive, i.e. susceptible to infection. This relatively high rate of positive reactors is the probable result of a general improvement of living conditions in this country and the decreased incidence of the disease, as a result of which chances of subclinical infection have been reduced.

IMMUNIZATION

Having detected susceptible individuals, several methods of inducing a relative immunity are available. The following preparations have been employed for this purpose:

(1) Formol-toxoid (F.T.)
(2) Toxoid-antitoxin floccules (T.A.F.)
(3) Alum-precipitated-toxoid (A.P.T.)
(4) Purified toxoid, aluminium phosphate precipitated (P.T.A.P.)

(1) Formol-toxoid (F.T.) has been extensively employed by Ramon in France. It is readily prepared, but although the specific toxicity of the toxin is greatly reduced by the action of the formalin, it tends to produce a marked non-specific reaction in sensitive individuals. For this reason it is considered to be unsuitable for general application in individuals over the age of 10 years.

(2) Toxoid-antitoxin floccules (T.A.F.) are composed of a saline suspension of the precipitate formed when equivalent amounts of toxin and antitoxin are mixed. With this preparation the risk of reaction is reduced to a minimum. It has a relatively high immunizing value.

(3) Alum-precipitated-toxoid (A.P.T.) is a turbid suspension containing the precipitate resulting from the addition of alum to the toxoid. The toxoid is liberated slowly from the precipitate and the antigenic stimulus provided thereby is maintained for a longer period. A.P.T. is now widely used for the immunization of children and two doses of 0·2 and 0·5 ml at 3–4 weeks' interval

are recommended for infants, with a third dose on reaching school-age.

(4) A more consistent and controllable product has recently been prepared by Holt—P.T.A.P. (purified toxoid aluminium phosphate precipitated). This has been shown to be a very potent antigen.

The choice of preparation depends to some extent on the age of the individual. Immunization should not be carried out under the age of six months; infants frequently possess some immunity transferred from the mother, and their antibody response at this age is poor. After this period most, if not all, children are susceptible and the most suitable time for immunization is at age 8–12 months. As children tend to react, the Schick test is unnecessary. The tissues of children are not so reactive as those of adults, and therefore in immunizing infants the selection of prophylactic is not important. Immunization against diphtheria, tetanus and whooping-cough is often combined with satisfactory results.

The results obtained by the use of these prophylactics have become highly satisfactory. Some 80–100 per cent of individuals become Schick-negative after the administration of the requisite course. The incidence of diphtheria in immunized individuals has been greatly reduced, but not entirely eliminated. It is interesting to note that many of the reported cases in Schick-negative individuals have been gravis infections, and that in most of these the severity of the attack has been reduced.

In this country little immunization had been carried out prior to 1940, but as a result of intensive propaganda campaigns much progress has since been made as indicated by the following statistics for England and Wales: 1901, approx. 10,000 deaths; 1941, 2,641 deaths; 1951, 30 deaths; 1956, eight deaths; 1967, no deaths. There has also been a corresponding drop in the incidence of the disease.

Passive immunization by the injection of antitoxin is now seldom used as the resulting immunity is of short duration, lasting only 2–3 weeks. It is used mainly as an emergency measure, particularly in hospitals and in dealing with delicate children. The usual dose is 1,000 units subcutaneously.

THERAPY

The essential feature of the treatment of diphtheria is the early and adequate administration of *antitoxin*. In all cases where the symptoms suggest diphtheria, serum must be given at once without awaiting the bacteriological report.

The antitoxin is produced by the immunization of the horse—usually with some toxoid preparation. Samples of blood are collected at intervals and the antitoxin content of the serum estimated. A reasonable titre, such as 500–600 units per ml may be obtained after immunization for 2–3 months. Blood is then collected in citrate solution, and from this the serum is prepared. The titration is carried out by comparing its action against a potent toxin with that of a standard antitoxin in the guinea-pig. The L+ and Lr doses are both employed for titration. In France the Lf dose is used, but owing to the possible event of irregularities this method is only employed as a preliminary process in this country to provide an index of combining power. The antitoxin is contained in the pseudo-globulin of the serum and is associated mainly with the beta globulin fraction. The antitoxin can be readily separated from the remainder of the serum by salting out with ammonium sulphate, and is obtained in a concentrated form and free from much extraneous protein. A recent development has been the preparation of serum "refined" by enzyme activity, which removes unwanted protein fractions. Concentrated serum is therefore a highly satisfactory therapeutic agent and may contain up to 2,000–3,000 units per ml.

In practice the dose and route of administration vary with the age and condition of the patient.

The intramuscular route, into the muscles of the thigh, is usually employed, but in severe cases where a rapid action is desired the serum should be given intravenously. In considering the dosage, it must be remembered that it is easy to give too little, but difficult to give too much. There is no generally accepted scheme of dosage, but as a working basis it may be taken that the older the patient and the severer the attack, the larger the dose. In dealing with mild cases the average dosage is 2,000–10,000 units; in the severe forms it is 10,000–50,000 units. In very severe cases doses up to 250,000 units have been given.

Serum-therapy has produced very satisfactory results, the incidence of complications and the mortality rate have both been reduced to a marked degree. The mortality rate is, however, still relatively high in gravis and intermedius infections. In these cases the administration of antitoxin not infrequently has little effect on the course of the disease. The nature of the offensive weapon of these strains has not been determined, but as it might be concerned with the bacterial protoplasm an antibacterial serum has been prepared. This has been given together with the antitoxin, but the results are not encouraging.

Penicillin has given good results in the severe forms of the disease when given to supplement the action of the antitoxin, for which it must be clearly understood that penicillin is not a substitute. Penicillin may arrest the growth of the diphtheria bacilli but it has no action on the preformed toxin. Attempts to clear carriers by the local administration of penicillin have not been very successful.

Chapter XXIII

Mycobacterium: Tuberculosis: Leprosy

MYCOBACTERIUM

Slender rods. Acid-fast. Variable morphology. Non-motile. Non-sporing. Aerobic. Several species pathogenic to man and/or animals. Pathogenic species generally slow growing; saprophytic species grow more rapidly. Type species: *Mycobacterium tuberculosis.*

The leprosy bacillus, *Myco. leprae*, was described by Hansen in Bergen in 1874. It was present in the large vacuolated cells of the characteristic leprosy nodules. This organism has not been successfully cultured in artificial media. In 1882 Koch described the *Mycobacterium tuberculosis* and showed its relationship to human tuberculosis. He later demonstrated that there were two types of tubercle bacilli, human and bovine, depending on the source from which they were isolated. Further types were isolated from birds and cold-blooded animals and were called respectively avian and cold-blooded or reptilian types. In 1895 Johne and Frothingham observed another member, Johne's bacillus, in cases of chronic enteritis in cattle known as Johne's disease. In 1903 Stefansky, at Odessa, isolated a Mycobacterium from rats now known as the rat-leprosy bacillus. During the period 1885–1906 a number of saprophytic mycobacteria were isolated and described. These organisms are of importance to the medical bacteriologist because they may be found in situations which might lead to a mistaken diagnosis of tuberculosis if reliance were placed on microscopical examination without investigation of cultural characteristics and pathogenicity. The most important members of this group are:

(1) *Myco. smegmatis*, frequently present in the smegma of both males and females
(2) *Myco. butyricum*, found in butter.
(3) *Myco. phlei*, present in timothy grass and dust and fodder.
(4) *Myco. stercoris*, isolated from the faeces of various animals.

Wells, in 1937, investigated an epidemic disease of voles and isolated an organism which is antigenically similar to human and bovine tubercle bacilli. It was known as the vole bacillus, *Myco. tuberculosis* var. *muris*, and now as *Myco. microti*.

Other mycobacteria of limited pathogenicity have been isolated from various skin lesions in man in recent years. *Myco. balnei* was originally isolated in Scandinavia from skin lesions contracted in swimming pools. *Myco. ulcerans* was isolated from skin ulcers in Australia. *Myco. fortuitum* was isolated from a cold abscess in a woman in Brazil, and has subsequently been found in many other countries.

THE TUBERCLE BACILLUS (MYCO. TUBERCULOSIS)

The tubercle bacillus is a strict parasite and is essentially pathogenic. There is a definite biological distribution of the various types; the human and bovine strains produce tuberculosis in mammals, including man, the avian strains in birds and the cold-blooded strains in cold-blooded animals and fish.

MORPHOLOGY

These organisms are usually straight or slightly curved rods with rounded ends and are arranged either singly or in small groups; clubbed and swollen forms are seen in cultures; branching is rare. The size is variable, the average dimensions are $0{\cdot}3\text{–}0{\cdot}5 \times 1{\cdot}2\text{–}4\ \mu$; organisms isolated from the body-tissues tend to be longer and thinner than those grown under artificial conditions. Granular or beaded forms are not uncommon.

Tubercle bacilli stain with difficulty owing to the presence in the cell of fatty and alcoholic materials; the ordinary staining methods are consequently of little value. They are Gram-positive, but the staining is not easy, and this method is not employed in the routine study of the acid-fast organisms. The organisms, when stained by the Ziehl–Neelsen method, are particularly resistant to the decolorizing action of acids and alcohol; this is stated to be due to the presence of a higher alcohol termed "mykol". By virtue of this property members of this genus are frequently referred to as "acid-fast" organisms.

CULTURAL CHARACTERISTICS

Growth under artificial conditions varies considerably with the different types of tubercle bacilli. The cold-blooded types behave as the saprophytic members of this group and grow readily on ordinary media in 2–3 days, for the human and bovine types, on the other hand, special media are required, and even then growth is very slow. The avian type occupies a position mid-way between these extremes.

The media commonly used for studying the mammalian types are Dorset's egg, made by mixing eggs with water and inspissating at 70°C, inspissated serum, glycerine agar or broth and glycerine-potato. It is important to note that the addition of glycerine to agar or broth enhances the growth of all members of this group with the exception of the bovine strains of the tubercle bacillus and the vole bacillus. Various synthetic media, such as that of Kirchner and of Dubos are also very useful.

Aerobic conditions and moisture are essential for growth. The optimum reaction of the medium varies with the types, and is about pH 7·0–7·6. The optimum temperature also varies with the different types: for the human and bovine bacilli it is 37°C, for the avian about 40°C, for the cold-blooded about 25°C, for *Myco. ulcerans* and *Myco. balnei* 30°–33°C.

On solid media, such as Dorset egg, growth occurs usually as dry, friable, confluent colonies with a granular or wrinkled surface. The bovine and human types grow slowly, 4–6 weeks being usually required to obtain an extensive growth. Evaporation of the medium is prevented by using screw-cap bottles.

In fluid media, such as glycerine broth, the human and bovine types grow as a greyish-white, dull, wrinkled, surface pellicle with no general turbidity; a slight granular deposit may be present. The avian and cold-blooded types show no surface growth or general turbidity, but give a granular deposit or a thick veil at the bottom of the flask. A peculiar fruity odour is given by cultures of all types.

A rapid and diffuse growth of the tubercle bacillus is promoted by the synthetic medium introduced by Dubos (p. 413). The Dubos medium contains bovine albumin (0·3 per cent) and a wetting agent, "Tween 80", which is a water-soluble ester of oleic acid. This medium is now widely used for testing the sensitivity of *Myco. tuberculosis*. In this medium *Myco. tuberculosis* grows as small aggregated chains of bacilli known as *serpentine cords*. The production of such cords is an important differential criterion between *Myco. tuberculosis* and other acid-fast bacilli.

The cultural reactions present features of importance in the differentiation of the human and bovine types. The human type grows more freely, particularly when the medium contains glycerine, and for this reason it is termed eugonic. The bovine type usually grows less luxuriantly, particularly on fresh isolation on glycerine media, when it may fail to grow, and is termed dysgonic. When growth is obtained on media containing serum of a rich yellow colour, the human type frequently forms a yellowish or orange pigment, the bovine type never does.

The cultural differentiation between the mammalian and the other types is not difficult. The avian type, after a few subcultures, usually grows readily at 40°–42°C, and in fluid media gives a granular deposit or a spreading growth at the bottom of the flask. The mammalian types form a surface pellicle in fluid media, and growth is slight or absent at 40°C. The cold-blooded types usually give an early and luxuriant growth when incubated at 25°C, at which temperature the other types do not multiply.

The various types appear to be stable. Attempts have been repeatedly made to demonstrate mutation of one type into another, but without success. Atypical forms have been isolated from a small percentage of lesions in man and animals. Some workers consider that these "atypical acid-fast bacilli" are intermediate between the human and the bovine types; their position in the classification of the mycobacteria is, however, far from clear. They appear to be pathogenic to man and may be differentiated from *Myco. tuberculosis* by biochemical tests. They are always resistant to isoniazid.

RESISTANCE

The tubercle bacillus possesses an unusual property in that it behaves as other non-sporing organisms to the action of heat, but is very resistant to chemical bactericidal agents. In a dried state the organisms may survive exposure to 100°C for 20 min, but as a saline suspension they are killed at 60°C in less than 30 min. In milk the tubercle bacilli are killed by this treatment provided it is contained in a closed vessel. In an open vessel a surface pellicle forms; this protects the organisms and renders their destruction more difficult.

Their resistance to chemical agents is similar to that exhibited by many sporing organisms and is doubtless due to the presence in the bacterial cell of fatty materials and alcohols, which resist the penetration of aqueous solutions. Marked resistance is shown to 5 per cent phenol, 15 per cent sulphuric acid, caustic soda and antiformin; the two latter solutions are frequently employed in the isolation of the tubercle bacillus from contaminated material, as they digest the solid matter and destroy the Gram-negative bacteria. The effect of the aniline dyes is variable; the tubercle bacillus can grow in the presence of dilutions of gentian-violet and malachite-green that prevent growth of many organisms. These substances have consequently been employed in the preparation of selective media, e.g. gentian-violet is added to coagulated egg in Petroff's medium, and malachite-green is employed in Löwenstein–Jensen's medium. The latter medium is now being used extensively, as growth occurs more rapidly than on the other varieties.

BIOCHEMICAL ACTIVITY

Little attention has been given to biochemical activities of the tubercle bacillus. There is some evidence that acid, but not gas, may be produced in some sugars, e.g. glucose and maltose, but these tests are seldom, if ever, used in routine practice. Unlike the saprophytic mycobacteria, *Myco. tuberculosis* possesses a peroxidase. A test for this enzyme is one useful and simple way to differentiate the saprophytic A.F.B. from the tubercle bacillus. Niacin is produced in the metabolism of the human type of *Myco. tuberculosis* but not in that of the saprophytic mycobacteria; this forms the basis of another differentiating test.

SEROLOGY

Agglutination, agglutinin-absorption and complement fixation have all been used in the study of the tubercle bacillus. By means of these tests it has been found that the tubercle bacilli fall into three serological groups—mammalian, avian and cold-blooded. The human, bovine and murine types constitute a homogeneous group and cannot be differentiated by serological methods.

PATHOGENICITY

The tubercle bacillus is one of the most widely distributed pathogenic organisms. It is essentially a pathogen and produces spontaneous tuberculosis in man and a large variety of animals. The susceptibility of the different animals varies, but cows, swine and monkeys are particularly sensitive. The various types of tubercle bacilli differ considerably in their pathogenicity. The mammalian types are pathogenic to man and or, mammalian animals, but not to birds or cold-blooded animals; the avian type is pathogenic for birds, but has little action on mammals except rabbits and pigs; the cold-blooded type produces tuberculosis in cold-blooded animals and fish.

Animals show marked differences in their susceptibility to experimental tuberculosis. Some species, e.g. guinea-pigs and monkeys, are highly susceptible to the human and bovine types; others, such as dogs, rats and mice, are very resistant. A third group, including goats, ferrets and rabbits, is relatively susceptible to the bovine bacillus but resistant to the human bacillus. This variation in susceptibility provides a further test for the differentiation of these two mammalian types; in practice the rabbit is the animal employed for this purpose.

The injection of tubercle bacilli into susceptible animals, such as the guinea-pig, produces a rapidly spreading fulminating form of tuberculosis ending fatally usually in 3–5 weeks. Following subcutaneous inoculation a well-marked local caseous lesion forms with caseating tubercles in the regional glands and most organs, particularly the spleen, liver and lungs. In relatively resistant animals, such as the rabbit for the human type, the subcutaneous inoculation of a small dose of the organisms produces a slowly developing local lesion; generalization of the infection and death may occur in some cases after several months, but healing lesions are usually found when the rabbit is killed.

TUBERCULOSIS

Tuberculosis has been and still is one of the most important scourges of civilized communities. It appears to be one of the disadvantages of civilization, as the disease is extremely rare in uncivilized communities. When the tubercle bacilli have been introduced into these communities a fulminating type of tuberculosis, with early death, has resulted. In this respect it is interesting to note that the disease is also practically unknown in monkeys in their natural surroundings, but when these animals are brought into captivity tuberculosis becomes very common and usually runs a rapid course.

Tuberculosis in man may exist in a great diversity of forms and may involve practically every organ of the body. Its distribution and form depend largely on the age and susceptibility of the patient and the route of entry of the organism. Some consider that genetic factors play an important role in the varying susceptibilities of different races. In resistant individuals a localized, indurated, chronic lesion usually develops, whereas in susceptibles the acute generalized forms predominate. Between these two extremes a large variety of lesions is found.

MODES OF INFECTION

In practically all cases infection takes place after birth. The old idea that infection might, in some cases, be hereditary is now largely discredited. Infants of tuberculous parents with active lesions frequently develop tuberculosis, but this is due to their greatly increased chance of infection after birth. The incidence of the disease in such infants has been decreased by their immediate removal to healthier surroundings. Tuberculous foetuses have been removed from cows but, while the infection might have occurred by the transplacental route, in many cases lesions have been demonstrated in the uterus and cotyledons of the parent.

In post-natal infection the tubercle bacilli may enter the body by any exposed surface. The three most important portals of entry are: (1) the respiratory tract, (2) the alimentary tract and (3) the skin. The infecting organism belongs to either the human or bovine type. Human infections with the avian type have been reported, but they are rare.

(1) **The respiratory tract.** The pathogenesis of pulmonary tuberculosis is an unsettled problem. Two main hypotheses have been formulated; one, advocated strongly by English bacteriologists, postulates that infection occurs primarily by the way of the respiratory tract; the other, advocated by French workers, maintains that infection results mainly from passage of the tubercle bacilli through the intestinal mucosa.

In the inhalation theory the tubercle bacilli are introduced into the lungs either by droplet infection, e.g. sneezing and coughing by patients with open tuberculous lesions, or in dust and dried sputum. Evidence supporting this theory is that respiratory tuberculosis is caused almost entirely by the human type, whereas the bovine type is largely responsible for abdominal tuberculosis. If infection occurred via the intestinal tract, one would expect the bovine type to be present in a greater number of respiratory lesions. The bacilli also pass through the mucosa of the upper respiratory tract and produce enlargement of the cervical or bronchial glands, from which they may spread to the lung.

The other hypothesis is based mainly on the observations of tuberculosis in animals. It is considered that, after absorption through the intestinal mucosa, the tubercle bacilli enter the lymph and blood streams and later pass to the lungs. The relative frequency of pulmonary tuberculosis in adults is explained by the reduced resistance of their lymph glands, as a result of which the bacilli are more liable to pass through and on to the lungs. This hypothesis appears less satisfactory than the former.

(2) **The alimentary tract.** Tubercle bacilli readily pass through the intact mucosa of the alimentary tract and reach the lymphatic system. The chief sites of entry are the mouth and naso-pharynx, giving rise to involvement and enlargement of the cervical glands, and the small intestine, particularly Peyer's patches, from which infection passes on to the mesenteric glands. The main vehicle of infection is milk; other sources include food and sputum.

(3) **The skin.** Infection via the skin is relatively infrequent; it may occur either by way of minute abrasions during the handling

of contaminated material and infected food or by direct inoculation, as by a contaminated knife or needle. Primary tuberculosis of the skin usually takes one of two forms, lupus vulgaris or verruca tuberculosa.

HISTO-PATHOLOGY

In all cases of tuberculosis the primary lesion, the tubercle, is essentially the same. This is a local lesion containing giant cells, lymphocytes and epithelioid cells; the initial step in its development is the formation of microscopical tubercles, which increase and coalesce to produce the characteristic macroscopic tubercle. Further developments depend on the virulence and number of the organisms and the resistance of the host, and whether it is a primary infection or a reinfection of the adult type. Under conditions favourable to the host, the progress of the tubercle is arrested and fibrosis with perhaps ultimately calcification may occur. In this manner the focus of infection becomes either quiescent or healed. If quiescent, it may become active at some later date when the resistance of the individual is lowered.

Under less favourable conditions the lesion progresses. This sequence of events is therefore commonly encountered in individuals in whom the general resistance has been lowered by prolonged fatigue, under-nourishment, the acute exanthemata, such as measles and whooping-cough, injury and chronic alcoholism. In certain industries, such as those involving the treatment of silica, and in miners, the inhalation of irritant dusts with the production of pulmonary lesions definitely predisposes to infection with the tubercle bacillus. Under such unfavourable conditions the tubercle increases in size, degenerates, caseates and may form a cavity, ulcer or sinus. General manifestations of the infection, such as sweating and pyrexia, are then observed and result from the absorption into the general circulation of toxic substances produced by the tubercle bacilli. Extension of infection may take place by three routes: (a) directly along continuous surfaces, (b) by the lymph stream and (c) by the blood stream.

(a) Spread along continuous surfaces is frequently seen in ulcerative pulmonary tuberculosis. The breaking-down of a lesion in the lung or bronchial gland may result by aspiration in a tuberculous broncho-pneumonia and by expectoration in tuberculous lesions of the upper respiratory tract and face. In a similar manner tuberculosis of the epididymis may give rise to an ascending infection of the prostate and bladder.

(b) Spread by the lymphatics to the regional lymph glands is an early feature and occurs very commonly in children. There may be no lesion at the primary site of entry. In many cases the course of the disease is arrested at this stage, but in others there is a progressive involvement of other glandular groups and perhaps passage into the thoracic duct and into the blood stream.

(c) The entrance of the tubercle bacilli into the blood stream is not uncommon; it may occur not only by the thoracic duct, but also from any local lesion by invasion and ulceration of the wall of a vessel. When a few organisms enter, they tend to localize in various organs, particularly the bones, joints and kidneys, and produce a chronic or subacute type of lesion. If, however, a large number of organisms invade the blood stream, there is a generalization of the infection with the production of miliary tuberculosis. In this condition small tubercles are present in many organs.

IMMUNITY

The mechanism of the immunity reaction in tuberculosis is still unsolved. There is no doubt that some immunity may be acquired, but this is always of a low grade of effectiveness. Available data suggests that an increased resistance can only be produced by the living organisms and only in virtue of a previous infection is an immunity obtained. There is thus a condition known as an infection-immunity. Evidence supporting this view is obtained from a study of both the spontaneous human disease and the experimental condition in laboratory animals. Antibodies, particularly agglutinins, opsonins and complement-fixing bodies, have been demonstrated, but there appears to be no definite relationship between their presence in the serum of an individual and his resistance to infection with the tubercle bacillus.

Clinical tuberculosis in civilized communities is relatively uncommon among the adult population considering the possibilities of infection and, when present, it tends to run a chronic course. In infants and children the disease, on the other hand, is usually more acute. This suggests that adults are less susceptible to infection than children. On post-mortem examination it has been found that a large percentage of the adult population in an urban community has some localized focus of infection, such as in a lymph node. Adults also often give a positive tuberculin reaction. These facts indicate that in virtue of the latent foci adults are endowed with an increased resistance to re-infection. On the introduction of tuberculosis into regions where the disease is uncommon, its course is the same in adults and children; an acute form of infection results. It is also recognized that non-specific factors, such as general health and good nourishment, play an important role in increasing the resistance to infection.

The immunity problem has been studied extensively in laboratory animals, particularly the guinea-pig, and it is from observations in these animals that much of our knowledge has been obtained. Koch (1891) found that after the subcutaneous injection of a small dose of tubercle bacilli into healthy guinea-pigs progressive lesions developed and the animals ultimately died. When, however, a similar injection was made into an animal which had also been injected with a small dose of tubercle bacilli some four weeks previously, a slight lesion developed and cleared up rapidly; there was no involvement of the regional glands. This observation was confirmed and the reaction is now known as "Koch's phenomenon". It was also found that if the second dose were given too soon after the first, i.e. before a definite lesion had developed, the usual type of lesion developed. There was thus a marked difference in the behaviour of the healthy and the infected animals. When infected, the guinea-pigs became hypersensitive to further infection; to this type of phenomenon the term bacterial allergy (delayed or tuberculin-type hypersensitivity) is now generally applied.

ALLERGY IN TUBERCULOSIS

Koch also found that the allergic reaction was manifested by guinea-pigs when, for the second inoculation, dead tubercle bacilli or extracts of the bacilli were employed instead of the living organisms. The bacterial extract was obtained by incubating tubercle bacilli in 5 per cent glycerine broth at 38°C for 6–8 weeks. After this it was evaporated in a water-bath at 100°C to 1/10th of its volume and passed through a filter. The filtrate was a brown, syrupy fluid containing about 40 per cent glycerine, and was termed tuberculin. It is now known as Old Tuberculin. It is standardized by its action on guinea-pigs sensitized by previous infection with the tubercle bacillus.

The action of tuberculin was tested in the human subject, and it was found that healthy infants gave no reaction, but in tuberculous patients, however, a very severe reaction was produced. The effect was thus analogous to that seen in guinea-pigs, in which the reaction has been studied in some detail. The reactions have been classified into: (1) local—formation of nodule with erythema, (2) focal—a short time after the second injection in the animals, it was observed that the tissue around the focus of infection becomes congested and oedematous, and (3) general—pyrexia, malaise, etc. The focal and general reactions are due to the absorption of tuberculin into the blood stream. In view of the striking reactions given by tuberculous patients, tuberculin, administered by various routes, has been widely applied as a diagnostic measure.

The precise mechanism of the reaction has not been determined. There appears to be little doubt that, in the presence of most forms of tuberculosis, a local tuberculin reaction can be elicited in any of the body tissues. It appears most probable that the reaction results from sensitization of the cells to some substance obtained from the tubercle bacilli and produced during the primary infection. It has also been shown that sensitization may follow the injection of killed tubercle bacilli. Some toxic substance is most probably produced by the interaction of tuberculin and the sensitive tissue-cells, and this gives rise to the general symptoms. The substance in the tuberculin responsible for the primary reaction is a protein; it is extremely heat-stable, but is destroyed by proteolytic enzymes. It has been prepared in a relatively pure form as P.P.D. or purified protein derivative. No significant difference has been detected in the tuberculins produced by the human and bovine types, but that formed by the avian type differs in several respects.

In tuberculosis the available evidence indicates that the hypersensitive state is indicative of previous infection, clinical or subclinical, with the tubercle bacillus and is usually associated with increased resistance to re-infection. A negative tuberculin test suggests absence of, and increased liability to, infection, and is found most commonly in infants; it may also occur in patients recently infected or in an advanced state of tuberculosis.

DIAGNOSIS

Methods employed in the bacteriological diagnosis of tuberculosis are: (1) microscopy, (2) culture, (3) pathogenicity, and (4) application of allergic reactions.

Serological tests, particularly complement-fixation, have been tried, but the results are not reliable. Middlebrook and Dubos introduced a form of haemagglutination test, in which the serum of the patient is tested against a suspension of sheep red blood corpuscles to which old tuberculin had previously been added. While many cases of tuberculosis give a positive reaction, false positive and contradictory results are obtained with some regularity; the test is therefore likely to prove of limited value in diagnosis.

MICROSCOPY

Due to the slow growth of the tubercle bacillus, and the urgency of making a diagnosis and starting treatment, particularly in cases of meningitis, some reliance must be placed on microscopical demonstration of acid-fast organisms. It is advisable in all cases to confirm microscopical findings by culture. Sputum should be carefully examined and smears prepared from thick nummular or purulent portions. When very few organisms are present, the bacilli may be concentrated by centrifugation after treatment of the sputum with antiformin. In chronic pulmonary tuberculosis, laryngeal swabs or gastric contents may give positive results.

In the examination of urine and faeces, other acid-fast bacilli such as the smegma bacillus or other saprophytes may be present.

Cerebrospinal fluid should be centrifuged, and smears made from the deposit. If the fluid is allowed to clot, the organisms are usually present in the clot, but it is difficult to prepare satisfactory films from the clot.

Pus aspirated from tuberculous abscesses has variable numbers of bacilli. There may be large numbers easily found, or they may be very scarce.

Biopsy specimens, such as endometrium, may require preliminary digestion in order to prepare a suitable film.

Smears should be stained by the Ziehl–Neelsen technique. When very few organisms are present it is best to use a very light counterstain such as dilute malachite-green, as the heavy staining with methylene-blue may obscure some organisms.

Smears may also be stained with auramine, and the acid-fast organisms then fluoresce in ultraviolet light, and are more readily detected. Some other particles may also fluoresce, and positive findings should be confirmed by the Ziehl–Neelsen method.

CULTURE

As various types of antibiotic and chemotherapy are now available, it is important to culture the tubercle bacilli and test their sensitivity. In chronic cases on prolonged therapy, cultures should also be made at intervals to detect the development of resistance.

The routine media should include Dorset egg, particularly to detect bovine types, Löwenstein–Jensen, and a fluid medium such as Kirchner. The average time for cultures to become macroscopically positive is three to four weeks, but cultures should be incubated for at least six weeks before being rejected as negative. Kirchner cultures are usually positive a few days earlier than Löwenstein–Jensen.

Specimens aspirated with aseptic precautions and free of secondary infection, such as C.S.F. or pus from a psoas abscess may be inoculated directly. Other specimens, such as sputum, are contaminated with saprophytes or secondary invading pathogens which would rapidly outgrow the tubercle bacilli. Such specimens must be treated to destroy the contaminants. The usual method is to mix the specimen with 4 per cent NaOH or 6 per cent H_2SO_4 for 15–20 min, centrifuge, and then neutralize the deposit before inoculation. This will destroy most vegetative forms except the mycobacteria. Some of these survive as their waxy structure makes them impermeable to aqueous solutions.

In Kirchner's medium the growth of tubercle bacilli is seen as characteristic small granules which swirl up on a quick twist of the tube. If the culture is left undisturbed, a pellicle will eventually form.

On Löwenstein–Jensen medium the human type gives rough, greyish to orange-yellow colonies. The bovine type on Dorset egg gives a discrete, smooth, whitish growth, never as luxuriant as the human type.

ALLERGIC TESTS

The tuberculin reaction has been widely applied in the diagnosis of tuberculosis in man and animals. Purified protein derivative (P.P.D.) or Koch's old tuberculin (O.T.) are used in this test: the subcutaneous method of Koch (1890), the cutaneous method of von Pirquet (1907), the conjunctival method of Calmette (1908), the patch test of Vollmer (1937), the jelly patch test, and the Heaf test. Of these tests, the Mantoux and the Heaf are the most reliable.

The Mantoux test is performed by the intradermal injection on the forearm of 0·1 ml of a 1/10,000 dilution of tuberculin. If no reaction follows, increasing concentrations of tuberculin, 1/1,000 and 1/100, are used. Readings should be made at two, three and four days. In a positive result an area of induration of at least 5 mm in diameter appears at the site of injection; the erythema, which is of less significance than the induration may spread over a wider area.

The Heaf test is simpler to apply, but requires a special apparatus, the Heaf gun. The skin of the forearm is cleaned with ether or spirit; a drop of concentrated old tuberculin or P.P.D. 2 mg per ml is spread over an area about 2 cm in diameter. The Heaf gun is then applied and the six needles prick the tuberculin into the skin. The reaction is read at two, four or seven days. The degree of sensitivity is judged by the degree of reaction. In a slightly sensitive individual there is a slight reaction, oedema and erythema, at each needle prick. In a very sensitive subject the reactions are confluent, oedematous and may blister. Intermediate grades of reaction are readily detected. The Heaf test has the advantage that it is easily applied; no accurate dilutions of tuberculin are required; the concentrated O.T. or P.P.D. retains its potency much longer than dilute solutions; and in surveys the result can be read in seven days, which is often a convenient interval.

The patch tests are useful for very young children but are less reliable than the Mantoux or Heaf tests.

PATHOGENICITY

Tubercle bacilli may also be detected by the guinea-pig inoculation test. Material is injected intramuscularly into the groin of one or two animals. Two animals give more satisfactory results than one, as one can be killed at the end of three weeks and the other after five weeks. In many, but not all, cases a positive report may be given in three weeks. By using two animals, the possibility of the test being nullified by death of the animal through intercurrent infection is also reduced. On post-mortem examination the distribution of the lesions varies considerably. When few organisms are present in the inoculum a local indurated or caseous lesion only may be produced; on the other hand if the specimen contains many organisms, a generalized infection develops. Smears are prepared from the caseous material and stained by the Ziehl–Neelsen method.

In determining the type of the infecting organism the following criteria are of value.

Human tubercle bacilli are eugonic, i.e. they grow relatively profusely on artificial media, particularly on those containing glycerine; they may produce a yellowish-orange pigment; they are of low virulence for rabbits as tested by the injection of 10 mg of the organisms subcutaneously.

Bovine bacilli are dysgonic, i.e. they grow poorly on artificial media, the growth is not enhanced by the addition of glycerine to the medium, and they do not produce a pigment; they have a high virulence for rabbits.

PROPHYLAXIS

General and specific measures have both played an important role in the prevention of tuberculosis. These have been directed both to the prevention of infection and to increasing the resistance of the community. Subclinical infection is common as shown by the fact that the majority of adults are tuberculin positive.

GENERAL MEASURES

In preventing infection with bovine tubercle bacilli attention has been given to the milk supply. By repeated inspection of the cows and bacteriological examinations of the milk this has been graded into different classes. Only those grades guaranteed free from tubercle bacilli should be consumed in an untreated condition. When contaminated, the milk should either be pasteurized in bulk or boiled in the home. By the increasing adoption of these measures the incidence of bovine tuberculosis has been greatly reduced.

Mass miniature radiography to establish an early diagnosis, and the segregation of open cases, have helped to reduce the spread of tuberculosis, but about 20,000 fresh cases of tuberculosis are found every year in England and Wales, in addition to the 300,000 cases already under treatment or supervision.

General and specific methods have been used to increase the individual resistance. The general measures include attention to the social condition of the community, particularly their nourishment and general health and access to fresh air and sunshine.

SPECIFIC MEASURES

Tuberculin. When Koch first introduced tuberculin he considered that he had a valuable therapeutic agent. The results were, however, extremely disappointing. Other preparations were later introduced; tuberculin O—a watery extract of tubercle bacilli disintegrated by grinding in an agate mortar; tuberculin R—the residue remaining after the removal of tuberculin O suspended in water, and tuberculin B.E.—a suspension of the organisms in 50 per cent glycerine. These were tried for prophylactic purposes, but the experimental results were unsatisfactory.

The Bacille-Calmette-Guérin (B.C.G.), was introduced by Calmette and his co-workers. It is a bovine strain of tubercle bacillus that has been rendered avirulent by cultivation through many generations on a glycerine bile potato medium. The avirulence of this organism was considered by Calmette to be permanent and was established by injecting a number of susceptible animals, including guinea-pigs and monkeys, with large doses both subcutaneously and by the mouth. Calmette also claimed that after receiving 50 mg of B.C.G. subcutaneously a number of calves were unaffected by the subsequent intravenous injection of 5 mg of virulent tubercle bacilli, which dose was sufficient to kill untreated animals in two months. After prolonged experimental work on laboratory animals, Calmette used the B.C.G. for human prophylaxis. Infants born of tuberculous parents were at once removed and were also given B.C.G. by the mouth. It was found that the incidence of tuberculosis in these infants was much below the average.

The application of the B.C.G. in human prophylaxis has increased enormously. It is now accepted that this vaccine is safe but minor lesions may be produced; there are retrogressive. The degree of immunity following the use of B.C.G. is uncertain, it appears to develop slowly and also to be of a low order. It is also well known that the incidence of tuberculosis can be reduced by the removal of infants from tuberculous parents; this fact prevented a ready acceptance of Calmette's early statistics.

B.C.G. has proved of great value for immunizing certain classes of the community, e.g. nurses and medical attendants, who are particularly liable to come in close contact with open cases of tuberculosis.

The vole bacillus, while highly virulent for the vole, is relatively avirulent for rabbits and guinea-pigs. Experiments have consequently been carried out to test the value of this organism as an immunizing agent and the preliminary results have been satisfactory. The subcutaneous inoculation of small doses of a pure culture of the vole bacillus afforded a high degree of protection in guinea-pigs against a later inoculation of virulent tubercle bacilli. A limited number of tests have been carried out on human subjects with encouraging results. Further tests are, however, necessary before a definite opinion can be given of the value of the vole bacillus as an immunizing agent against tuberculosis for general application.

THERAPY

The original attempts of Koch at the therapy of tuberculosis by means of tuberculin were particularly disappointing in view of the hopes expressed. The doses employed were too large; serious and even fatal reactions developed and the practice fell into disrepute. Other tuberculin preparations were introduced without any striking success. The use of minute doses in the treatment of chronic localized lesions has, however, given more satisfactory results. Some workers have consequently suggested that the use of tuberculin in minute doses should be more widely applied in the treatment of tuberculosis. The initial dose should be extremely minute (0·1 ml of a 1/1,000,000 dilution), subsequent doses may be gradually increased in the absence of a marked reaction; great caution must, however, be exercised. In all instances specific therapy should be supplemented by general therapeutic measures. The object of tuberculin therapy combined with antibiotic and chemotherapy is to stimulate a mild focal reaction to improve the access of the drugs to the lesion through the blood supply.

Streptomycin, an antibiotic obtained from *Act. griseus*, was the first satisfactory therapeutic agent for tuberculosis, in particular the meningitic and miliary forms. Prolonged dosage, for many months, was necessary and good results were obtained in spite of the tendency of tubercle bacilli to develop resistance and the appearance of toxic symptoms, particularly vestibular disturbances, due to the streptomycin.

Other agents, particularly salts of para-aminosalicylic acid (P.A.S.) and the isoniazids, have proved very active. In view of the liability of the tubercle bacillus to develop resistance when a single agent is used, it is now the accepted practice to use at least two of the drugs in combination. Daily isoniazid plus streptomycin or daily isoniazid plus P.A.S. have been used. It has been found that daily isoniazid with streptomycin twice a week may give rise to isoniazid-resistant tubercle bacilli. When streptomycin is given with P.A.S. the dose of the latter is very important, a daily dose of 20 g is necessary to prevent the development of streptomycin-resistant strains. Treatment may need to be continued for up to two years. Bacteriological control of therapy is essential, especially when the clinical response is poor.

LEPROSY

Leprosy is still a widely distributed disease. The World Health Organization has estimated that the incidence of leprosy in the world population is between 10 and 12 million. The majority of cases are found within the tropics, but the great increase in the number of emigrants from those territories where leprosy is endemic make it necessary for bacteriologists in this country to be familiar with the techniques for the laboratory diagnosis of this condition.

Leprosy is a human disease of marked chronicity which affects particularly the skin and the peripheral nervous system. In advanced cases with little natural resistance to the infection, mycobacteria may be found throughout the reticulo-endothelial system of the body, in the blood stream, and in large numbers in the nasal mucosa.

It is not known with certainty how the disease is transmitted. There is undoubtedly a wide variation in susceptibility and the clinical manifestations of the disease may not appear for many years after the organism has gained entry into the body. Attempts to infect healthy individuals by the injection of mycobacteria from known cases of leprosy into the skin have always proved unsuccessful.

Until recently it has not been possible to reproduce this condition in experimental animals; during the last decade Shepard in Atlanta, Georgia, found that inoculating the mouse footpad with a suitable dose of *Myco. leprae* obtained from an advanced case of lepromatous leprosy led to a multiplication of the organisms after a lag period. The optimum inoculum contained 5,000 mycobacteria and after 90–120 days the organisms began to multiply until a ceiling of 10^6 mycobacteria was reached. Whatever the size of the inoculum, this ceiling is not normally exceeded and thereafter the mycobacteria die and disappear presumably as a result of macrophage activity.

More recently, Rees, in London, has extended this work and has been able to reproduce a condition very similar to human leprosy in mice which have been thymectomized and exposed to whole body irradiation and infected by the footpad or intravenous route. In these mice whose natural immune mechanisms had been depressed, mycobacteria were found in the dermis of the skin, in skin nodules, in Schwann cells of the peripheral nervous system and particularly in the nose—in fact distributed in a way very similar to that found in advanced cases of human leprosy. Mice infected in this way can now be used to study the action of various anti-leprosy drugs which previously could be studied only in patients suffering from the disease.

In man the essential lesions of leprosy are of two kinds and are dependent upon the immunological state of the patient. In the one termed *tuberculoid leprosy*, areas of the dermis of the skin and certain of the peripheral nerves are infiltrated with lymphocytes, epithelioid cells and giant cells. The cutaneous nerves of the affected areas are destroyed and only very occasional mycobacteria can be found in skin biopsies. The patient with tuberculoid leprosy responds very strongly to an intradermal injection of a suspension of killed *Myco. leprae*, the lepromin test; and this is taken as an indication of a well-developed delayed hypersensitivity response to the invading organism. In the other type of leprosy termed *lepromatous leprosy* there is a generalized infiltration of the dermis of the skin over the whole body, with large vacuolated macrophages, many of which contain numbers of mycobacteria; these large vacuolated cells were first described by Virchow and are known as Virchow cells. Mycobacteria are also found in the

Schwann cells throughout the peripheral nervous system, and in the naso-pharyngeal mucosa. In advanced cases the whole of the reticulo-endothelial system is involved and mycobacteria may be found even in the blood stream. No peripheral nerve destruction is found in the majority of cases of lepromatous leprosy. A greater than normal cutaneous sensitivity to tactile stimulation is present. Large numbers of mycobacteria may be found in skin smears, skin biopsies, and scrapings from the nasal mucosa. The lepromin response is invariably negative in cases of lepromatous leprosy.

Between the high resistance, localized anaesthetic, tuberculoid leprosy cases and the low resistance, generalized and hyperaesthetic lepromatous leprosy cases there is a whole range of cases which present with some or other of the characteristics of both tuberculoid and lepromatous leprosy. These cases have been termed *dimorphous leprosy* because of this combination of characteristics, or *borderline leprosy* for it has been found that the great majority of these cases, in time, and without treatment, gradually lose their natural resistance and become cases of lepromatous leprosy—hence they are considered cases of borderline lepromatous leprosy from the onset. The greater the degree of natural immunity as shown by the lepromin response, the more the borderline case will resemble a case of tuberculoid leprosy, but any weakening of the lepromin response should indicate the development of lepromatous characteristics.

Leprosy, *per se*, is seldom fatal; death frequently results from some intercurrent infection. Spontaneous regression may occur at any stage as a result of some increase in the general resistance of the body. Therapeutic measures involve attention to the general health and chemotherapy. The administration of such drugs as chaulmoogra oil gave disappointing results but the sulphone compounds are now being used with much success. These drugs appear to have a bacteriostatic rather than a bactericidal action on the leprosy bacillus; in consequence clinical and bacteriological response is slow and prolonged treatment is required.

BACTERIOLOGY

The leprosy bacillus was first described by Hansen in 1874, who saw large numbers in the granulation tissue-cells of leprous lesions. These observations have been repeatedly confirmed. The bacilli are usually situated in large numbers in the "lepra" cells and are often present on the naso-pharyngeal mucosa. They stain with difficulty; they are straight or slightly curved, Gram-positive, acid-fast rods, 1·5–8 μ in length, usually arranged in clusters. They are non-motile and non-sporing.

The organism has never been satisfactorily cultivated. Numerous attempts on a great variety of media have been made. Successful results have been claimed, but conclusive evidence of multiplication of the organisms under artificial conditions has not been obtained. Under such conditions it is not unnatural that other organisms, such as other acid-fast bacilli and certain diphtheroids, have been incriminated as the aetiological agents of leprosy. Their relationship to the disease has, however, never been conclusively substantiated.

The clinical diagnosis may be confirmed by the demonstration of acid-fast bacilli, particularly inside the "lepra" cells, either in smears or sections stained by the Ziehl–Neelsen method. If doubt exists, tissue containing the organisms should be ground up and injected into guinea-pigs. The absence of lesions suggests the leprosy bacillus.

THERAPY

Treatment with sulphones has proved very satisfactory, particularly in cases of lepromatous leprosy. Care must be taken to avoid toxic side effects by initiating treatment slowly and using low weekly doses of the drug. Where toxicity to the sulphones has developed, other anti-leprosy drugs such as B.663 Geigy, also known as G.30.320 Geigy, may prove effective.

Chapter XXIV

Vibrio: Cholera: Spirillum

VIBRIO

Short bent rods, sometimes almost straight; motile by means of a single polar flagellum; aerobic, growing well on ordinary media. Frequently liquefy gelatin; do not form spores; Gram-negative. Some species are pathogenic to man and animals.

The first member of this group to be described was *V. cholerae*, or the cholera vibrio, which was isolated by Koch in Egypt in 1884 from the faeces of cholera patients. He was unable to isolate this organism from patients who had died of other diseases and so concluded that it was the causative organism of cholera.

Organisms resembling the cholera vibrio were isolated from other sources. Some were obtained from water and others from the faeces of patients suffering from intestinal conditions other than true cholera. These organisms were termed paracholera vibrios and the conditions to which they gave rise, paracholera. The El Tor vibrio was first isolated from six pilgrims who died of paracholera in 1905 at the Tor quarantine station in the Sinai peninsula. Later this organism was found to be responsible for endemic paracholera in Indonesia. Recently the El Tor vibrio has spread widely throughout Asia and it has been decided that the El Tor infection should be regarded as essentially identical with classical cholera and treated as such.

Vibrios are also responsible for diseases in birds and animals. One was isolated by Gamaléia in 1888 from cases of enteritis in fowls and was designated *V. metchnikovi*. Another, obtained from the uterine discharge of sheep suffering from infectious abortion and also from the foetuses of aborting cattle, was termed *V. fetus* by Smith (1918).

Many saprophytic members exist, and they have added confusion to the study of this genus. A number of vibrios have been isolated from water, which appears to be their normal habitat. Other sources include the nasal secretion, carious teeth, cheese, hay-infusion and manure.

V. CHOLERA (Cholera vibrio)

MORPHOLOGY

The cholera vibrio, on first isolation from the tissues, tends to be in the form of a curved rod (0·3 × 1–5 μ in size) and bears a close resemblance to a comma; it has therefore been termed the "comma" bacillus.

It may occur singly, in S-shaped pairs or spirals. After prolonged cultivation pleomorphism is marked and straight rods and involution forms are common. It is actively motile, due to the presence of a single terminal flagellum. It is non-sporing, non-capsulated and Gram-negative.

CULTURAL CHARACTERISTICS

V. cholerae grows readily on the ordinary media. It prefers an alkaline medium, the optimum reaction being about pH 8. For isolation from contaminated material, e.g. faeces, special selective media with a high pH, such as that introduced by Dieudonné, are therefore employed. The optimum temperature is 37°C, but growth occurs over a wide range, 16°–42°C. The cholera vibrio is a doubtful facultative anaerobe, growth being very slight in the absence of free oxygen. On potato a yellowish-brown pigment is formed by many strains.

On solid media, such as agar, after incubation for 24 hr at 37°C there is an abundant growth; the colonies are moist, translucent, greyish, round and small (1–2 mm).

In fluid media, such as broth, multiplication rapidly takes place. After incubation for 24 hr there is moderate turbidity with a thick surface pellicle and a slight deposit.

RESISTANCE

V. cholerae behaves as most vegetative organisms and is readily destroyed by bactericidal agents; heat at 55°C kills the vibrios in 15 min. The life of the cholera vibrio outside the body is short; varying from a few days to a few weeks according to the environmental and atmospheric conditions. This observation is based on several experiments carried out to determine the viability of the organism in faeces, fresh water, sea water, etc.

BIOCHEMICAL ACTIVITY

V. cholerae has marked biochemical activity. This is seen particularly in its proteolytic actions which are most pronounced when the organism is grown in an alkaline medium. Indole is rapidly produced and nitrates are reduced to nitrites. These changes are produced simultaneously and form the basis of the "cholera-red" reaction, which was originally considered to be given only by *V. cholerae* and was used extensively for identification. It has since been found that the reaction is produced by many other vibrios. In carrying out this test a few drops of concentrated sulphuric or hydrochloric acid are added to a 24-hr peptone-water culture; a pink coloration results from the action of the acid on the indole and nitrites produced from the peptone by the cholera vibrio. Gelatin and coagulated serum are slowly liquefied; NH_3 and H_2S are formed.

The saccharolytic activity is also important; acid, but not gas, is formed from glucose, laevulose, maltose, mannite, saccharose and starch. Lactose is sometimes fermented late (5–12 days). In litmus-milk acid or acid and clot may be formed.

As a general rule *V. cholerae* does not produce a soluble haemolysin capable of lysing suspensions of sheep red cells. This property is, however, possessed by the El Tor vibrio. The El Tor vibrio also differs from *V. cholerae* in producing a true thermolabile exotoxin which is fatal to mice on intraperitoneal injection; in agglutinating chicken red cells and being resistant to polymyxin B.

SEROLOGY

Serological reactions play an important role in the identification of the cholera vibrio. The tests commonly employed are agglutination and the bactericidal reaction known as "Pfeiffer's phenomenon".

In demonstrating Pfeiffer's phenomenon, the vibrios may be injected intraperitoneally into either an immune guinea-pig or a normal animal after mixture of the organisms with an immune serum. Lysis and disintegration of the organisms occur as a result of the action of complement on the sensitized vibrios. This action can also be demonstrated *in vitro*.

In the agglutination test several factors are concerned; being motile, *V. cholerae* possesses both the heat stable O or somatic component and the heat-labile H or flagellar component. As *V. cholerae* is monoflagellate the H factor is less pronounced than in the case of the salmonella organisms. Gardner and Venkatraman have shown that the H factor is non-specific, being common to all strains of *V. cholerae* and the paracholera vibrios, while the heat stable O antigen contains two factors, one of which is specific for *V. cholerae* and the other non-specific.

The various phenomena associated with the smooth → rough (S → R) change may also be found. A complex polysaccharide which behaves as a haptene has been demonstrated and is considered to be associated with the somatic antigen, most probably the specific factor. The agglutination test shows a degree of specificity and is now used extensively in diagnosis in place of the more complicated bactericidal reaction. There is, however, no doubt that the specificity of the agglutination tests will be considerably increased by the use of standard sera prepared against the specific O antigen.

PATHOGENICITY

V. cholerae produces the spontaneous disease, cholera, only in man. It is pathogenic to most laboratory animals; guinea-pigs and rabbits have been usually used for experimental work. Following intraperitoneal inoculation a peritonitis rapidly develops, followed by septicaemia and death in 24–48 hr. Infection also results from the introduction of the vibrios *per os* after gastric juice has been previously neutralized by the administration of bile.

There appears to be no doubt that a toxin is produced by *V. cholerae*; the profound collapse shown by cholera patients indicates very strongly the generalization of some toxic substance, but marked discrepancies are found in the reports of different workers on its formation and nature. The available evidence suggests that it possesses many of the properties associated with the so-called "endotoxins", and is liberated by the disintegration of the vibrios.

CHOLERA

Cholera has been an endemic disease in the East for centuries. Since the early part of the nineteenth century a series of world-wide pandemics has occurred and almost every country has at some time been affected. The spread of infection has invariably been along the trade routes and the source of infection is the faeces of either cases or carriers. Snow was the first to show that water could be a source of infection; and his careful documentation of the Golden Square cholera outbreak in London in 1855 led to the incrimination of the Broad Street pump whose water was contaminated. Since that time there has been a gradual reduction in the severity and frequency of cholera epidemics. No cholera has been observed in North or South America in this century, and since 1923 Europe has been free of it. An epidemic occurred in Egypt in 1947 but thereafter no epidemic has occurred in Africa. By 1950 cholera appeared to be retreating to its endemic foci in Bengal, and to be on the point of disappearing spontaneously. Since 1955 a series of outbreaks has occurred, and since 1961 cholera El Tor has spread from its endemic foci in Indonesia to re-establish cholera in South East Asia in territories where classic cholera had long disappeared. In fact, cholera El Tor would appear to be invading those areas where classical cholera was once endemic. In its pathology and physiology the disease caused by the El Tor vibrio is indistinguishable from that caused by the vibrio of classical cholera. It is now generally accepted that the El Tor infection should be regarded as essentially identical with classical cholera and treated as such.

The onset of the condition usually commences some 24–48 hr after the ingestion of the cholera vibrios and is characterized by pyrexia, abdominal pain and diarrhoea. The symptoms rapidly become more severe; a subnormal temperature, profuse diarrhoea with rice-water stools, suppression of urine, cramps and toxaemia are found. There is marked dehydration of the tissues, and this, more than anything else constitutes the most dangerous aspect of the disease. The infection usually remains localized in the intestines; invasion of the blood-stream is extremely rare. The mortality rate varies from 30–80 per cent. Recovery is slow, but is not usually accompanied by complications. The vibrios are excreted in the faeces during convalescence by most cases, and in a few this excretion continues after recovery. These carriers do not, as a rule, excrete vibrios for more than a few weeks. Healthy carriers, such as attendants of cases, may also occur but the excretion of the organisms by these individuals is of short duration.

In fatal cases the most important lesions are found in the lower half of the small intestine, the mucosa of which is congested, necrotic and desquamated. The intestinal contents are usually clear and fluid, containing a large number of vibrios and flakes of mucous membrane, which give the characteristic "rice-water" appearance. The vibrios are frequently present in the bile and gall-bladder, and a cholecystitis may be present.

DIAGNOSIS

The bacteriological methods used in diagnosis of cholera are (1) the isolation of the vibrio from the faeces or vomit, and (2) the detection of agglutinins or bactericidins in the blood-serum.

(1) When vibrios are present in large numbers they can be readily recognized on microscopical examination; their isolation is then relatively simple. When the numbers are small, care is required. Satisfactory procedures are the use of special alkaline media, such as Dieudonné's, or serial cultivation in peptone-water (pH 8·4) with subculture from the surface growth after a short

time, usually some 4–6 hr. The organism is finally identified by its biochemical and serological reactions. The isolation of *V. cholerae* from sporadic cases is sometimes a matter of considerable difficulty and requires much experience.

(2) The serological test now employed is agglutination, which has replaced the bactericidal test in most laboratories. Agglutinins appear early, but owing to the low titre their presence is usually of little diagnostic value before the sixth day. A titre of 1/100 or more is usually taken to be diagnostic.

PROPHYLAXIS

The prophylactic measures adopted in the prevention of cholera are similar to those employed in dealing with enteric fever. Great care is taken at ports to prevent the entrance of cases or carriers into countries free of cholera. In all suspicious cases examinations should be performed for the presence of agglutinins in the serum and of vibrios in the faeces. In infected regions strict attention should be paid to the food, drinking-water and general sanitary arrangements.

Many attempts at specific immunization have been made since Ferran used a living culture in 1885, but an entirely satisfactory vaccine has not been produced. Attenuated, phenolized, etherized and heat-killed cultures have all been tried. Bili-vaccines, given by the mouth, have also been used. The ordinary vaccine, prepared by heating a culture at 56°C for 1 hr, has proved as satisfactory as the others. The immunizing substance is associated with the somatic and not the flagellar antigenic component. The disadvantages of vaccines are that reactions may follow their injection, and even after several doses the increased resistance is of short duration, probably less than one year, and further courses may be required. Statistics indicate, however, that vaccines have proved of some value in prophylaxis. The usual dosage is 4,000 and 8,000 million vibrios at an interval of 7–14 days.

THERAPY

Cholera is an acute self-limiting disease in which the most important part of treatment is the replacement of fluid lost from the body and the prevention of circulatory collapse. Any drug to be effective would have to be given during the early stages of the disease before the massive multiplication of the vibrio in the intestine had occurred. It would have to be non-toxic and to act with great rapidity. Such a drug has yet to be discovered.

GENUS SPIRILLUM

The generic characteristics of this group are: rigid, spiral rods with great variation in length, number and breadth of the spirals; usually motile by means of polar flagella; frequently Gram-positive; grow readily on simple media, some species producing a yellowish pigment.

These organisms have been isolated frequently from water and are practically all non-pathogenic to man or animals. They may, however, be mistaken for spirochaetes or vibrios on morphological examination. One member, *Sp. minus*, is responsible for rat-bite fever, which is a febrile disease produced in man by the bite of an infected rat after an incubation period of several weeks. The organism was originally considered to be a spirochaete, but it has since been proved to belong to the *Spirillum* genus. It was first described by Japanese workers, but it has since been isolated in many countries, including Great Britain.

Chapter XXV

Loefflerella: Glanders: Meliodosis

Loefflerella: small, slender, Gram-negative rods. Pleomorphism common. Slight biochemical activity. Pathogenic to animals and man.

The taxonomic position of this genus has not yet been generally agreed. Various authorities have used the terms Pfeifferella or Malleomyces but the International Committee on Bacteriological Nomenclature has rejected both names. As the first organism of this genus, the glanders bacillus, was isolated by Loeffler and Schütz in 1882, recent authorities have adopted the name *Loefflerella.*

Lf. mallei or the glanders bacillus.

MORPHOLOGY

Lf. mallei is usually a slender rod, straight or slightly curved, about $0{\cdot}5 \times 2\text{–}5\,\mu$. It may be arranged singly or in irregular clusters. In culture it tends to be shorter and less uniform; in old cultures pleomorphism is marked and long filaments with true branching may be present. It is non-motile, non-sporing and has no capsule. *Lf. mallei* does not stain intensely and in old cultures may be beaded; it is Gram-negative and non-acid-fast.

CULTURAL CHARACTERISTICS

The glanders bacillus grows readily but slowly on the ordinary media. The optimum temperature is 37°C, range 20°–42°C; the reaction of the medium is perhaps optimal when slightly acid, but multiplication occurs over a wide range of reaction around pH 7·0. It is a strict aerobe.

In fluid media, such as broth, after incubation at 37°C for two days there is moderate turbidity with a slight deposit. No surface growth occurs until 7–14 days, when it is of a slimy, ring-like nature; the deposit is viscid and difficult to disintegrate.

On solid media, such as agar, after two days' growth at 37°C the colonies are small, translucent, butyrous and greyish-yellow, becoming later opaque, granular and yellowish-brown.

When grown on potato a characteristic pigment is formed; at first it is slightly yellowish or brownish but later becomes reddish-brown to chocolate.

RESISTANCE

Lf. mallei is easily killed by destructive agents; it is destroyed by heat at 56°C in 15 min.

BIOCHEMICAL ACTIVITY

Biochemical activity is very limited. Glucose is fermented by some strains with the production of acid, but not gas; proteolytic activity is slight; indole is not formed, and gelatin is not liquefied. Litmus-milk is rendered slightly acid. No haemolysin is produced.

SEROLOGY

The serological reactions of *Lf. mallei* have not been extensively studied. By means of agglutination, agglutinin absorption and complement fixation two main sub-groups have been found. One of these is closely related to *Lf. whitmori.*

PATHOGENICITY

Lf. mallei is primarily a pathogen of equine animals, in which it produces two clinical conditions—glanders, with involvement of the lungs, and farcy, where lesions are present in the skin.

Experimental infection may be produced in a number of laboratory animals, guinea-pigs being the most susceptible. In the guinea-pig subcutaneous injection of the organism is followed first by abscess formation locally and in the regional glands, then by inflammatory changes in the testicles with a generalized infection and usually by death in 3–4 weeks. Following intraperitoneal injection, progressive swelling of the testicles, commencing in the tunica vaginalis in 2–3 days and becoming purulent, is followed by death usually about the tenth day. This sequence of events is characteristic, and is known as the "Straus" reaction.

GLANDERS IN MAN

The human disease is almost invariably acquired by direct contact with an infected animal; infection from man to man is exceedingly rare. The disease is therefore found chiefly in the attendants of the equine species. In Great Britain the introduction of the Glanders and Farcy Order of 1920 has eradicated the disease; no notification has been made since 1926. The disease is prevalent in other countries, particularly in Russia and Asia.

Several cases have been reported in laboratory workers, and the organism should therefore be treated with the greatest respect.

Infection occurs either by the cutaneous route or the nasal mucosa, and the course of the disease may be either acute or chronic. In the acute form there is usually a local nodular swelling with a spreading pustular skin eruption and foci in the lungs and other organs. There is high fever, muco-purulent nasal discharge and extreme prostration; death occurs in 7–10 days. In the chronic form subcutaneous and intramuscular nodules, becoming purulent, with enlargement of the regional glands and foci in the respiratory tract and other organs are usually present. The course is long, and complete recovery is rare; cases of apparent recovery, in which there have been latent periods of 1–5 years, have been reported, but in practically all cases a recurrence eventually takes place.

DIAGNOSIS

If the organism can be isolated from the lesions, its cultural reactions, particularly its reaction on potato, should be determined. Pathogenicity tests, with demonstration of the Straus reaction is important. An allergic test with mallein has been widely employed in veterinary practice. Mallein is prepared in a way similar to the preparation of tuberculin, and is usually injected subcutaneously; if glanders is present, definite signs of local, focal and general reaction are observed in 10–24 hr. Complement fixation tests may also give useful information.

PROPHYLAXIS AND THERAPY

The principal prophylactic measures have been directed to the elimination of the disease in horses. The various orders introduced in this country have been particularly satisfactory. Attempts to produce active immunity have not been very successful.

Mallein has been used in therapy of both human and animal disease, but in view of the spontaneous retrogressions in the chronic form the results are difficult to evaluate.

LF. WHITMORI; MELIODOSIS

Lf. whitmori is the aetiological agent of a rare tropical disease, meliodosis, which has been recognized only in Burma, Indo-China, Ceylon and British Malaya. The disease runs an acute course and appears to be of a septicaemic or pyaemic nature; purulent foci are present in many organs. It is usually fatal in 3–4 weeks. Epidemics have been reported in rodents, particularly guinea-pigs, rats and rabbits. No evidence has been obtained of direct infection;

it appears that the disease is primarily one of rodents, from which infection passes to man who is relatively resistant.

Lf. whitmori resembles *Lf. mallei* in many respects; it is Gram-negative, grows readily on the ordinary media, giving a creamy growth which later deepens to a chocolate colour. It does not produce indole, but liquefies gelatin. Acid is formed in glucose, lactose and mannite. *Lf. whitmori* is motile in young cultures.

Strains of *Lf. whitmori* are serologically homogeneous, and are antigenically related to one of the groups of *Lf. mallei.*

Chapter XXVI

Actinomyces: Nocardia: Actinomycosis: Madura Disease

ACTINOMYCES

Mycelial organisms which break up into rods of various sizes. No aerial hyphae produced. Non-motile. Gram-positive; non-acid-fast. Microaerophilic or anaerobic. Parasitic in man and animals; some species are pathogenic.

NOCARDIA

Microscopically similar to actinomyces. Aerobic; produce aerial hyphae. Some species weakly acid-fast.

ACTINOMYCES

MORPHOLOGY

Bollinger (1877) observed a fungus-like organism in the lesions of "woody-tongue" in cattle. This was investigated by Harz who termed it *Actinomyces bovis*. Wolff and Israel (1891) observed a similar organism in human lesions which is now known as *Actinomyces israeli*.

In culture, branching filaments are thickly matted together with rods of varying lengths. The staining is granular and irregular. In actinomycotic lesions the appearance is similar, but the peripheral ends of the filaments are thickened and appear as "clubs" radiating from the mycelial mass. The term "ray fungus" describes this microscopic picture. The clubs are probably formed by the deposition of protein and insoluble salts from the serous fluids on the dead ends of the filaments. The clubs are Gram-negative, and may be weakly acid-fast. They are more prominent in bovine than in human lesions.

CULTURAL CHARACTERISTICS

A. bovis is able to grow on ordinary media under anaerobic or micro-aerophilic conditions, but growth is poor and slow; it can, however, be improved by enrichment of the media with blood, serum or glucose.

In a glucose-agar shake-culture maximum growth is found in the second centimetre below the surface; this indicates that the organism is a micro-aerophile rather than a true anaerobe. The optimum temperature is 37°C and the optimum reaction of the medium pH 7·2–7·6.

In fluid media, such as broth, after several days' incubation at 37°C there is poor growth in the form of a white, nodular deposit resembling puff-balls; there is neither turbidity nor surface growth.

On solid media, such as agar, incubated anaerobically at 37°C for seven days a poor growth is seen in the form of small, greyish-white, opaque, convex colonies.

RESISTANCE

A. bovis behaves as other non-sporing organisms to the action of destructive agents. It is killed by heat at 60°C in 10–15 min.

BIOCHEMICAL ACTIVITY

The biochemical activity of the actinomyces is very slight. Glucose and a few other sugars may be fermented slowly with the production of acid. Gelatin is not liquefied and indole is not produced.

SEROLOGY

Erickson has claimed that the bovine strains of *A. bovis* can be differentiated from human strains of *A. israeli* by agglutination and agglutinin absorption tests.

PATHOGENICITY

Actinomycosis is a disease of cattle and man. Laboratory animals are resistant to infection.

ACTINOMYCOSIS

Actinomycosis is a relatively common disease of cattle. The first detailed account of the disease was given by Bollinger (1877), who examined cases of the so-called "woody-tongue" in cattle and found the lesions were of a granulomatous nature and contained a number of yellowish granules. This condition he termed "actinomycosis". In cattle the lesions are frequently confined to the jaw and soft parts of the head and neck.

In man actinomycotic lesions are usually situated in one of three regions—head and neck, the neighbourhood of the caecum and the lungs. The mouth is a common primary site. A chronic abscess with marked proliferation of the connective tissue forms and spreads locally by direct extension; the infection later may generalize by way of the lymphatic or blood streams; the lymph glands are, however, seldom attacked. Thick viscid pus containing the yellowish "sulphur" granules is usually discharged from the broken-down central portion of the abscess.

The epidemiology and pathogenesis of the disease are unknown. The source of infection has not been determined, but it is ultimately the saliva and nasal discharges of cases, probably animal. It is, however, frequently impossible to trace a human infection to an animal source, and there is no evidence that the disease is contagious in man. Carriers may possibly be responsible for the transmission of the infection in many cases; supporting this theory is the fact that organisms closely resembling *A. bovis* have been isolated from the alimentary tract of healthy individuals. While the method of transmission is uncertain it appears probable that the infectivity for man of *A. bovis* is slight and that some predisposing factor, such as trauma, is necessary to enable infection to take place.

DIAGNOSIS

Examination should be made for the presence of the "sulphur" granules in the thick pus; on rolling the tube containing the pus the granules usually adhere to the sides and can be readily detected. The granules are removed and examined, stained and unstained, after crushing. In early lesions the granules are small, soft, and white or creamy; later they may become yellow or brown, and hard, as calcium and iron compounds are deposited in them. The interlacing Gram-positive mycelia with Gram-negative clubs are seen.

Cultures should be prepared from the granules after they have been washed in sterile saline. A useful method is the glucose-agar shake-culture, in which the zone of maximal growth appears about 1 cm below the surface of the medium. The colonies should be removed and examined microscopically. There is often found in association with *A. bovis* a small, Gram-negative, non-motile cocco-bacillus which has been termed *B. actinomycetum comitans*.

Material sent for examination is generally grossly contaminated and may contain saprophytic actinomyces; care must be taken to differentiate these from *A. bovis*, which is essentially microaerophilic.

THERAPY

Good results have been given in some cases by large dosage with sulphonamides. Penicillin has also proved useful in large and prolonged dosage.

MADURA DISEASE

Madura disease is a chronic granulomatous disease bearing a close resemblance to actinomycosis and chiefly affecting the foot; it is frequently referred to as Madura foot. It is caused by *A. madurae* (*Nocardia madurae*) and is found mainly in India. Irregular nodular swellings are found in the infected region; these break down and form abscesses. Later thick yellowish pus is discharged and this contains greyish or yellowish granules, which are larger than those found in actinomycotic pus. The granules consist of interlacing mycelia, which are Gram-positive and non-acid-fast. Growth in culture is pinkish and occurs readily under aerobic conditions; these points differentiate *A. madurae* from *A. bovis*. The disease has not been successfully reproduced in experimental animals.

There is another variety of the disease in which the granules are black and consist of interlacing segmented mycelia; cultures, however, give a growth of a fungus, which is quite distinct from *A. madurae*.

ACTINOBACILLOSIS

A condition in cattle has been reported which occurs as an epizootic and presents a distribution of lesions which is in the main similar to that found in actinomycosis; this is termed "actinobacillosis". The aetiological agent is a Gram-negative, non-motile cocco-bacillus, to which the name *Actinobacillus lignieresi* has been given. It is an aerobe and prefers enriched media. Granules are formed in infected tissues. This condition is exceedingly rare in man, but one case has been reported from South America.

Chapter XXVII

Bacillus: Anthrax

BACILLUS

Aerobic spore-bearing rods. Usually Gram-positive. Usually motile. Catalase positive. Normal habitat is soil, dust, water, vegetation. Mostly saprophytes. *B. anthracis* highly pathogenic.

There is a large number of these saprophytes which are frequently encountered as contaminants in the bacteriology laboratory, and as contaminants in pathological specimens. The most common species are *B. subtilis*, *B. mesentericus*, *B. mycoides* and *B. megatherium*. They are occasionally found as primary agents in mild localized lesions such as conjunctivitis, but are otherwise non-pathogenic. They are differentiated from *B. anthracis* by cultural and microscopic characteristics. They are motile and do not have capsules. They liquefy gelatin rapidly without producing the characteristic inverted fir tree growth of *B. anthracis*. They are not pathogenic for laboratory animals.

B. anthracis or the anthrax bacillus, is an important pathogen of animals and man. It was first described in detail in 1877 by Koch, who definitely established its relationship to anthrax. This organism had, however, been observed by several earlier workers, including Davaine, who suspected this pathogenic role. Many other species have also been isolated, but these have little pathogenic activity and are usually present as harmless saprophytes. A number of these have interfered with the isolation and indentification of *B. anthracis* and have been referred to as "pseudo-anthrax bacilli". The original classification of the "pseudo-anthrax bacilli" does not appear to be of any particular value, and many workers consider that all organisms which might be confused with the anthrax bacillus should be included. The most important members of this group are *B. subtilis*, *B. mesentericus*, *B. mycoides* and *B. megatherium*. These organisms usually lead a saprophytic existence, being present in dust, soil, water, hay, etc.; they are consequently frequently found in the body-tissues as contaminants, secondary invaders and, in a few instances, as primary agents in mild localized lesions such as conjunctivitis.

B. ANTHRACIS OR THE ANTHRAX BACILLUS

MORPHOLOGY

B. anthracis. Usually straight or slightly curved rods, $0{\cdot}4 \times 1\ \mu$, with truncated ends; single forms are common but, under artificial cultivation, chains or unjointed filaments may be found. Spores are formed in culture, but not in the animal tissues, and are central or subterminal and do not bulge. A well-marked capsule is formed during growth in the animal body, but it is lost on cultivation on artificial media unless these contain much animal protein. *B. anthracis* is non-motile and Gram-positive. The capsules, when present, may be demonstrated in films by McFadyean's method; an old 1 per cent aqueous solution of a polychromatic methylene-blue stains them very well.

CULTURAL CHARACTERISTICS

B. anthracis grows readily on the ordinary media and growth is not greatly improved by the addition of blood, serum or glucose. The optimum temperature is about 37°C, range 14°–43°C, and the optimum reaction is pH 7·5–7·8. It is a facultative anaerobe.

In fluid media, such as broth, after an initial slight turbidity a flocculent deposit forms and the supernatant fluid becomes clear; there is no pellicle-formation.

On solid media, such as agar, the colonies present a characteristic irregular appearance, being opaque, greyish-white, with a reticular structure. The surface is uneven and has a "roughish" appearance; on examination with a low magnification lens interlacing chains of bacilli are found. The colonial appearance has been likened to the coils of a barrister's wig. When the growth is confluent a greyish-white pellicle is formed with irregular edges and an uneven surface. On prolonged cultivation on artificial media, smooth, glistening, regular colonies may be formed; these "smooth" types are relatively avirulent.

On blood agar some slight haemolysis is frequently present. Spores are not formed in the body tissues, as free oxygen is required for sporulation. They also are not produced when growth takes place at the extremes of temperature, i.e. above 42°C or below 16°C; when substances such as phenol (1/800) or potassium bichromate (1/2,000) are added to the media a permanent asporogenous strain may be produced.

RESISTANCE

The resistance offered to bactericidal agents by the vegetative forms is similar to that exhibited by the non-sporing organisms;

they are killed by heat at 56°C in 1 hr. The spores offer a much higher degree of resistance, but are usually killed by boiling at 100°C for 10 min or by a 4 per cent solution of potassium permanganate in 15 min. Of the chemicals employed as bactericidal agents the oxidizing substances have proved to be most efficacious.

Stock-cultures are conserved without difficulty at room temperature; spores in the dry state remain alive for years.

BIOCHEMICAL ACTIVITY

Glucose, maltose, sucrose and salicin are fermented with the production of acid, but not gas. Indole is not produced and gelatin is slowly liquefied. The type of growth in a gelatin stab-culture is a useful differential point; filaments are formed at right angles to the line of the stab, particularly at the top of the culture, and an inverted fir-tree appearance is thus produced. This is lost as liquefaction progresses downwards from the surface. Litmus-milk is first coagulated and then decolorized and digested.

SEROLOGY

Serological tests have not yet been widely employed in the study of this genus. Precipitins, opsonins, agglutinins and bactericidins have been demonstrated, but only precipitation is used to any extent for identification. The antigenic structure of *B. anthracis* is complex; more than two somatic antigens are present together with a capsular antigen. The presence of antibody to one of the somatic antigens—the "protective" antigen appears to be related to immunity to infection.

PATHOGENICITY

The anthrax bacillus has a high degree of pathogenicity for all classes of animals, but the carnivorous species are usually the most resistant. Birds are relatively insusceptible. It gives rise to spontaneous disease mainly in cattle and sheep, from which man becomes infected.

Laboratory animals such as guinea-pigs, mice and rabbits, are readily infected by subcutaneous or intra-cutaneous injection of the bacilli. Following the subcutaneous inoculation of a small dose of bacilli guinea-pigs die in 24–48 hr; on autopsy there is usually a localized haemorrhagic oedema with enlarged, congested regional glands and enlargement and congestion of the spleen, which is dark and friable. The blood is dark and usually contains large numbers of bacilli.

It has been found that there is a definite relationship between capsule formation and virulence, which is quite independent of spore formation. Capsulated strains, which give the typical irregular or roughish colonies, are the most virulent. Avirulent strains tend to form small smoother colonies, in which chains of bacilli or filamentous forms are not found.

ANTHRAX

Anthrax occurs primarily in animals and secondarily in man as an occupational disease. It has long been recognized and has played an important part in the development of bacteriology and immunology. Pasteur's early attempts at immunization in the field were aimed at protecting sheep against this disease.

Anthrax in animals is a widespread disease, being particularly prevalent in Russia, Africa and parts of Asia; it is encountered in the herbivora, particularly in cattle, sheep, goats, horses and pigs; Algerian sheep are, however, relatively resistant. In view of the frequent enlargement of the spleen, the disease has been called "splenic fever". The infection may be spread directly from animal to animal or may enter the body via the alimentary tract, being ingested with the food. A septicaemia develops and the bacilli are excreted in the saliva, urine and faeces; death supervenes in the majority of cases a few hours to several days after the onset of symptoms. Flies have occasionally transmitted the disease following contamination with the blood or excreta of an infected animal. It is interesting to note that in this country single cases are usually reported; infection is considered to occur by way of the intestinal tract, the organisms, as spores, being present in imported food. The course is generally short, death occurring in 12–48 hr after the onset.

In man there are three portals of entry—the skin, the respiratory tract and the alimentary tract; the latter form is, however, a rarity.

The cutaneous form occurs mainly in people handling infected meat or some product of an infected animal such as hair, hide, wool, etc. It is the commonest form of the disease and is usually encountered in butchers, farmers, dock workers, tanners, brushmakers. It may also arise from the use of a shaving brush with contaminated bristles. The spores are implanted directly on to the skin, through which they pass by way of cuts, minute abrasions or the hair follicles. A local lesion is produced which has been termed a "malignant pustule". This term is actually a misnomer, as pus formation is not usually a prominent feature and also the condition is not particularly malignant. The incubation period is short, about 24 hr, and is followed by the appearance of a local papule, which increases in size, and by the third or fourth day becomes a red indurated mass with a central, necrotic black area which is surrounded by an elevated zone of small vesicles. Localized oedema surrounding the pustule is frequently present. The mortality rate is low and depends to some extent on the site of the lesion. It is highest where the tissues are loose, as in the neck; in fatal cases the oedema spreads and septicaemia develops.

The pulmonary form is essentially an industrial disease and is often referred to as "wool-sorter's disease". It is seen in individuals concerned with the treatment of animal products and in this country cases are found mainly in the Bradford district, where the worsted industry is carried on. When the hair or wool is contaminated with anthrax spores, these are inhaled with the dust produced during the sorting and combing of these products. A generalized infection with septicaemia develops. The mortality rate is extremely high; Legge reported 54 cases, of which 53 ended fatally. On autopsy the principal changes are discoloration of the skin, widespread cellular oedema and congestion, particularly involving the trachea, the bronchi, where the primary lesion is usually situated, the lungs and the mediastinum. At the primary focus of infection the mucosa is necrotic and ulcers are formed. Multiple subcutaneous haemorrhages and serous effusions into the pleura, peritoneum and pericardium are also common features.

DIAGNOSIS

Bacteriological methods are required not only to diagnose anthrax in man and animals, but also to detect the presence of the bacilli or spores in suspected material, such as shaving brushes, wool, hair, etc. When anthrax is suspected in animals an autopsy is illegal in this country, as by such a practice the bacilli are liberated and become a source of infection for other animals. A swab soaked in the blood of the animal and collected with due precaution or an ear of the animal should be submitted for examination.

In the human cutaneous form, the bacilli are present in greatest numbers in the vesicles surrounding the black central eschar. The vesicular fluid should be collected either in a capillary pipette or on a swab. In the pulmonary type the sputum and blood should be examined, and in the intestinal form the vomit.

The methods employed in identifying the organisms are microscopy, culture, pathogenicity and sometimes serology. In microscopy the presence of capsules and absence of motility are points of most significance. Cultural tests are carried out on agar, blood-agar plates and in a gelatin-stab to obtain the inverted fir-tree appearance. For the pathogenicity tests guinea-pigs or mice are usually employed. With pure cultures subcutaneous injection is used, but with grossly contaminated material the most valuable procedure is cutaneous scarification. As the penetration of a few anthrax bacilli is sufficient to produce infection, the organism can frequently be isolated by this technique. The most useful serological test is the thermo-precipitin test of Ascoli. In this test extracts of material or infected tissue are tested against an immune serum. Although the reaction is not absolute specific, the specificity becomes serviceable by suitable dilution of the reagents.

In the demonstration of anthrax bacilli on shaving brushes, wool or hair, the suspected material is soaked in a weak solution of KOH (3–5 per cent), incubated for a few hours and then heated at 80°C for several minutes; after this treatment cultures are prepared and the material should also be injected subcutaneously or rubbed on the scarified skin of a mouse or guinea-pig.

As saprophytic members of this genus are frequently present along with the anthrax bacilli, the identification of the latter requires great care.

PROPHYLAXIS

Human anthrax is dependent on the presence of the animal disease, and is controlled mainly by attention to industrial hygiene. The incidence of the condition has decreased to a marked degree in this country following the introduction of regulations to prevent the infection of animals and to ensure the disinfection of material likely to be contaminated, e.g. the Anthrax Order 1928. The disease is notifiable, animals with anthrax are immediately destroyed and post-mortem examinations are illegal. The carcases are either buried in a deep pit surrounded by lime or burned. If through a mistaken diagnosis an autopsy is carried out, the carcase should be burnt without delay and the buildings thoroughly treated with disinfectants.

Suspected material, such as wool or hair, introduced into this country, is subjected to treatment at Liverpool by the "Duckering" process. After preliminary treatment with alkalies, the material is acted on by a 2 per cent solution of formaldehyde at 102°–105°F for 20 min; it is then washed in water and dried by air at 160°F. The treatment of shaving brushes and hides is more difficult and no generally accepted method is available.

Further preventive measures include the early notification and prompt medical attention of all suspicious lesions and the education of the workers to the potential dangers of the work.

Various vaccines have been employed extensively in combating the disease in sheep and cattle on the Continent. Organisms attenuated by cultivation at 42°C are used for the first injection, and less attenuated ones for the second dose, as originally recommended by Pasteur. The results are not always satisfactory, as it is impossible to standardize the virulence of the organisms. Another disadvantage is that the induced immunity is of relatively short duration.

A protective antigenic substance in the oedema fluid from infected animals has been demonstrated by Bail (1904) and several subsequent workers. Gladstone produced the protective antigen *in vitro* in media containing plasma or serum and showed it to be a protein. Belton and Strange have produced the antigen in synthetic media, and have prepared a non-toxic immunizing agent which provides long-lasting immunity in experimental animals. It has also been used to immunize veterinary surgeons, agricultural workers and others who may be exposed to anthrax infection.

THERAPY

B. anthracis is sensitive to penicillin and clinical trials have proved satisfactory. This is now the usual treatment; large doses are necessary.

Before antibiotics were available, arsenical compounds such as salvarsan were used extensively, particularly in South Africa. In Italy, Sclavo's serum, an antiserum prepared in the ass, was given intravenously in large doses.

In the cutaneous form surgical intervention is usually unnecessary and is perhaps dangerous, as it may facilitate the introduction of the bacilli into the blood stream and so lead to a generalization of the infection.

Chapter XXVIII

Clostridium: Tetanus: Gas Gangrene: Botulism

CLOSTRIDIUM

Anaerobes. Spore-bearing rods; diameter of spores larger than the rods. Usually motile. Gram-positive. Many species are pathogenic, producing potent toxins. Natural habitat is probably the soil. The type species is *Clostridium butyricum.*

Members of this group were recognized in the early days of bacteriology. Pasteur and Koch both described pathogenic and non-pathogenic species, but the methods of examination were not very satisfactory; fluid cultures only were used and the inherent difficulties involved in the study of this group were not fully appreciated. It is, therefore, not surprising that results of this early work soon reached a state of chaos. Species were isolated, subjected to an incomplete examination and, in many instances, unjustifiable claims for the discovery of fresh species were made. During the First Great War, when cases of gas gangrene were seen with great regularity, an added impetus was given to the study of this group. Greatly improved technical methods were introduced and the use of solid media became practicable. Several valuable reports resulted, and the knowledge of the genus was gradually freed from confusion. It was found that different strains of one species had frequently been given quite distinct names and were considered as separate species; e.g. *Cl. welchii* had been styled by different bacteriologists—*B. welchii*, *Bacillus of Achalme*, *B. aerogenes capsulatus*, *B. phlegmonis emphysematosae*, *B. perfringens* and *B. enteritidis sporogenes.*

The early discrepancies are easily explained by the difficulties involved in obtaining pure cultures, particularly with the methods then in use. These organisms appear to grow as well, if not better, in mixed as in pure cultures; the spores also tend to adhere, and when different varieties are present the individual forms may be exceedingly difficult to separate. Irregularities were obtained mainly in the study of gas gangrene, in which condition gross contamination of the wound is invariably found. Pathogenic and non-pathogenic species are generally present and the identification of the individual members is frequently a matter of great difficulty.

The pathogenic members of this genus have been described in some detail, but the study of the saprophytic species is far from complete. Some have been accurately defined, but it appears highly probable that many others have not yet been recognized as separate species.

The natural habitat of most, if not all, members of this genus is considered by the majority of workers to be the soil. They are widely distributed in nature and some are isolated with great frequency from the faeces of man and animals. Their presence in the faeces is due to the ingestion of contaminated foods.

The species important to a medical bacteriologist are those concerned either with the causation of, or present in the lesions of gas gangrene, tetanus and botulism. In the former group are *Cl. welchii*, *Cl. septicum*, *Cl. oedematiens*, *Cl. histolyticum*, *Cl. fallax* and *Cl. sporogenes*. Some of the more pathogenic members of this group, and *Cl. chauvoei*, are also responsible for important diseases in animals, e.g. braxy, blackleg and lamb dysentery. In the tetanus group are *Cl. tetani*, *Cl. tetanomorphum* and *Cl. tertium*, and in botulism the various types of *Cl. botulinum*.

MORPHOLOGY

The *Clostridia* are usually large, straight or slightly curved rods, $0{\cdot}5 \times 8\,\mu$, with rounded or truncated ends. Pleomorphism is a common feature and a great variety of forms may be seen—filaments, clubs, navicular and citron shapes. In old cultures irregularly staining involution forms are frequent. Motility is shown by most species, but it is frequently difficult to demonstrate; an important non-motile member is *Cl. welchii*. Capsule formation is exhibited only by *Cl. welchii* and *Cl. butyricum*.

All members form spores which may be demonstrated, in the case of the pathogens, in the body tissues. Some members, such as *Cl. sporogenes*, spore readily, but others are less consistent, e.g. *Cl. welchii* forms spores only in media free from a fermentable carbohydrate. Variation is encountered in the position and shape of the spore which, when mature, is almost invariably wider than the vegetative bacillus—hence the term "clostridium". The spores may be terminal, as with *Cl. tetani*, or more commonly equatorial or subterminal.

The anaerobes stain readily with the ordinary dyes. They are Gram-positive, but in old cultures they tend to be decolorized more readily than most Gram-positive organisms. The spores are not stained by these methods and the special spore stains are required for their demonstration.

METHODS OF CULTIVATION

The clostridia will grow on the ordinary simple media if oxygen is excluded from the environment. Thus in stab or shake cultures in agar, growth will occur in the depth of the medium, but not near the surface.

In liquid media, some reducing agent is required; cysteine hydrochloride, sodium thioglycollate, ascorbic acid, strips of iron, or pieces of tissue may be used. Before inoculation the medium should be heated to expel dissolved oxygen, and then quickly cooled. Air may also be excluded by a layer of oil or vaseline over the surface.

To isolate the clostridia from the usually heavily contaminated pathological specimens, it is necessary to culture the material on the surface of a solid medium to obtain isolated colonies in the usual way. As the clostridia will not grow in atmospheric oxygen, it is necessary to absorb or remove the oxygen. The most satisfactory apparatus for this purpose is the McIntosh and Fildes jar. This consists of a glass or metal jar with a tightly fitting metal lid. Two metal tubes, with taps, lead through the lid into the jar. On the inner surface of the lid is fixed a small packet of palladium black enclosed in a wire gauze with a heating coil of resistance wire connected to two terminals on the outside of the lid. Inoculated cultures are placed in the jar. The lid is firmly clamped down, being sealed with wax or vaseline if necessary. The air is then exhausted by a suction pump, and replaced with hydrogen. A current is then passed through the coil to heat the palladium black, which acts as a catalyst to combine any oxygen remaining in the jar and any which may diffuse out of the medium, with the hydrogen. With this apparatus, even the strictest anaerobes will produce a surface growth and isolation is greatly facilitated.

CULTURAL CHARACTERISTICS

The optimum temperature is 37°C but growth may occur over a fairly wide range. The optimum pH is 7·0–7·4. Growth on solid media is not profuse; colonial forms are variable. Some types tend to spread in a thin effuse film over the surface, particularly if moist. This spreading can be inhibited to some extent by increasing the concentration of agar to 6 per cent, and drying the plates before inoculation.

In fluid media under anaerobic or reducing conditions growth is heavier. Robertson's meat media is particularly useful. Proteolytic types digest and blacken the meat, while saccharolytic types frequently turn it pink and produce gas.

BIOCHEMICAL ACTIVITY

The clostridia may be predominantly proteolytic, such as *Cl. sporogenes*, *Cl. histolyticum*, *Cl. bifermentans* and *Cl. botulinum*, or predominantly saccharolytic such as *Cl. welchii*, *Cl. septicum*, *Cl. chauvoei* and *Cl. oedematiens*. *Cl. tetani* has very slight proteolytic activity and does not ferment sugars.

TOXIN PRODUCTION

The pathogenic properties of the clostridia are due almost entirely to their capacity for producing exotoxins.

Cl. tetani produces tetanospasmin which attacks the central nervous system with characteristic clinical effect. It also produces a tetanolysin which is of lesser importance. All serological strains of *Cl. tetani* produce the same toxin, although there are some non-toxigenic strains.

Cl. welchii. There are six different serological types of *Cl. welchii* designated A to F. At least 12 different toxins are produced by various strains. These toxins are designated by letters of the Greek alphabet, alpha, beta, gamma, delta, etc. All strains produce the alpha toxin and varying numbers of the other toxins. Type A is generally found in human gas gangrene infections. Type F is found in *enteritis necroticans* and occasionally in gas gangrene. Types B, C and D are responsible for various diseases of sheep, and Type E produces enterotoxaemia in calves. The alpha toxin is a lecithinase, and is demonstrated by the Nagler reaction, in which the lipoprotein of serum or egg yolk is split. An agar plate containing 20 per cent human serum is dried and a few drops of alpha antitoxin are spread over half the plate. The cultures for test are inoculated across both sides of the plate. After incubation there is an area of opacity surrounding the growth of *Cl. welchii* on the side of the plate without antitoxin. This reaction is prevented by the antitoxin on the other side of the plate which remains clear.

Cl. oedematiens, produces eight different toxins, also designated by Greek letters. The alpha and beta toxins have necrotic and lethal activity.

Cl. septicum, produces five different toxins, at least one of which has a lethal action on animals.

Cl. botulinum produces the most potent toxin known. There are five different serological types of *Cl. botulinum*, and each produces an antigenically distinct toxin, so antitoxin to one of them does not neutralize the toxins of the others.

PATHOGENICITY

While members of this genus are important pathogens of man and animals, the anaerobes are not strictly parasitic; they are actually unable to multiply in living tissues and only produce infection when implanted in tissue which has been damaged or destroyed in some way, as by trauma. In man the main diseases produced are tetanus, gas gangrene and botulism; in animals they give rise to braxy, blackleg and lamb dysentery. In all instances the clinical picture is mainly due to the activity of the toxins produced by the causative organisms. Botulism in fact is entirely an intoxication; the toxin is produced outside the body and is absorbed from the intestinal tract after ingestion with food. Certain heat-resistant strains of type A *Cl. welchii* have been incriminated with increasing frequency as the organisms responsible for outbreaks of food poisoning.

Experimental infection can be produced by the more pathogenic members of the gas gangrene group by the injection of whole cultures; washed spores or bacilli have no such effect. The most satisfactory animals are guinea-pigs, mice and pigeons; death follows the intramuscular injection of a culture in 24–72 hr. The post-mortem appearance given by the different species varies, but local swelling, oedema, haemorrhage and gas production with necrosis, redness and oedema of the local muscles are frequently found.

Most laboratory animals are susceptible to both *Cl. tetani* and *botulinum* toxins; mice and guinea-pigs are those generally employed in experimental and routine work.

TETANUS

Tetanus is a disease of antiquity and is widely distributed throughout the world. Its incidence is, however, relatively low considering the great possibility of infection with *Cl. tetani*, which has been isolated with frequency from soil and the faeces of man and animals. The spontaneous disease is encountered in man and various animals, particularly the horse. The relationship of *Cl. tetani* to tetanus was definitely established in 1889 by Kitasato, who isolated the organism in pure culture from the tissues of the wound and reproduced the condition experimentally with the pure cultures. Earlier workers, including Rosenbach (1886), had reproduced the disease experimentally by the injection of material collected from the local lesion, but although organisms resembling *Cl. tetani* were seen microscopically this organism was not isolated in culture.

In man the majority of cases have followed wounds contaminated with soil and manure, such as those occurring in war or accidental traumatic lesions. Cases have also occurred in newly-born infants following infection of the umbilical cord, and after such operations as criminal abortion, the haemostatic injection of gelatin, vaccination, surgical operations and teeth-extraction. In some cases the initial injury is relatively trivial, e.g. prick from a sharp object such as a splinter, nail or rose thorn, and is neglected. In post-operative tetanus the catgut is generally held responsible for the introduction of the tetanus spores; great care is therefore necessary in the preparation of catgut, and the sterilization process requires special attention.

The tetanus bacillus or spore gains entrance at the time of the injury, and the subsequent events depend largely on the local environmental conditions. If there is little tissue damage and no contamination with foreign material, the spores may either undergo phagocytosis or remain dormant for some time. The wound heals and the host does not develop tetanus unless the general or local resistance is lowered by some other infection or local trauma. This type of infection has been termed "idiopathic tetanus". When, however, the implantation of the organism or spore is accompanied by tissue damage and gross contamination of the wound the organism is able to multiply locally and toxin is produced. Under such conditions the development of tetanus may be rapid, within 24 hr of the initial tissue damage and contamination in some cases.

The pathogenesis has been investigated in some detail in experimental animals, particularly rabbits, guinea-pigs and mice. The results indicate that the toxin, produced locally, passes to the nerve-cells of the anterior horn of the cord by way of the axones or perineural lymphatics of the regional motor nerves. This gives rise to the tonic muscular contractions characteristic of tetanus. The nerve-cells corresponding to the area of infection are usually the first to be involved, after which the toxin passes to other parts of the cord. The toxin enters the nerve by the exposed nerve-endings or by breaches in the sheath caused by the trauma. Some toxin also passes into the blood stream, but this is not the usual route of transport to the cord. The bacilli remain localized and their main activity is to produce the toxin.

In laboratory animals two types of tetanus have been described —ascending and descending—but these are frequently combined. The ascending type is demonstrated by injecting mice or guinea-pigs subcutaneously into one of the hind limbs. The toxin passes by the local motor nerves to the central nervous system, through which it subsequently diffuses. The spasms are first observed in the injected limb, then in the corresponding limb and later in the upper limbs and head.

The descending form is seen after intravenous inoculation of the toxin, which gains entrance to the central nervous system probably via the motor nerve endings. The spasms quickly develop and the upper limbs are usually affected before the lower extremities. It may follow the injection of the toxin by any route in large animals, e.g. the horse.

The production of the infection does not depend solely on the introduction of the tetanus bacilli or spores into the tissues. This must be accompanied by other factors, such as trauma or the presence of gross contamination of the wound. The role played by these factors has been investigated by various workers, who found that infection did not follow the injection of the washed bacilli or toxin-free spores into mice or guinea-pigs. It occurred, however, on the addition to the washed organisms of such substances as a small amount of toxin, the ionizable salts of calcium, soil or other bacteria. The importance of these additional substances is that they render the tissues of the host favourable for the multiplication of the organisms. Fildes considered that the failure of the washed organisms to grow was due to the presence of oxygen in the normal undamaged tissues. When local necrosis was produced the oxygen tension was reduced and germination of the tetanus bacilli or spores proceeded. The function of the foreign bodies was to produce this local necrosis.

A similar sequence of events is also found in the spontaneous human disease. It thus follows that the incubation period is variable and depends mainly on the nature and site of the local lesion. The average time is about eight days, but this has been increased by the introduction of antitoxin as a prophylactic measure. The disease is characterized by a series of tonic muscular contractions which are local at first, but later become general. As the muscles of the jaws are frequently involved, the disease has been popularly termed "lock-jaw". The mortality of the disease in the absence of serum prophylaxis or previous active immunization is high, 80–90 per cent. The most severe cases are those in which the incubation period is short, while tetanus resulting from lesions of the head and upper extremity tends to be more severe than that following injuries of the lower extremities.

DIAGNOSIS

The bacteriological methods used in the diagnosis of tetanus are microscopy, culture and pathogenicity tests.

The important points of the microscopical examination are the presence of motility and the demonstration in the local lesion of rods with terminal spores, giving the so-called "drum-stick" appearance to the organism. This appearance is, however, only suggestive of the presence of *Cl. tetani*, as it is also given by other closely related but non-pathogenic *Clostridia* (e.g. *Cl. tetanomorphum*).

In cultural tests as much tissue as can be obtained from the wound should be submitted for examination. This is added to tubes of Robertson's meat medium, some of which are then heated at 80°C for 20 min to destroy vegetative organisms, and all are incubated under anaerobic conditions at 37°C. The tetanus bacillus produces a general turbidity and gas, but does not blacken or digest the meat. Another satisfactory and rapid method of isolating *Cl. tetani* is to inoculate the water of condensation of an agar slope with the tissue heated at 80°C for 20 min. *Cl. tetani* produces a thin effuse growth which spreads upwards on the surface of the medium; in 24–48 hr the upper edges of the filamentous growth are examined by means of a hand-lens and subcultured on to various media. *Cl. tetani* liquefies gelatin, but neither digests serum nor ferments any sugars.

Pure cultures or suspensions of the original material are finally injected subcutaneously into experimental animals, usually mice or guinea-pigs; *Cl. tetani* gives rise to characteristic spasms resembling tetanus in man. Conclusive proof of the identity of the organism is obtained by demonstrating the neutralizing action of the specific antitoxin.

PROPHYLAXIS

The value of antitoxin as a prophylactic measure was conclusively demonstrated during World War I. It was introduced during the last quarter of 1914 as a routine procedure in the treatment of wounded soldiers. The development and progress of the disease were influenced in several ways: (1) The incidence was greatly reduced; figures collected from the British hospitals in France indicate that in September 1914, tetanus developed in 9 per 1,000 wounded, but by December 1914, following the introduction of antitoxin, the incidence was only 1·4 per 1,000. (2) The incubation period was lengthened. The average figure of 10 days in the unprotected was increased to an average of 45 days in the protected individuals. (3) The severity and type were changed; the disease became more sub-acute and local tetanus was a more prominent feature.

While there was no doubt about the beneficial effects of antitoxin in prophylaxis, the results were less satisfactory than those obtained in diphtheria. Several factors are responsible for this: (1) the relatively low concentration of the antitoxin in the blood after injection and the possible high toxin production of the local lesion. (2) The shortness of the immunity conferred by the antitoxin and the possibility of tetanus spores lying dormant for some time and subsequently germinating and producing toxin.

The main prophylactic application of antitoxin was seen during war-time conditions, but in civil practice all individuals with wounds, in which contamination with soil or manure has taken place, should receive antitoxin unless they have been actively immunized. The dosage employed during the 1914–18 war was four subcutaneous injections of 1,000 international units at intervals of seven days. In all cases surgical measures are necessary. The wound must be thoroughly cleaned and all necrotic tissue removed; this renders the focus less favourable for germination of the tetanus spores. Antitoxin (1,500–3,000 units) should always be given before surgical treatment is commenced. Further dosage, at weekly intervals, may be necessary in certain civilian cases where active immunization has not been carried out.

In the case of individuals liable to receive contaminated wounds, e.g. soldiers, active immunization with tetanus toxoid is now carried out; this procedure was officially introduced into the British Army as a routine measure in 1938. Immunity is produced slowly, but once developed it appears to be of relatively long duration; two doses at an interval of 6–8 weeks were originally given. Further doses at intervals of 10–12 months are now recommended, as it has been found that the administration of a third dose is followed by a rapid and marked increase in the antitoxin titre of the serum and that this increase is maintained for some time. The results have been striking; tetanus in the British Army during the 1939–1945 War was negligible. In the American Army it is interesting to note that toxoid, and not antitoxin, was given to wounded who had been previously immunized, while in the British Army a single dose (1,500 international units) of antitoxin (A.T.S.) was used to reinforce the active immunity.

It is recommended that civilians, whose occupation renders

them liable to tetanus, e.g. agricultural and horticultural workers, should be actively immunized.

Tetanus toxoid is now frequently included in the immunization schedules for children. It may be given together with diphtheria toxoid. This has the great advantage that if the possibility of tetanus infection should arise at a later date through some injury, a further dose of toxoid can be given to stimulate active antitoxin production, and the disadvantages of serum therapy are eliminated.

THERAPY

The value of antitoxin as a therapeutic agent is undecided. It is accepted that once the toxin has reached the central nervous system it is fixed by the nerve cells and antitoxin has little effect. Serum must therefore be given as early as possible in order to neutralize the toxin in the blood, the site of infection and other tissues before it reaches the C.N.S. Large doses are necessary, and the intrathecal route, supplemented by intravenous or intramuscular injections, should be used. It is therefore important that cases should be diagnosed in the initial stages and that large doses of serum should be administered without delay. An immediate dose of 200,000 units intravenously has given promising results. Anticonvulsant therapy, penicillin and perhaps blood transfusions are also required.

The sulphonamides and penicillin have given disappointing results in the treatment of tetanus.

Preparation of antitoxin. Antitoxin is prepared by the injection of the horse with increasing doses of alum-toxoid. The selection of a toxin-producing strain of *Cl. tetani* does not present any difficulty, although it is important to note that non-toxic strains are not uncommon. Eight serological types have been demonstrated by agglutination, but they all produce the same toxin. Broth is seeded and incubated anaerobically at 37°C for 5–14 days after which it is passed through a Seitz or Berkefeld filter.

The activity of the toxin is determined by estimating the M.L.D. on subcutaneous or intramuscular injection of mice or guinea-pigs. For mice of 20 g this is, on an average, 0·00005 ml, and for guinea-pigs of 350 g ten times greater. The tetanus bacillus produces two distinct substances—(1) tetanospasmin and (2) tetanolysin.

Tetanospasmin is the more important of the two substances, and is the one usually referred to in discussions of the tetanus toxin. It has an affinity for nervous tissue, and is directly responsible for the symptoms of tetanus. It is extremely active when injected into the tissues (a quantity as small as 0·000005 ml of a broth culture filtrate may kill a mouse), but is inert when given by the mouth; it is readily destroyed by moderate heat (60°C), oxidizing agents, acids and strong alkalies.

Tetanolysin has a lytic action on red blood corpuscles, but probably has little pathogenic significance. It is very labile, being readily destroyed by heat at 50°C. The so-called "non-toxic" strains of *Cl. tetani* may produce abundant tetanolysin while virulent strains may not produce any.

After its activity has been determined the toxin is formolized to give the toxoid, which is treated with potash alum. Immunization is slow and may take several months. The antitoxin is contained in the pseudo-globulin fraction of the serum and may be readily concentrated to approximately three times its original value. The standardization in various countries of the antitoxin content of a serum has lacked uniformity, owing to the existence of three different units and four methods of assay. Attempts are being made to overcome this, and the method now advocated is one in which a satisfactory toxin is standardized against a dried standard antitoxin by determining a dose analogous to an "L+" dose in a 350-g guinea-pig (1/10 of a unit of antitoxin is used and the animal must die in just 96 hr); from the data available the titre of the freshly prepared antitoxin can be determined. This method was originally worked out in America and it was later adopted by the League of Nations Standards Commission; the adopted international unit is, however, one-half the original U.S.A. unit.

GAS GANGRENE

Gas gangrene, a common disease in the days of pre-antiseptic surgery, is now a rarity in civil practice and is essentially a war disease. The anaerobes not infrequently give rise to wound infection and may also be responsible for cases of puerperal fever and mild types of enteritis, but the classical cases of gas gangrene are seldom seen.

Our knowledge of this condition is due largely to observations made during World War I by British and French bacteriologists. Prior to 1914 the bacteriology of the disease was in a state of chaos. Widely divergent reports were made and disagreement was common. This was not surprising, as the technique employed in these investigations was of a primitive nature. The great improvement in technique, particularly the introduction of the McIntosh and Fildes jar, simplified the bacteriological investigations of the disease. The report of the Anaerobic Committee of the Medical Research Council (1919) and the monograph of MM. Weinberg and Séguin (1918) give comprehensive details of some of the various observations carried out on gas gangrene during the war and indicate the factors which caused previous work to be inaccurate and unreliable. These findings have been largely confirmed by observations in World War II.

The incidence of gas gangrene is closely related to the liability to contaminated wounds in which extensive tissue destruction and imperfect drainage are present. The anaerobes are able to multiply only in devitalized tissue. It thus follows that war-time conditions are more favourable than those prevailing in ordinary civil life. In civil practice the disease is now usually seen as a complication of street or other accidents, in which contamination of the wound with soil or faeces has taken place. In the pre-antiseptic days, where the tissue destruction at the operation was invariably followed by infection of the wound, most surgeons had much unpleasant experience of the condition.

During the 1914–18 war it was noted that cases were most numerous during active operations, when the wounded were not collected for some time and when the possibility of the wounds becoming contaminated with soil was increased. The incidence was also highest following battles fought over highly cultivated and manured ground and in wounds caused by bombs, shrapnel and fragments of high-explosive shells.

The onset of gas gangrene may occur within a few hours of the receipt of the wound or may be delayed for several days. An early sign is the discharge from the wound of a foul-smelling, brownish material which later increases and contains gas; there is also local swelling of the tissues and the neighbouring skin becomes purplish. The fluid and gas extend in the subcutaneous and intermuscular tissues. Early diagnosis may be made by radiology, when small bubbles of gas may be seen on the X-ray film of muscles extending from the site of the wound. When the muscles are involved they become black, friable and diffluent locally; the necrotic area of the muscle is bordered by a red zone. General symptoms are also found and are usually those of a profound toxaemia. In the late stages septicaemia may develop. The mortality rate is high (20–80 per cent).

The isolation of *Clostridia* from a wound does not necessarily indicate the presence of gas gangrene, the diagnosis of which must be based on clinical grounds. In many instances the organisms are merely harmless contaminants, in others they produce a local infection, termed anaerobic cellulitis, and in only few cases muscle is involved with the development of true gas gangrene. The condition may clinically be similar to streptococcal myositis, in which the causative organism is an anaerobic streptococcus.

Bacteriology. Gas gangrene is not usually a single bacterial infection; in practically all cases gross contamination with anaerobic and aerobic organisms is present. The isolation and incrimination of the main pathogenic species have consequently been problems of considerably difficulty. The incidence of the various anaerobes in the diseased tissues was the subject of investigation during the 1914–18 war. Similar results were obtained during World War II.

Various anaerobes together with aerobic organisms are usually present in the wound. *Cl. welchii* is found in about 80 per cent of all cases. *Cl. septicum* and *Cl. oedematiens* are also possible causative organisms in gas gangrene. *Cl. sporogenes*, *Cl. histolyticum* and *Cl. fallax* are frequently present. These are probably less pathogenic, but they may produce tissue damage and thus prepare suitable conditions for the development of the more pathogenic species. Aerobic species such as staphylococci, coliforms and pseudomonas also help to produce reducing conditions which favour the anaerobic species.

DIAGNOSIS

The isolation of the individual organisms may be extremely difficult. The material that should be submitted for examination is principally the exudate and necrotic tissue collected from the wound. In the late stages blood cultures may be positive.

The methods of examination are (1) microscopy, (2) culture, and (3) pathogenicity tests.

(1) The examination of smears, stained by Gram's method and for spores, gives some indication of the number and nature of the organisms present.

(2) Numerous cultural methods have been employed in the investigation of gas gangrene. The most valuable procedure is to add the tissue to several tubes of Robertson's meat medium, heat at 80°C for 20 min to kill off the vegetative forms, subculture into meat medium, selective media (saccharolytic and proteolytic) and on to plates of solid media and incubate under anaerobic conditions.

(3) For the pathogenicity tests mice or guinea-pigs are used. In most instances normal animals may indicate the presence of gas gangrene pathogens but are of little value in identification of species in a mixed inoculum. The most useful test is one in which

groups of animals are used, some being previously immunized by inoculation with antitoxins of the three important pathogenic species. Five groups of mice are usually prepared: (1) normal, (2) immunized against *Cl. welchii*, (3) immunized against *Cl. septicum*, (4) immunized against *Cl. oedematiens* and (5) immunized against all three. The animals protected against the homologous organism survive, whereas the controls and those protected against the other organisms are killed. In the case of a mixed infection, group (5) animals survive and the others die; the survival time of the animals protected against one of the infecting organisms would, however, probably be longer than in the unprotected. This thus provides a means of isolating and identifying the individual species from the original mixture.

The Nagler reaction is useful in identifying *Cl. welchii* and *Cl. oedematiens* and depends on the lecithinase activity of the toxin which splits the lipo-protein complex of serum. This test can be carried out in the following manner: 20 per cent human serum and 5 per cent Fildes' extract are added to 1·5 per cent agar at 50°C, plates are poured and dried. To the surface of one-half of a plate a few drops of *Cl. welchii* antitoxin are added and allowed to dry; material under examination is seeded evenly on both sides of the plate; after incubation, colonies of *Cl. welchii* produce an opacity of the surrounding medium on the half of the plate without anti-toxin; the other half of the plate does not show this reaction owing to the presence of antitoxin and serves as a control. Egg-yolk suspension can be used instead of the serum. Specific results can also be obtained with *Cl. oedematiens* antitoxin.

The following are other criteria which are useful in the identification of the various *Clostridia* present in gas gangrene.

Cl. welchii (*Cl. perfringens*) is non-motile, capsulated when freshly isolated from the tissues, grows readily in glucose-agar with much gas production, liquefies gelatin, gives rise to stormy fermentation in litmus-milk, forms an active haemolysin; gives a specific Nagler reaction; on intramuscular inoculation into the guinea-pig gives rise to a massive crepitant swelling containing much gas, and muscles are pale and necrotic, while the suprarenals are often congested.

Cl. septicum is motile, grows readily in the meat medium with the production of gas, but does not digest the meat; liquefies gelatin; ferments glucose, maltose, lactose and salicin with the production of acid and gas; forms acid and clot with some gas in litmus-milk; produces a haemolysin; and on agar gives a filamentous growth; on intramuscular inoculation of the guinea-pig gives rise to a local haemorrhagic, gaseous, oedematous mass, the muscles are soft and deep red, while the suprarenals may be slightly congested.

Cl. oedematiens (*Cl. novyi*) is motile and behaves culturally as *Cl. septicum*: ferments glucose and maltose and produces a haemolysin; gives a specific Nagler reaction. On intramuscular inoculation of the guinea-pig it gives rise to a gelatinous mass, the local muscles are congested, purplish-red and infiltrated with small bubbles of gas; the suprarenals are not congested.

Cl. sporogenes is motile, spores readily, grows well and is markedly proteolytic, meat being digested and blackened; non-pathogenic to laboratory animals, but its presence tends to increase the infectivity of small doses of the pathogenic species.

Cl. histolyticum is motile and spores readily, actively proteolytic —digests and blackens meat with the production of tyrosin crystals, ferments glucose and maltose with the production of acid only; produces a weak exotoxin, which in large doses produces in guinea-pig local necrosis and may even kill the animals.

Cl. fallax is motile and does not spore readily, has little proteolytic activity but ferments glucose, lactose and maltose with the production of acid and gas; when freshly isolated is pathogenic for guinea-pigs; this action is soon lost in culture.

PROPHYLAXIS AND THERAPY

Two factors were mainly responsible for the greatly reduced incidence of gas gangrene after the early days of the 1914–18 war: surgical measures and the introduction of antitoxic sera. Of these, adequate and immediate surgical treatment was the more important; in all cases the wound must be cleaned and drained, while necrotic material must be removed. When the disease develops the affected tissues must be removed *en bloc*; in the severe cases amputation of the affected limb may sometimes be necessary. These observations were fully confirmed in World War II.

Specific sera, prepared by the immunization of the horse with the various toxins, were available for general application in the late days of the 1914–18 war and, while their value could not be definitely determined, they appeared to serve a useful purpose both in prophylaxis and therapy. In the 1939–45 war polyvalent gas-gangrene antitoxin was used extensively. All severely wounded cases received a prophylactic dose as soon as possible after wounding. The standard dose was increased threefold in the later stages and now contains: 10,000 units of *Cl. welchii* antitoxin, 5,000 units of *Cl. septicum* antitoxin and 10,000 units of *Cl. oedematiens* antitoxin.

Sulphonamide and penicillin powders were also widely used in the 1939–45 war and undoubtedly played an important role in preventing the infection of wounds. There is, however, no doubt that the most important factor in the prophylaxis of gas gangrene is early and radical surgery with particular attention to puncture wounds. Antitoxin, sulphonamides and antibiotics are useful but they play a secondary role to good surgery.

In the therapy of the established disease, drastic measures are necessary; these include surgery, intravenous antitoxin, antibiotics, sulphonamides and blood transfusion. In spite of these, the mortality rate is high, 20–80 per cent.

BOTULISM

Botulism in man is a comparatively rare disease, which follows the ingestion of contaminated food, such as sausages (hence the name, *botulus*—a sausage), prepared meats, canned foods, vegetables and fish. The aetiological agent, *Cl. botulinum*, was first isolated in 1896 by v. Ermengem, who investigated a small outbreak in Belgium following the consumption of uncooked ham. From this a sporing anaerobe was isolated, and both the organism and the ham produced the same series of symptoms when fed to animals.

Cases have been extremely uncommon in this country; one outbreak at Loch Maree was reported in 1922, and a few cases were notified in 1935. At Loch Maree eight individuals developed botulism after eating potted duck paste, and all died. Outbreaks have been more frequent in America and Germany, where prepared and home-canned foods are consumed on a larger scale.

Symptoms appear a short time (18–72 hr) after the consumption of the affected food, and although some gastro-intestinal disturbance may be present they are due mainly to the action of the toxin on the central nervous system. Paralysis of accommodation, ptosis, general weakness, aphonia and dysphagia pass on, in fatal cases, to bulbar paralysis, coma and death. Sensory symptoms are seldom marked.

The mortality has varied in different outbreaks from 25 to 100 per cent, and is highest when the incubation period is shortest. On post-mortem examination there is little to be seen, the cerebral vessels are congested, and some of the nerve-cells are swollen and show signs of degeneration. The exact mode of action of the toxin is uncertain, but experimental observations suggest that the central nervous system is involved peripherally rather than centrally.

Cl. botulinum is responsible for spontaneous disease in animals, giving rise to the condition known as "limber-neck" in chickens. It has also been incriminated as the aetiological agent of a paralytic disease in cattle, but the evidence on which this relationship has been based in unconvincing.

Bacteriology. Botulism is essentially a toxaemia and not an infection. The symptoms are caused by a preformed toxin ingested with food and absorbed from the alimentary tract. It is in fact doubtful whether *Cl. botulinum* is able to multiply in the human alimentary tract.

Cl. botulinum is widely distributed in nature, being present mainly in virgin soil. It has also been isolated occasionally from samples of animal faeces, but its presence is probably due to the consumption of soil-contaminated food. Two types (A and B) were originally described, each produces a distinct toxin; type A is encountered most frequently in America and type B in Europe. Further types (C, D and E) have since been isolated. Members of the various types have been subdivided by serological tests, e.g. agglutination and complement-fixation.

Cl. botulinum grows quite well on the ordinary media under anaerobic conditions, the optimum temperature being about 30°C. Its proteolytic activities are more marked than the saccharolytic; gelatin is liquefied by all types, but the action on coagulated egg-white is variable, type A strains being the most active; acid and gas are produced in glucose, maltose and sometimes salicin.

Cl. botulinum produces the most active bacterial toxin known; mice, guinea-pigs and monkeys are very susceptible, a dose of 0·00001 ml subcutaneously is sufficient to kill a mouse. Rats and dogs are very resistant.

The toxin is readily obtained by cultivation in broth (pH 7·4–8·0) at 35°C, the maximum amount being present on about the eleventh day. It is relatively unstable, being rapidly destroyed by light and alkalies, but it exhibits a marked resistance to weak acids; in virtue of this resistance it escapes destruction by the acids of the gastric juice; it is more resistant to heat than most other exotoxins, requiring exposure to a temperature of 80°C for 30 min to effect complete destruction. A toxoid is produced by treatment with formalin.

Botulism in man invariably results from the ingestion of the preformed toxin that has been produced by the growth of the organism in food which has been contaminated either directly or indirectly with soil. Many foods, including meat and vegetable products, particularly string beans, corn, olives and spinach, are suitable media for *Cl. botulinum*, and conditions are especially favourable for growth in canned foods. The spores of *Cl. botulinum* are relatively resistant to the action of heat and may resist the preliminary sterilization involved in the canning process; they multiply later and produce the toxin. A temperature of 20°C has been found suitable for growth and toxin production; growth is, however, relatively slow at this temperature. If the food is consumed in an unheated state the toxin is ingested and botulism is produced.

The food sometimes manifests some sign of contamination, there may be a rancid odour and some alteration in its physical appearance; in the case of canned foods, the cans may be blown.

DIAGNOSIS

The clinical diagnosis may be confirmed by the demonstration of the toxin in the suspected food or by the isolation of *Cl. botulinum* from the food, vomit or faeces.

The examination of the suspected food for the presence of the toxin gives the more satisfactory results. Suspensions are made in saline and injected subcutaneously into a series of mice, some of which have been protected by the inoculation of antitoxins A and B. If the toxin **is** present in the suspension, the unprotected animals die.

The food is seeded into glucose-broth, incubated anaerobically at 35°C for 10 days, filtered and injected into a similar series of mice. Pure cultures may be obtained by heating the broth at 80°C for 30 min and seeding on glucose-agar plates.

PROPHYLAXIS

Measures adopted in the prevention of botulism are (1) attention to the preparation and condition of the food; spoilage is not always detectable, but all suspected food should be discarded, and (2) adequate precaution and control in the sterilization of canned foods. Dormancy and heat-resistance are features of botulinum spores, and these must be considered in the process of heat sterilization. It is interesting to note that in the majority of botulism cases following the consumption of canned foods the food was home-canned and had not been subjected to the factory processes.

THERAPY

Gastric lavage should be used at the earliest possible time to remove any remaining toxin. The stomach should thereafter be kept alkaline with sodium bicarbonate solution. Antitoxin should be given, but it will not reverse damage already done at the myoneural junction. Failure of antitoxin therapy is probably due to late diagnosis in a rare and unfamiliar disease.

Chapter XXIX

Lactobacillus, Fusiformis and Miscellaneous Genera

LACTOBACILLUS

Gram-positive rods. Non-motile. Non-sporing. Ferment carbohydrates to acid, chiefly lactic acid. Micro-aerophilic on first isolation.

The *lactobacilli* have been isolated from the alimentary tract of man and animals, carious teeth and milk which has undergone fermentation. *L. acidophilus* and *L. bifidus* are frequently present in the faeces of breast-fed infants. Döderlein (1892) isolated a member from the acid vaginal secretion of pregnant women; this was termed Döderlein's bacillus, but it is now accepted as a strain of *L. acidophilus*. Another member, *L. caucasicus*, was originally isolated from "kefir", the fermented milk of the Caucasus, while *L. bulgaricus* was obtained from the fermented milk of Bulgaria. *L. odontolyticus* was recovered from carious teeth, and the Boas-Oppler bacillus was observed by Oppler (1895) in the stomach contents of an individual suffering from a gastric carcinoma.

These organisms are Gram-positive, slender, non-motile and non-sporing rods, which may exhibit pleomorphism. Growth is never profuse on the usual media, but is increased by enrichment with glucose and blood. On first isolation they do not, as a rule, grow under aerobic conditions; micro-aerophilic or strictly anaerobic environment may then be necessary, but after a few subcultures most species grow aerobically. The optimum reaction of the medium is about pH 7·0.

The *lactobacilli* behave in most respects as the non-sporing organisms to bactericidal agents, being readily killed by heat, but an important distinction is their high resistance to acids. For this reason they are referred to as "acidophilic" or "aciduric" bacilli. This resistance is employed in the isolation of these organisms from contaminated material. A commonly employed method is to incubate the material in 0·5 per cent acetic acid broth for 1–3 days and then subculture on 2 per cent glucose agar.

Little proteolytic activity is displayed, indole is not formed and gelatin is not liquefied. The saccharolysis is more marked, though variable; acid is usually produced in glucose, maltose, lactose and sucrose. Litmus-milk is acidified and frequently clotted.

These organisms appear to have no pathogenic action in man or animals. They are normal members of the intestinal flora and are particularly numerous in individuals fed on a carbohydrate diet.

Living cultures, particularly of *L. acidophilus*, administered by the mouth have been recommended by some workers, particularly in America, for various intestinal disorders, e.g. to suppress the growth of putrefactive organisms and to relieve constipation and the so-called "auto-intoxication" of intestinal origin. The results, have, however, been far from satisfactory. This may be due to the use of unsatisfactory preparations, as it has been found that only certain strains of *L. acidophilus* are able to multiply in the human intestines. *L. bulgaricus*, which has been used therapeutically in this respect, is unable to establish itself in human intestines.

FUSIFORMIS

This genus has been defined in the following manner: obligate parasites; anaerobic or micro-aerophilic; cells frequently elongated and fusiform, staining somewhat unevenly. Filaments sometimes formed; non-branching. Non-motile. No spores. Reaction to Gram variable. Growth in laboratory media feeble.

This group has received relatively little attention; it is considered by some workers to be related to the corynebacterium and pfeifferella groups. While some members belong to the normal bacterial flora of the alimentary tract, others have been isolated from ulcerative conditions of the throat and various suppurative lesions, including appendicitis. One member, *F. fusiformis*, is frequently associated with a spirochaete in the lesion of Vincent's angina, but its role in the causation of this condition is uncertain.

These organisms tend to be large fusiform rods, but pleomorphism is a common feature. Some are Gram-positive. Growth occurs at 37°C under anaerobic or micro-aerophilic conditions. In some cases there appears to be little doubt that they are either wholly or partially responsible for the lesions from which they have been isolated. In others they are either harmless parasites or secondary invaders. Necrotic lesions and death have been produced experimentally by some members in laboratory animals.

BACTEROIDES

Members of this group are responsible for the condition termed "necrobacillosis". This condition is often fatal and consists essentially of multiple putrid abscesses, frequently found along the pleural margins of the lungs and associated with a foul-smelling empyema. The organisms may be present in large numbers in the pus or blood stream but they are difficult to culture as they are very sensitive to oxygen; strict anaerobic conditions are essential and Brewer's medium gives excellent results. They may also be isolated from abscesses in relation to the intestinal tract; particularly from chronic appendix abscesses. They are Gram-negative and non-sporing and, while they may become very pleomorphic on culture, they seldom, if ever, present a fusiform appearance. "L" forms are common. In consequence *F. necrophorus* does not seem an appropriate title for these organisms and it has been suggested that these organisms belong to a separate genus, which has been tentatively named bacteroides.

LISTERELLA

Listerella organisms are Gram-positive, non-sporing rods; coccobacillary forms are usual although filaments may occur. They are motile; motility is, however, most satisfactorily demonstrated by incubation at 26°C. Growth occurs on the usual laboratory media, including MacConkey, under aerobic conditions; there is a narrow zone of haemolysis around the colonies on blood-agar plates. Acid, but no gas, is produced in glucose, maltose, salicin and, after five days, sucrose; lactose dulcitol and mannitol are not usually fermented. Indole is not formed but catalase is produced. The organisms tend to be sensitive to the routine antibiotics.

Listerella monocytogenes is a common animal pathogen which has been isolated from 50 different species. It is also an important human pathogen which has been increasingly recognized in the last few years. In rodents and in man the infection is characterized by a monocytosis in the blood. In man, the infection may occur as meningitis, lymphadenopathy, septicaemia, and metritis. The majority of cases occur during pregnancy and the foetus rarely survives.

ERYSIPELOTHRIX

Members of the group are responsible for infection in a wide range of animals and they are closely related to the *Listerella* group. One member, *E. rhusiopathiae*, is responsible for a localized cutaneous condition known as "erysipeloid", which is seen mainly in persons handling animals, e.g. abattoir workers, butchers, veterinary officers. Very rarely a generalized infection develops which may be fatal. *E. rhusiopathiae* is usually a slender, Gram-positive, non-sporing, non-motile rod but filaments may be seen. Growth is poor on ordinary media and, for primary isolation, the addition of serum or blood is required. *E. rhusiopathiae* is aerobic or microaerophilic with little biochemical activity; acid formation is usual in glucose, lactose and galactose but not in maltose, sucrose or mannite. A generalized infection can be produced in mice. The organisms can be isolated from swabs obtained from the local skin lesions or from biopsy material. *E. rhusiopathiae* is resistant to the sulphonamides but sensitive to most antibiotics.

STREPTOBACILLUS MONILIFORMIS

Streptobacillus moniliformis (Act. muris) is a Gram-negative, aerobic organism which exhibits marked pleomorphism, "L" forms are common. Enriched media are necessary for growth and little biochemical activity is exhibited. The organism is normally present in the upper respiratory tract of rats and is responsible for one type of rat-bite fever. It is also responsible for a febrile disease characterized by a rash and polyarthritis, known as Haverhill fever or *Erythema arthriticum epidemicum*. The organism is usually isolated from the blood stream. Treatment with sulphonamides has proved of little value but penicillin has given good results.

THE MORAX-AXENFELD BACILLUS

This organism was described independently by Morax (1896) and Axenfeld (1897) and is responsible for a type of subacute conjunctivitis in man. It usually occurs as a short Gram-negative rod arranged in pairs or short chains, but in old cultures pleomorphism is frequently seen. It is non-motile, non-sporing and non-capsulated.

Growth occurs under aerobic conditions on media enriched with serum, ascitic fluid or blood.

On solid media, such as serum agar, small, greyish, translucent colonies appear in 24 hr; serum is liquefied. In fluid media a general turbidity is at first formed, but later a greyish deposit appears. It is non-pathogenic in animals, but on instillation into the healthy human conjunctiva a definite inflammatory reaction develops in 4–5 days.

Chapter XXX

Spirochaetes

Syphilis, Yaws, Weil's Disease, Seven-day Fever, Relapsing Fever and Vincent's Angina

The term *spirochaete* was first used by von Ehrenberg (1883) to indicate certain elongated, spiral, motile organisms present in stagnant water. Since then many varieties have been described and *spirochaete* now indicates pathogenic and non-pathogenic spiral organisms which are motile without flagella.

GENERAL PROPERTIES

The *spirochaetes* have a spiral, flexible form. They are motile without flagella. Motility is produced by flexion of the body, by rotation around the long axis, i.e. corkscrew motion, or by a general movement of the protoplasm termed translation. Movement takes place equally well in a forward or backward direction; there is no antero-posterior polarity.

There is considerable variation not only in size of the different spirochaetes, but also in the number and regularity of the spirals. Most spirochaetes are not easily stained, and special methods are required for their demonstration. They are best seen in wet preparations by dark-ground illumination.

The multiplication is by transverse division only; there is no sexual phase of reproduction. The spirochaetes do not grow readily *in vitro*, and the ordinary media are of little value for culture. Noguchi (1911) introduced a satisfactory medium, composed of sterile ascitic fluid arranged in long narrow tubes and containing a piece of sterile rabbit-kidney. This is particularly useful for the cultivation of the anaerobic members and may be solidified by the addition of agar for the isolation of spirochaetes from contaminated material. For the more aerobic spirochaetes, such as the blood spirochaetes and *Lept. icterohaemorrhagiae*, success has followed the use of semi-solid agar to which fresh rabbit blood is added immediately before use. In many cases isolation from the tissues is extremely difficult, but when growth occurs subcultures are usually obtained without much trouble.

The resistance offered to destructive agents is similar to that exhibited by the vegetative bacteria. Most spirochaetes are killed by temperatures over 50°C, but can withstand low temperatures. They are very susceptible to drying; this usually accounts for the early death of the pathogenic members outside the animal body.

The serology is similar to that of the bacteria; agglutinins, spirochaeticidins and protective bodies have all been demonstrated. A so-called "adhesion phenomenon", the *Rieckenberg reaction*, is also found; in the presence of the specific antiserum the spirochaetes are altered in such a way that small particles, such as platelets and bacteria, adhere to them.

The spirochaetes have been loosely divided into three groups: (*a*) pathogenic, (*b*) commensal, and (*c*) the free-living. The free-living types are found mainly in fresh- and sea-water containing organic matter. A large number of spirochaetes inhabit the bodies of man and animals as commensals. It appears probable that a number may become pathogenic and be responsible for certain ulcerations of the alimentary tract. Some pathogenic members have been isolated from stagnant water and may also exist as commensals; others, however, are strict parasites. There is thus much overlapping of the various groups.

Various classifications have been suggested, all of which are based mainly on morphological criteria and are therefore open to question. As our knowledge of the spirochaetes is so incomplete a final method of classification is not yet available. A useful and convenient division, which is now widely used, is one in which four main groups are included:

(1) **Spirochaeta**—free-living members with an axial fibre and an unsegmented protoplasm; non-pathogenic to higher animals.

(2) **Cristispira**—free-living members with a band-like membrane or crista running spirally along the length of the body; the body is divided by transverse septa; non-pathogenic to higher animals.

(3) **Treponema** (including the spironema class)—members with a number of primary spirals, an indefinite axial fibre and sharply pointed extremities; some are pathogenic, e.g. *Trep. pallidum*, the causative agent of syphilis.

(4) **Leptospira**—members with closely set, regular, primary spirals, and an indefinite axial filament; the ends are very flexible and are frequently arranged at angles to the main axis; some are pathogenic, e.g. *Lept. icterohaemorrhagiae*, the causative agent of Weil's disease.

The main diseases of man produced by the spirochaetes are:

Syphilis . . .	*Trep. pallidum*
Yaws . . .	*Trep. pertenue*
Relapsing fever . .	*Trep. recurrentis*
Leptospirosis (Weil's disease) . .	*Lept. icterohaemorrhagiae. Lept. canicola*
Vincent's angina . .	*Trep. vincenti*—(doubtful).

SYPHILIS

Syphilis is a disease of some antiquity. The origin of the disease is uncertain, but it appears probable that it was introduced into Europe by Columbus and his crew on their return from discovering the New World. There are several records of the alarming and widespread outbreaks of the disease in Europe during the last decade of the fifteenth century. The disease then presented a clinical picture quite different from that seen at the present time. It ran an acute course with symptoms of marked severity and a high mortality rate in the early stages. This suggests the introduction of a pathogenic organism into a virgin soil. The name "Syphilis" originated at this time as a result of a poem by Fracastor in which the hero, a shepherd named "Syphilis", was smitten with the disease for disrespect shown to the Gods.

Although many reports of the condition had been published in the meantime, the aetiological agent was not detected until 1905. Then Schaudinn and Hoffmann discovered a spiral organism in fluid from chancres and the associated enlarged inguinal glands. Lesions had, however, been previously produced in the rabbit by Haensell (1881), and in apes by Metchnikoff and Roux (1903). Following the demonstration of the aetiological agent, then termed "Spirochaeta Pallida" and now known as *Treponema pallidum,* much progress was made in the study of the disease. Wassermann and his colleagues introduced the serological reaction, universally known as the "Wassermann test", in 1906; Ehrlich (1910) published his results of treatment by salvarsan; and in 1913 Noguchi cultivated the organism from the brain of cases of general paralysis.

Bacteriology. *Trep. pallidum* is approximately 8–14μ in length and contains some 10 regular spirals, the depth and length of each being about 1μ. The spirals tend to be more open and shallower at the ends, which terminate as fine, tapering, pointed threads. Division takes place by transverse fission, Motility is sluggish.

Trep. pallidum is difficult to stain, and is examined most satisfactorily either unstained by dark-ground illumination or by methods which stain the background of the smear but leave the organism unstained, e.g. mixing with indian-ink or Congo-red, spreading and drying. For stained smears, Giemsa's method, which is prolonged for 16–24 hours and colours the organisms pinkish, or special silver methods such as Fontana's, by which the organism is stained black, are used. For sections Levaditi's silver method gives good results.

Cultivation is not easy and some still doubt whether *Trep. pallidum* has actually been cultivated *in vitro*. Spirochaetes morphologically resembling *Trep. pallidum* have been isolated in pure culture from syphilitic lesions, but these have generally proved avirulent on injection into rabbits and monkeys. Anaerobic conditions and an enriched medium containing serum or ascitic fluid are necessary, and growth occurs most favourably at 37°C and with a reaction of *p*H 7·2–7·5; subcultures should be made every 10–14 days.

Trep. pallidum is readily killed by bactericidal agents and is particularly susceptible to drying.

Experimental infection can be produced in monkeys, rabbits and mice by intratesticular, corneal or intradermal inoculation of tissue containing the spirochaetes. The rabbit is the animal of choice, as infection is readily obtained and its course is comparable to that found in the human disease. On intratesticular inoculation, after an incubation period of 2–3 weeks, the testicles become swollen and indurated, the inguinal glands are later involved and a generalized infection follows, i.e. there is a stage analogous to the secondary stage in man. It is, however, important to note that another allied treponema, *Trep. cuniculi*, gives rise to spontaneous lesions in rabbits and, as it is morphologically similar to *Trep. pallidum*, the failure to differentiate these organisms may be a source of irregularities in experimental work.

Pathogenesis. Syphilis in man may be either congenital or acquired.

Congenital syphilis results from syphilis in the mother. The organisms pass from the maternal to the foetal blood by way of the placenta. A syphilitic mother does not necessarily produce a syphilitic infant, as if she is infected very late in pregnancy, or if the disease has reached the latent stage, a healthy infant may result. The explanation is that at these particular stages of the disease the organisms are not usually present in the blood stream and consequently have not the opportunity of reaching the foetus. It is doubtful whether transmission from the father occurs other than by producing infection in the mother primarily and in the foetus secondarily. In congenital syphilis the infection is generalized and the organisms are widely disseminated; lesions may be present in most tissues, particularly the liver, spleen and bones. Death of the foetus not infrequently occurs *in utero*.

In **acquired syphilis** the organisms gain entrance as a rule through the mucous membrane of the urino-genital tract; in some cases the primary infection may occur in the mucosa of the mouth or the skin. In a few instances infection has followed blood-transfusion, in which event the organisms are implanted directly into the blood-stream.

Syphilis is usually transmitted by direct contact during sexual intercourse. It may also be contracted either directly as a result of kissing, wet-nursing and circumcision, or indirectly through the agency of towels, handkerchiefs, pipes, etc. As the life of *Trep. pallidum* outside the body tissues is probably short, cases of indirect transmission are relatively few.

Following the entry of the organisms into the tissues, there is usually an incubation period of some three weeks, but this may vary from 10–90 days. The course of the disease is characterized by its chronicity and is divided clinically into three stages—primary, secondary and tertiary; in many cases these stages are not clearly demarcated.

The "primary stage" consists initially of the appearance of a local lesion or chancre at the site of infection. It is usually single, painless, indurated and tends to ulcerate. When the chancre is well developed there is an associated enlargement of the regional glands.

The "secondary stage" sets in some 6–12 weeks after the appearance of the primary lesion and represents a generalization of the infection. The chief symptoms are sore-throat, cutaneous rash, joint and bone pains, headache, general enlargement of the lymph glands and eye lesions with some pyrexia during the initial phase. This stage may last from a few months to several years and merges imperceptibly into the tertiary stage. In some cases the lesions are so mild that they may escape notice.

The "tertiary stage" is usually observed after a so-called "latent period", in which the infection has localized in some organ where it has been slowly developing. It is manifested by the appearance in any tissue of chronic, progressive and destructive lesions or gummata, which are of a granulomatous nature. When healing occurs marked scarring is produced. This stage lasts indefinitely. In the late phases changes in the nervous system, such as tabes dorsalis or general paralysis, may appear. These late changes are considered by some people to represent a "fourth or a quaternary stage". This is the most disabling stage of the disease, and frequently gives rise to insanity or premature death.

Observations made in both the human and the experimental disease indicate that, while the disease may remain localized clinically during the early stages of the primary stage, the infection actually soon becomes generalized. In the very early stages the organisms can be detected only in the local lesion, but in the later part they can also be obtained from the regional glands and the blood stream. In the secondary stage the organisms are widely disseminated, being present in the blood stream, mucous patches of the throat, rash, semen and lymph glands. As this stage subsides the organisms become increasingly difficult to demonstrate and, in the tertiary stage, they tend to remain localized in the specific lesions. They occasionally generalize and have been isolated from the blood stream during the late stages of the disease.

It is thus important to note that individuals are most infectious during the early stages of the disease when the lesions are relatively mild and may escape notice. In the late stages, when the lesions attract attention, the infectivity of the individual is relatively low. These are points of great importance in the control of the disease.

Immunity. Man has little natural immunity to syphilis, although the marked change in the clinical picture of present-day syphilis from that seen in the Middle Ages suggests that a certain degree of general resistance has developed in man. Nevertheless, all uninfected individuals are susceptible to infection. There is, however, much evidence to indicate that during the course of the disease some immunity of a low grade is acquired against re-infection. There is thus an "infection-immunity" as found in tuberculosis, but while there appears to be increased resistance to re-infection, spontaneous cure of the primary infection does not take place. It has been found in rabbits that with the development of a well-marked primary lesion super-infection of the animal with *Trep. pallidum* becomes increasingly difficult. Infected rabbits, successfully treated in the early stages of the disease with arsenical preparations, can later be re-infected without difficulty, whereas animals treated in the later stages are usually resistant. In man the rarity of re-infection during the course of the disease is well known. It is, however, important to note that the intradermal inoculation of the mouse gives rise to infection without the production of a primary lesion and the spirochaetes have been demonstrated in various tissues, e.g. liver and spleen, several months after the original injection. If such a phenomenon followed the re-inoculation of infected rabbits or man, re-infection would be extremely difficult to detect. It is considered by many workers that the termination of the secondary stage, with the great reduction in the numbers of the organisms in the tissues, is brought about by some form of immunity response by the host. The search for antibodies has, however, yielded little information about the nature of this response in syphilis.

DIAGNOSIS

The bacteriological methods used in the diagnosis of syphilis are (1) microscopy and (2) serology.

(1) Microscopy is the most satisfactory method in the early stages of the disease. The organisms can usually be seen without difficulty in the exudate of the primary or secondary lesions. In the collection of the exudate the surface of the lesion should be cleaned to remove contaminants and the serous fluid obtained; haemorrhage should be avoided. Closed lesions should be punctured and the regional glands aspirated.

The examination is made either under dark-ground illumination or on stained preparations. Under dark-ground illumination *Trep. pallidum* presents a silvery appearance with regular spirals and pointed ends and exhibits rather sluggish motility. Other spirochaetes may be present, particularly if the specimen has been collected from the mouth, and some difficulty may be experienced in identifying *Trep. pallidum.*

Specimens may be stained by Giemsa's or Fontana's method, but a more satisfactory procedure is to mix the exudate with indian-ink, prepare smears and dry. *Trep. pallidum* stands out unstained against a dark background.

(2) Serological tests are of little value in the early stages of the disease, but after the development of the secondary stage they have proved of great clinical importance. Two main tests have been used extensively in the sero-diagnosis of syphilis—complement-fixation (Wassermann test) and flocculation (Kahn test). Tests involving the *Trep. pallidum,* e.g. T.P. Immobilization Test and T.P. Complement-Fixation Test, are useful in doubtful cases.

Wassermann Test. The rationale of this test may more easily be understood by a preliminary recapitulation of the essential features of the complement-fixation test. Certain antigens in the presence of the homologous antiserum are able to fix complement; red blood corpuscles (R.B.C.) in the presence of the inactivated homologous antiserum are unchanged, but when complement is also added lysis takes place. Sensitized R.B.C. consequently constitute a satisfactory indicator for the presence of complement. Antigen, inactivated patient's serum and complement are incubated at 37°C for one hour, sensitized R.B.C. are then added, the mixture is re-incubated for a further hour and the presence or absence of lysis determined.

While the above principles hold for the Wassermann test, it is important to note that the reaction is not specific in the immunological sense. When Wassermann and his colleagues (1906) first applied complement fixation to the diagnosis of syphilis, they employed as antigen a suspension of a syphilitic liver. It was, however, later found that satisfactory results were obtained by the use of an extract of normal liver, which naturally did not contain *Trep. pallidum.* Further observations indicated that alcoholic extracts made better "antigens" than the watery extracts and that the reacting substance was contained in the lecithin fraction of the lipoids and was soluble in alcohol. Some time later it was shown that the "antigen" could be rendered more sensitive by the addition of an alcoholic solution of cholesterol. The "antigen" in general use at the present time is prepared by adding a 1 per cent alcoholic solution of cholesterol to an alcoholic extract of normal human heart muscle and diluting rapidly with saline to give a turbid suspension.

The antigen is thus non-specific, but the test exhibits a high degree of disease-specificity. The explanation of this is not known. The nature of the reacting substance in the serum has not been determined; it is uncertain whether it is a true antibody or a specifically changed property of the serum. The reacting substance may be a tissue disintegration product resulting from the multiplication and activity of *Trep. pallidum* or it may be a true antibody stimulated by the presence of these disintegration products. Another theory suggests an analogy to the stimulation of the heterophile antibodies.

The original test has been modified by many workers, and at the present time it is improbable that the exact technique of any two laboratories is identical. It is usual to carry out a quantitative as well as a qualitative test. In order to do this it is possible to vary the quantities of either antigen, complement or patient's serum. In practice the patient's serum is generally varied and the other factors standardized.

Despite the limitations of our knowledge concerning the mechanism of the reaction, in the hands of experienced workers the results are extremely reliable. The test has consequently been one of the most widely used of the serological diagnostic reactions. In the primary stage of syphilis the reaction is frequently negative. With the development of the secondary stage, and in the absence of treatment, a positive result is obtained in 90–100 per cent of cases. A high percentage of positive results is also obtained when active lesions are present in the tertiary stage, but negative results may be given during the so-called "quiescent" or "latent" period. In cases of neurosyphilis the reaction is usually given by both serum and cerebrospinal fluid. It is, however, accepted that the reactions given by the serum and the C.S.F. may not run parallel; this is particularly so in treated cases.

Cases rigorously treated in the early stages may never give a positive Wassermann reaction. It thus follows that while a positive diagnosis is strongly indicative of the presence of syphilis, a negative test does not exclude it. A persistent negative reaction in a suspected case of secondary syphilis would, however, almost certainly exclude the disease. In all instances it is essential to correlate the clinical and laboratory evidence before making a diagnosis.

A positive Wassermann reaction is obtained in yaws and pinta (both Treponemal diseases) and may be given in other diseases, e.g. febrile conditions, especially chronic ones such as malaria and leprosy, after vaccination with calf-lymph, and also in "collagen diseases". A positive result is sometimes given by a healthy person with no history or evidence of infection; this is known as a biological false positive, the explanation of which is still unknown and it may be associated with pregnancy. The results are therefore sometimes difficult to interpret and other tests such as the treponemal immobilization test are then valuable.

Flocculation Tests. Many flocculation tests have been introduced for the diagnosis of syphilis: Sachs-Georgi, Sigma, Kahn, Meinicke, etc. Of these, the Kahn test is used most widely at the present time. In all two reagents only are required. This is an obvious advantage over the complement-fixation test. Several slide techniques have been introduced but these tend to lack specificity and are used mainly for preliminary screening purposes.

Kahn Test. This is used extensively in America. The antigen consists of an alcoholic extract of the ether-insoluble lipoids of beef heart. The inactivated serum is added to the antigen-suspension and the mixture is shaken vigorously for three minutes. After standing for three minutes saline is added and readings taken; the presence of flocculation indicates a positive reaction.

The test is probably as reliable and sensitive as the Wassermann reaction, and, as it is simpler and quicker to carry out, the latter test has been superseded in some laboratories; the main disadvantage of the Kahn test is that doubtful results are sometimes difficult to read. It has been recommended that both complement-fixation and flocculation should be used in the sero-diagnosis of syphilis. Although the two tests agree in a high percentage of cases, there is not complete overlapping and it is possible for cases to be missed if only one test is used. In practice the use of both tests is not always possible, but it is certain that, when unexpected or unsatisfactory results are obtained by one method, the other should always be tried.

The Treponema Immobilization Test (T.P.I. Test). This test requires live spirochaetes which are maintained by intratesticular passage in rabbits. When required for the test they are carefully teased out to give a saline suspension, which is then mixed, in definite proportions, with the patient's serum and complement. After incubation at 37°C for 18 hours under anaerobic conditions, wet preparations are examined under the microscope for the presence or absence of motility. A normal serum control is also prepared.

This is a specific treponemal test and in the presence of syphilis, immobilization of the spirochaetes takes place. It is therefore being demanded with increasing frequency to determine the specificity of a positive Wassermann Reaction when the clinical diagnosis is obscure.

This test is technically difficult and potentially dangerous; therefore in order to obtain a more useful test for routine purposes, attempts have been made to elaborate a "Treponemal Complement Fixation Test". A satisfactory technique has been developed using a treponemal protein as antigen. Like the T.P.I. test this is of value in the study of biological false positive reactions.

Prophylaxis. There are no specific measures of prophylaxis. The prevention of syphilis is an important and difficult problem of public health administration. Such questions as the regulation of prostitution, education of the public and the compulsory notification of the disease have been advocated, but unanimous opinion has not been obtained. There was a big increase in the incidence of syphilis during the late war as a result of war-time conditions and consequent moral relaxation, but there has been a steady and definite fall during recent years. *Trep. pallidum* is harboured almost exclusively by the actual cases of the disease, early and adequate treatment of which would naturally decrease the chances of dissemination of the spirochaetes.

THERAPY

Arsenical therapy, introduced by Ehrlich (1910), has given good results in the early stages of the disease, but it was necessary to enforce prolonged courses of treatment. In the later stages the results were less satisfactory. Salts of mercury and bismuth have also proved useful therapeutic agents. In neurosyphilis, particularly general paralysis, malarial infection has given beneficial effects, mainly the result of the induced pyrexia.

Treatment with penicillin is now the routine practice; large doses are essential. The slow release products, procaine or benezathine penicillin, are frequently used; a total of 4–6 million units, in divided doses for 8–10 days, by the intramuscular route is required. Treatment may be combined with arsenic or bismuth but this is rarely necessary. Other antibiotics are not so effective; tetracycline or chloramphenicol is sometimes used for patients who are allergic to penicillin.

It is important to note that the rendering of a positive serum-reaction negative by treatment does not necessarily indicate a cure; treatment should be continued until all signs of the disease have cleared up and the serological tests remain persistently negative. A complete cure is extremely difficult to determine. There is, however, no doubt that syphilis is curable if proper treatment is carried out in the early stages of the disease.

YAWS

Yaws is a tropical disease found mainly in the children and young adults of rural districts. The lesions are found on the skin, and consist of papules which develop into an encrusted granulomatous eruption; apart from the spleen, the internal organs are not generally affected. Yaws is not usually a venereal disease; it is transmitted by direct contact, the organism gaining entrance by open wounds or slight abrasions.

The causative organism is a spirochaete, *Trep. pertenue*, first described by Castellani in 1905. It is morphologically indistinguishable from *Trep. pallidum*, and it has been suggested that the two organisms are related, i.e. yaws is a modified form of syphilis. Conclusive evidence of this relationship has not, however, been produced. It is well known that patients suffering from yaws give a positive Wassermann reaction, but cross-immunity tests indicate that the organisms are not absolutely identical. Patients with yaws have developed syphilis and *vice versa*. Experiments in monkeys have yielded the same results; a definite immunity was not produced by one organism against the other. It has, nevertheless, been noted that syphilis is uncommon in individuals who have suffered from yaws in childhood, e.g. syphilis is rare in the Fiji Isles where yaws is rife. Further work is required before the relationship between yaws and syphilis can be definitely determined.

Penicillin has proved a useful therapeutic agent.

LEPTOSPIROSIS (WEIL'S DISEASE)

Leptospirosis or Weil's disease is a widespread disease occurring mainly in sewage workers, miners, agricultural workers in damp regions and in soldiers during war-time conditions; it may, however, occur in any individual coming in contact with stagnant water in rat-infested districts and it has been contracted by bathing in canals and rivers. The disease has a world-wide distribution and cases have been reported in various parts of this country; several cases have been reported among sewage workers in London, while an outbreak at Aberdeen was recorded in 1934 among fish cleaners working in rat-infested buildings. Further investigations by Smith and Davidson have shown that the condition is particularly prevalent among the fish workers in Aberdeen; serological tests were carried out on 210 workers in the fish-cleaning trade, and some 25 per cent gave a positive result, indicating previous infection. A large number of controls were tested and all gave a negative result. A common aetiological agent, *Lept. icterohaemorrhagiae*, was first discovered in cases of the disease by Inada and his colleagues in Japan during 1915.

Some cases have also been reported of human infection with allied spirochaetes. Some 40 varieties have been identified by serological methods, and they are unevenly distributed throughout the world. In this country *Lept. icterohaemorrhagiae* and *Lept. canicola* are the types usually seen. *Lept. canicola* is primarily a pathogen in dogs, giving rise to nephritis leading to uraemia. *Lept. grippo-typhosa* is a pathogen of rodents and was responsible for a few cases of Weil's disease during the Normandy campaign of World War II. In man the clinical signs of infection with these organisms are similar to, but, milder than, those found in infection with *Lept. icterohaemorrhagiae.*

Bacteriology. *Lept. icterohaemorrhagiae* varies in length, the average being 7–14μ, and contains a number of closely wound spirals, each about 0·25–0·5μ in length. It has tapering ends which are frequently set at an angle to the main axis, giving to the organism the form of the letter C, L or S. It exhibits active motility.

Growth is obtained without much difficulty on artificial media, such as semi-solid blood agar. Multiplication occurs at or near the surface in 6–8 days when incubated at a temperature of 30°C. Cultures can be kept at room temperature for 4–6 weeks.

TRANSMISSION

Lept. icterohaemorrhagiae has not only been isolated from cases of Weil's disease, but it has also been established that the organism is a frequent pathogen of rats, and other rodents, by which it is commonly excreted in the urine. The number of infected rats, *R. norvegicus* (the grey sewer rat), caught during examinations of the wild-rat population in different countries has varied from 10 to 30 per cent. The disease in the rat is comparatively mild, jaundice is not found. Another interesting point is that *Lept. icterohaemorrhagiae* can survive in stagnant water, under favourable conditions, for several months.

These observations indicate the probable mode of transmission of the infection in man, which is generally considered to take the following course. The disease is primarily one of rodents, particularly the grey rat, which excretes the organism in the urine. In this way the organism is widely disseminated, and in badly drained, rat-infested areas stagnant water becomes readily contaminated. Individuals working in these regions become infected as a result of the contaminated material coming in contact with cuts, scratches or minute abrasions on the exposed skin, the nasal mucosa or conjunctiva.

After an incubation period of 5–7 days, there is an initial febrile stage of some 7 days, during which the main signs are headache, prostration and muscular pains, with marked irregularity of the temperature; meningitic symptoms are sometimes found. The condition may then settle down with complete recovery; in a minority of cases it passes on to an icteric stage, which usually lasts from 7–13 days. There is either a gradual subsidence of the jaundice with the commencement of convalescence or death supervenes as a result of haemorrhage or renal damage. The mortality rate among the diagnosed cases has varied in different outbreaks from 6–32 per cent.

It is important to note that the appearance of jaundice is not an essential feature of the disease, and that in exposed workers many influenza-like illnesses are probably caused by a leptospiral infection. This is strongly suggested by the findings of Smith and Davidson (1936), who consider that infection with *Lept. icterohaemorrhagiae* occurs in three grades: (*a*) severe infection with jaundice, (*b*) mild pyrexial infection without jaundice, and (*c*) latent infection without clinical manifestations. It is now accepted that meningeal and eye symptoms may occur and occasionally pure meningeal forms of the disease may be encountered, these may simulate meningitis of virus or tuberculous origin.

The condition can be reproduced experimentally in guinea-pigs, which are highly susceptible. After intraperitoneal infection of the organism the main symptoms are fever and jaundice; the temperature falls about the fourth day when the jaundice appears; the disease terminates in death after 5–10 days. Infection may also be obtained by rubbing infected material on the skin or conjunctiva. On autopsy there is generalized jaundice with multiple haemorrhages and enlargement of the spleen; the spirochaetes are most numerous in the liver. Other animals are relatively resistant but hamsters are useful in the case of *L. canicola*.

DIAGNOSIS

The distribution of the organisms in the body depends on the stage of the disease; during the first stage they are widely distributed and can be readily demonstrated in the blood stream. After seven or nine days they become increasingly difficult to find in the blood, but can be demonstrated in the urine, in which they are most numerous about the third or fourth week. They then gradually disappear and are rarely found after the fifth week. In individuals dying during the first week spirochaetes can be found in many tissues, afterwards they are most numerous in the kidney.

The bacteriological methods employed in the diagnosis of Weil's disease are (*a*) microscopy, (*b*) pathogenicity tests, and (*c*) serology.

(*a*) Microscopical examination is usually carried out on unstained material under the dark-ground illumination. The spirochaetes can be demonstrated in the blood during the first week, but the most satisfactory results are obtained with the urine after the initial stage of the infection. Attention is directed to the presence of the closely wound spirals and hooked ends, but atypical forms are frequently seen, particularly with urine having a strongly acid reaction.

Tissues collected at autopsy are most satisfactorily stained by Levaditi's method.

(*b*) Pathogenecity tests are carried out in the guinea-pig or the hamster. During the first week blood and during the later stages the urine are injected subcutaneously or intraperitoneally; 2–5 ml are injected, and if the spirochaetes are present a typical infection is produced.

(*c*) Various serological tests have been devised—agglutination, complement-fixation, lysis, neutralization and the demonstration of the adhesion phenomenon. An agglutination test introduced by Schüffner has proved very successful; dilutions of the patient's serum and a suspension of the leptospira are mixed in a small sterile container and then incubated at 32°C for three hours. A drop of the mixture is then examined by dark-ground illumination for lysis and agglutination of the leptospira. Agglutination is usually seen in the higher and lysis in the lower dilutions of the serum. The responsible antibodies have been demonstrated after the seventh day of the illness. This method of examination is particularly valuable in those cases settling down after the febrile stage and not manifesting jaundice. A macroscopic agglutination test, in which young actively growing cultures of leptospirae are used, has also been tried with success. The mixture of serum and culture is incubated for three hours at 37°C, followed by 30 minutes at 55°C. The results are read with a hand lens. A complement fixation test is now available with a leptospiral suspension as antigen.

If negative results are obtained in suspected cases similar tests should be carried out with *Lept. canicola* and *Lept. grippo-typhosa*, which have antigenic differences from *Lept. icterohaemorrhagiae*.

The identification of *Lept. icterohaemorrhagiae* in samples of stagnant water, slime, etc., is more difficult, as many closely allied but non-pathogenic forms may also be present in these specimens. Isolation is most readily carried out by the injection of guinea-pigs, in which *Lept. icterohaemorrhagiae* produces the characteristic infection.

PROPHYLAXIS

Important measures in the prevention of Weil's disease are the suppression of rats, attention to badly drained areas, use of protective clothing and the treatment and protection of wounds, scratches, etc., in individuals likely to be exposed to infection. Attention to sick dogs is important in the case of *Lept. canicola*; infected animals may excrete the organisms in the urine for many months after recovery.

THERAPY

Arsenical compounds have proved unsatisfactory as a therapeutic measure, but favourable results have been reported following the use of immune horse-serum. All types of *Leptospira* are sensitive to most antibiotics and it has been claimed that penicillin in large doses gives satisfactory results, but the figures so far produced are not very convincing.

SEVEN-DAY FEVER OF JAPAN

A febrile condition lasting about a week is encountered in Japan chiefly in agricultural workers; jaundice is uncommon, and the death-rate is nil. The aetiological agent, *Lept. hebdomadis*, is closely related to *Lept. icterohaemorrhagiae*, from which it can be distinguished by serological tests. It is present in the blood during the first week of the illness, after which it passes to the kidneys and can usually be obtained from the urine from the eighth to the fortieth day after the onset. The organism is pathogenic to young guinea-pigs, producing enlarged glands, slight jaundice and a mortality rate of about 60 per cent. The vector of the organism appears to be the field-mouse.

Spirochaetes have also been isolated from an allied febrile condition in Japan known as "autumnal fever". Two serological types have been distinguished—one is identical with *Lept. hebdomadis*, the other is more virulent and has been termed *Lept. autumnalis*.

RELAPSING FEVER

Relapsing fever is the name given to a group of closely related diseases which are widespread in tropical and sub-tropical regions. The disease is characterized by the sudden onset, severe frontal headache, fever and muscular pains. The incubation period may vary from 3–10 days, and the fever generally persists for some five days, when it falls by crisis. After remaining normal for about a week there is a recurrence of the symptoms; there may be as many as ten relapses before the condition finally settles down. The mortality rate is usually low, about 5 per cent. In fatal cases the post-mortem findings are not particularly characteristic.

The aetiological agent of European relapsing fever was first observed by Obermeier (1873) in the early days of bacteriology, during an outbreak in Berlin. He considered the organism to be a spirochaete; this is now termed *Trep. recurrentis*, and is responsible for European relapsing fever. Subsequently a number of similar but serologically distinct organisms have been described in outbreaks of relapsing fever in different countries. These are considered to be distinct species and have been given separate names, such as *Trep. carteri* of the Indian disease, *Trep. duttoni* of the African, *Trep. novyi* of the American, and others.

All present a similar morphological picture. The average length is about 14μ, and each organism contains some five to seven loose spirals, which have a wavelength of 2–3μ. The ends are pointed and motility is active. *Trep. recurrentis* were first cultivated by Noguchi (1912) in ascitic fluid with a small piece of fresh rabbit kidney. Growth takes place most readily at 30°C under micro-aerophilic conditions and at a *p*H 7·4.

TRANSMISSION

Relapsing fever is transmitted by the agency of blood-sucking insects, particularly ticks and lice. Infection may be introduced directly into the tissues by the bite, but generally it follows contamination of wounds, caused by scratching or biting, with the excreta of the insect. The tick, once infected, may remain so throughout life, and the infection is transmitted through several generations. Monkeys are susceptible to infection, and a condition similar to that seen in humans may be produced by intraperitoneal, subcutaneous or intradermal inoculation of the organisms. Rats and mice are more resistant, while guinea-pigs and rabbits are quite refractory.

The spirochaetes are present in the blood in large numbers during the febrile stages, but as these subside the organisms become increasingly difficult to demonstrate. During the quiescent period they apparently lie dormant in the tissues, but they reappear in the blood during the relapses. The blood can be examined either as a stained film or by dark-ground illumination, when the actively motile spirochaetes, in some cases straight and in others bent and resembling the letter S or C, are seen. For stained preparations Giemsa's method is satisfactory—the spirochaetes are stained blue, but in some cases the staining is irregular.

IMMUNITY

That the immunity following an attack of relapsing fever is only of a low order of efficiency is indicated by the fact that second attacks with the same treponema are not infrequent after 1–2 years. Antibodies—agglutinins, complement-fixing bodies and lysins—have, however, been demonstrated in the serum of recovered individuals. These antibodies are specific for the different treponemata responsible for the various types of relapsing fever; there is no marked cross-immunity.

The development of antibodies during the attack is considered to be responsible for the termination of the febrile periods. As antibodies are produced in sufficient quantities the spirochaetes are rapidly agglutinated, disintegrated and ingested by leucocytes. Within a short time they disappear from the blood stream and the temperature falls by crisis. Experimental work indicates that during the quiescent period the spirochaetes are present in such tissues as the brain, where, owing to the relatively feeble blood supply, there is a low concentration of antibodies. At the end of the apyrexial period the antibodies in the blood have decreased and the spirochaetes re-infect the blood and a similar cycle of events follows. The number of spirochaetes in the blood tends to decrease with each successive relapse until the immunity produced is sufficient to prevent further invasion. It is, however, uncertain whether a true spontaneous cure actually takes place.

PROPHYLAXIS

Measures adopted to prevent relapsing fever are attention to the general sanitation of the environment, personal cleanliness, and the avoidance of bites by ticks, lice, etc.

TREATMENT

Relapsing fever responds very well to the administration of arsenical compounds. A suitable dose may cure the disease. Some strains of the relapsing fever spirochaetes have become arsenic-resistant, and for such infections potassium bismuth tartrate is recommended. Penicillin and other antibiotics have given encouraging results in some cases.

VINCENT'S ANGINA

Vincent's angina is an inflammatory condition of the mouth and throat, in which extensive ulceration and necrosis of the mucous membrane are generally found. The aetiology is uncertain, but an organism described by Vincent (1896), and now known as *Trep. vincenti*, is usually demonstrated, often in considerable numbers, in the lesions, together with a large fusiform bacillus (cf. p. 317). *Trep. vincenti* is a thin organism (about 5–10μ in length) containing some three to eight irregular spirals; it is actively motile. It is Gram-negative, but does not stain very deeply or evenly by this method. It can be cultivated anaerobically at 37°C on serum-agar, forming small, tenacious colonies. The pathogenicity of these two organisms is doubtful, and their *rôle* in the aetiology of Vincent's angina has not been definitely established. Similar organisms may be present in the mouths of normal individuals but never in the great numbers which are found in cases of Vincent's angina. In the tonsillar lesions Gram-negative spirochaetes and fusiform bacilli may be the only organisms seen. In cases of acute ulcerative gingivitis, unless the swab is carefully collected, these organisms may be masked by the numerous organisms usually associated with the gum margins. Lesions are not produced on subcutaneous injection into guinea-pigs. It has been suggested that the spirochaete and the fusiform bacillus represent different phases in the life cycle of one organism, but this does not appear probable on the evidence available. They appear to be distinct organisms which thrive in symbiosis.

TREATMENT

Cases tend to clear up satisfactorily with local treatment in the form of mouth washes and application of chromic acid solutions to the lesions, combined, in severe cases, with intravenous arsenical injections. The local application of penicillin in the form of lozenges has given good results when accompanied by adequate surgical treatment.

Chapter XXXI

Viruses

Many diseases of man, animals, insects and plants, some of which exhibit a marked degree of infectivity, are caused by agents which cannot be studied by the ordinary bacteriological methods. These disease-incitants have certain properties which distinguish them from the ordinary bacteria; they are not visible by ordinary microscopy, they do not grow under the usual conditions of cultivation and they tend to pass through the bacterial filters. They thus constitute a distinct group of disease-agents and they have been designated filterable viruses, or, more simply and correctly, viruses.

Viruses have long been associated with the problems of immunity. In fact, active immunization was first widely used for the prevention of human disease in the case of two virus infections, viz., small-pox by Jenner (1800) and rabies by Pasteur (1885). Although the practice was for many years empirical, Pasteur had indicated the probable nature of the rabies virus when he postulated that he was dealing with an organism which he was unable to see with the microscope then available. Further development in the knowledge of the viruses was uneventful until Iwanowski (1892), a botanist investigating the mosaic disease of the tobacco plant, discovered the filterable nature of the responsible agent. These observations were confirmed later by Beijerinck (1899), who considered that he was dealing with a "Contagium vivum fluidum".

About the same time a virus was discovered to be responsible for a disease of animals, when Loeffler and Frosch (1898) found that foot-and-mouth disease could be reproduced by a bacteriologically sterile filtrate of the vesicular lymph from the specific lesions. These observations stimulated the interest of many bacteriologists, and viruses have since been incriminated in the aetiology of many human and animal diseases. Nevertheless, it is only in recent years that any definite advance has been made in our knowledge of the nature of these agents, and this is still incomplete. Many methods have been used in their study, but only those yielding useful information need be considered here.

FILTRATION

As the original designation of these agents indicates, filtration has been extensively employed in their study. It was early recognized that an important criterion of the viruses was their ready passage through bacterial filters. This criterion still holds for general application, but certain reservations are necessary. The process of, and the factors involved in, filtration have already been discussed. It is sufficient now to point out that filtration is not a simple mechanical action governed only by the relative sizes of the pores of the filter and the particles to be filtered. Many factors are concerned and the most important are: the composition and electrical charge of the filter; the pH and electrical charge of the material undergoing filtration; the amount of extraneous material in the suspension; the duration of filtration; the pressure employed; the temperature at which it is carried out. Viruses may be adsorbed on to the filter or on to albuminous material in the suspension and so fail to pass through a filter. Irregular results have been due to some variation in the filtration technique. The value of filtration in the study of viruses is therefore limited. The process cannot be used to determine with any accuracy the size of the virus particles, but by the use of the various grades of filter a means of obtaining a rough estimate of their dimensions is available.

Many viruses readily pass through the usual bacterial filters, such as Berkefeld "N" and Seitz "EK" types; others, e.g. the vaccinia and herpes viruses, only traverse with difficulty the coarser types, such as the Berkefeld "V", when special precautions are taken. Some smaller bacteria and various spirochaetes may pass through the coarser filters more readily than the larger viruses. It consequently follows that filtration alone is not an absolute criterion for the identification of a virus.

ULTRA-FILTRATION

More precise studies of the size of different viruses have been made by the use of ultra-filtration, which was first employed for the study of colloidal suspensions by Bechhold (1907). This process involves the use of collodion membranes, which are less influenced by outside factors than the bacterial filters. Much of the early application of these membranes to the study of the viruses was unreliable owing to the use of faulty membranes, but in recent years a successful technique has been introduced by Elford. Some of the main points of this are: the grading of acetic acid collodion membranes in terms of their average pore diameters, constant area of filtering surface, use of constant pressure and membranes supported on a perforated plate.

Many viruses have been studied by this technique, and marked variation has been found in their approximate sizes:

psittacosis virus—0·275μ*	= 275mμ*
vaccinia virus—0·125–0·175μ	= 125–175mμ
influenza virus	= 100mμ
poliomyelitis virus	= 8–12mμ
foot-and-mouth virus	= 8–12mμ

The minute size of the smallest viruses is more easily appreciated when it is realized that the size of a molecule of oxyhaemoglobin is 3–5mμ, while the average size of the staphylococcus is 1μ= 1,000mμ.

The particulate nature of viruses is also indicated by the results obtained from centrifugalization—viruses have been deposited from virus suspensions by prolonged high-speed centrifugalization. Various types of centrifuge are available for this purpose and estimations of the particle dimensions of several viruses have been made. These figures agree very closely with results obtained by ultra-filtration, ultra-violet light photography and electron microscopy.

MICROSCOPY

The small dimensions of viruses indicate that they are beyond the ordinary range of microscopic vision. The limit of resolution using dark-ground illumination and visible rays is approximately 0·2μ; under the usual laboratory conditions, however, the figure is probably never less than 0·25μ. Smaller objects may, however, be rendered visible by using photography with ultra-violet light and a special optical system of quartz lenses. As the results of ultra-filtration have shown that the greatest dimension of most viruses is no more than 0·15–0·175μ, it follows that the use of ordinary microscopical methods is of little avail.

Microscopy has three main applications in the study of viruses: (1) to observe the nature of the lesions produced in the various diseases, (2) to detect the presence of inclusion bodies, and (3) to determine the presence of elementary bodies.

(1) The nature of the lesions produced by the viruses may indicate the identity of the causative agent. Some viruses exhibit a marked affinity or tropism for special tissues, e.g. the affinity for nerve-cells or neurotropism of the viruses of rabies and poliomyelitis. During the reaction between the virus and the tissue-cells characteristic changes are frequently produced. In some cases, e.g. warts and molluscum contagiosum, there is a distinct

* μ or 1 micron = 1/1,000 mm. or a 1/25,000 inch. mμ or millimicron = 1/1,000 μ.

tendency to cell proliferation; in others, e.g. small-pox and poliomyelitis, there is an inflammatory reaction with accompanying degeneration and destruction of the tissue-cells.

(2) The presence of abnormal bodies in the affected cells of many virus diseases has long been observed; these bodies have been termed inclusion bodies. They may be present in either the cytoplasm or nucleus of the cells and are called respectively cytoplasmic or intranuclear inclusions. Inclusion bodies can frequently be demonstrated by ordinary staining methods, such as haematoxylin and eosin, but in a few cases a special staining technique is required. They vary considerably in size and shape. They are often of great diagnostic value, e.g. the Negri bodies in rabies, but their nature and composition remain unsettled. There is now much evidence to indicate that in some cases at least, they are composed of large numbers of elementary bodies held together by a lipoidal matrix.

(3) In recent years attention has been re-directed to the presence of minute bodies in the exudates formed in certain virus diseases; these had been observed by Borrel (1903) in fowl-pox, and by Paschen (1906) in vaccinial lesions, and were considered by these workers to be aetiologically related to the respective diseases. These elementary bodies have since been demonstrated in the lesions of many conditions, such as vaccinia, varicella, herpes zoster, fowl-pox and psittacosis. The bodies are agglutinated by the serum of animals recovered from the homologous disease. Pathogenicity tests indicate that they are the aetiological agents of the various diseases and there is little doubt that elementary bodies are the actual viruses. Elementary bodies are stained with difficulty; it is necessary to use either the prolonged Giemsa's method or a special technique. When stained they come within the limits of microscopical resolution, probably through the deposition of the stain on the virus (Fig. 1.XXXI). Their relation to the inclusion bodies has already been mentioned.

The electron microscope, with which magnifications of 1–200,000 diameters can be obtained, is now widely used for the study of viruses. Observations indicate that viruses have a definite but heterogeneous structure, e.g. the vaccinia virus is rectangular, resembling a brick, the influenza virus is spherical, but filaments may also be seen under certain conditions, while the tobacco mosaic virus is rod-shaped, approximately $280 \times 15\text{m}\mu$.

CULTIVATION

The methods of cultivation used for obtaining growth of the bacteria have proved unsuitable for the multiplication of viruses.

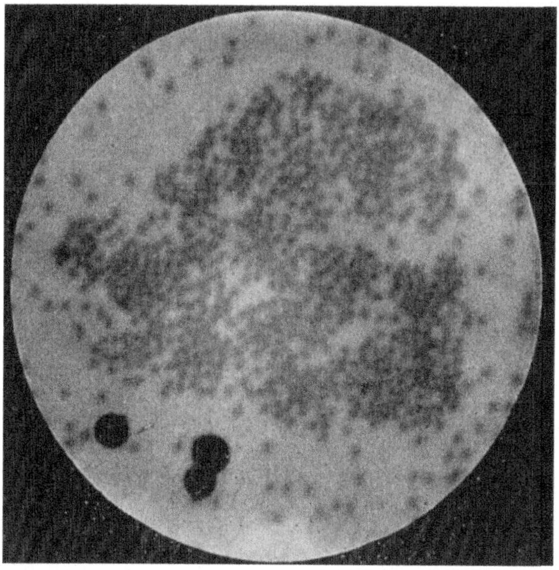

Fig. 1.XXXI. Elementary bodies.—The Borrel bodies of fowl-pox with three staphylococci; note that the diameter of the staphylococci, approximate 1 μ, is about 5 × that of the Borrel bodies (Giemsa). × 4,000.

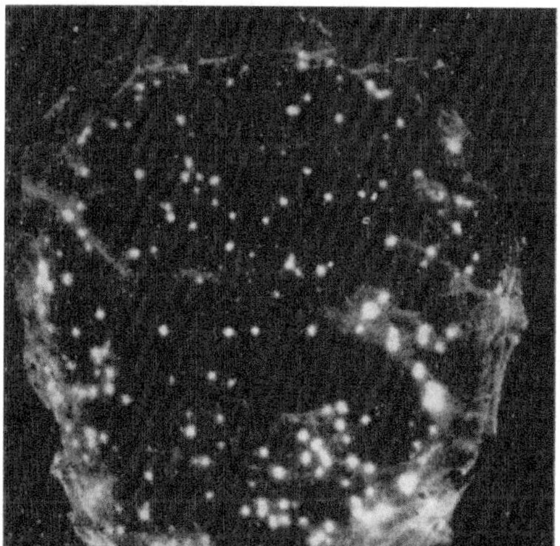

Fig. 2.XXXI. Growth of the variola virus in a chorio-allantoic membrane; note the discrete, opaque lesions. × 1·5.

[*facing page 342*

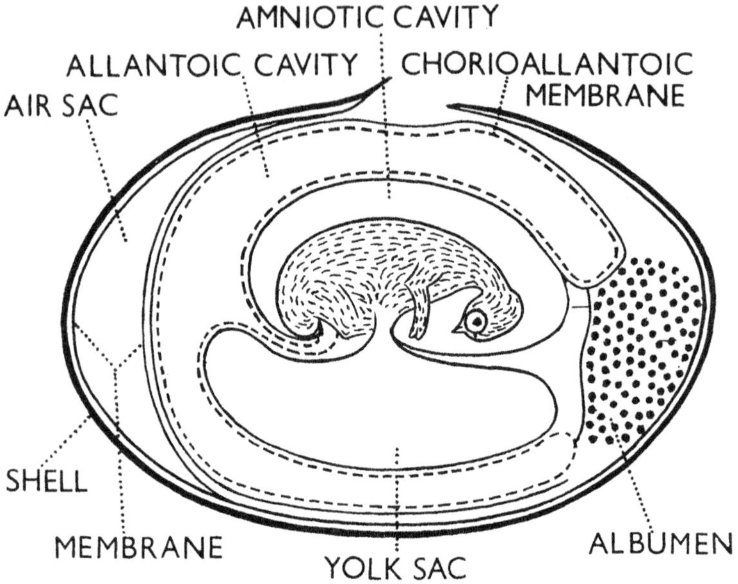

Fig. 3.XXXI. Diagrammatic representation of the developing chick embryo (10th day) prepared for the inoculation of the chorio-allantoic membrane.

At one time it was thought that actively growing cells were always necessary for virus growth, however Maitland in 1929 found that growth of the vaccinia virus took place in a medium composed of minced rabbit kidney and serum; no evidence of multiplication of the tissue cells was obtained. This indicated that growth of the tissue-cells was not essential for virus multiplication. Many viruses have since been cultivated in this, or a similar, medium.

A further advance was the cultivation of viruses on the chorioallantoic membranes of the developing chick-embryo (10–12 days old) (Fig. 3.XXXI). The position of the membrane is determined by means of trans-illumination in a darkened room, and a flap is made in the shell over the membrane by means of a file or other suitable instrument. The flap is turned back and the virus-suspension dropped on the membrane; the flap is then replaced and sealed with melted paraffin wax. After incubation at 37°C for 2–3 days, distinct opaque lesions can be seen and the membrane is dissected out, ground up in saline and tested for the presence of virus by infectivity tests. This method is simple and has proved successful with the viruses of influenza, vaccinia, variola, herpes, psittacosis, Rift Valley fever and many animal diseases (Fig. 2.XXXI).

Other developments have been the cultivation of viruses in the yolk-sac and amniotic sac of the chick-embryo. Inoculation of the yolk-sac is a relatively simple operation which is now widely practised. A burr hole is made through the shell of a 6-days' fertile egg, over the air sac, and the material is injected by means of a sterile syringe, into the yolk-sac. The eggs are candled daily and when dead, the sacs are harvested. Inoculation of the amniotic sac is a difficult technical procedure and requires much experience; it is consequently not used for routine work.

One of the most important developments in virus work has been the great advance in culture techniques. In 1948, Enders and his colleagues reported that they had grown the poliomyelitis virus in tissue culture. This stimulated considerable interest and culture techniques are now widely employed; as a result many new viruses have been isolated.

There are several essential requirements for the successful isolation of viruses by the new culture methods—(1) a basic salt solution to maintain a suitable pH, (2) an indicator to determine alterations in pH, (3) serum and special synthetic salt solutions to promote the growth of the tissue cells, (4) antibiotics, particularly penicillin and streptomycin, to prevent the growth of contaminating bacteria and (5) suitable cells—the cells normally employed are either fresh monkey or human kidney tissue or human amnion, or special "Hela" cells, which are a special strain of epithelial cells originally obtained from a case of cancer of the cervix and have since been maintained by continuous subculture.

The cells are broken down by trypsinization into small clumps and are then added to roller tubes or other containers. Under suitable conditions, cell-growth occurs in 7–10 days as a confluent sheet layer (monolayer) and the tubes can then be seeded with the virus. The multiplication of the virus is detected by the cytopathic changes in the cells, sometimes specific for a particular virus, and/or by changes in the pH. The cytopathic changes can be readily detected under the low power of the microscope—often gross changes are present.

Tissue cultures are now widely used for diagnostic purposes, the production of virus for vaccines and antibody assays.

It is important to note that definite multiplication of viruses has only been obtained in the presence of living cells and that cultivation *in vitro* has not been successful with all viruses. Some viruses appear to produce infection only in man and their study is rendered extremely difficult.

RESISTANCE

The viruses tend to behave as the vegetative bacteria in their resistance to destructive agents. As a rule they are readily destroyed by heat and oxidizing agents, but exhibit a marked resistance to drying, low temperatures, glycerine and weak concentrations of phenol (0·5 per cent). In many cases viruses may be conserved in 50 per cent glycerine at 0°C for considerable periods. Under these conditions the poliomyelitis virus has survived over eight years.

IMMUNITY

Recovery from most virus diseases is accompanied by a high degree of immunity; second attacks are very uncommon. This resistance to re-infection is frequently associated with the presence of antibodies in the serum. These antibodies can also be produced by experimental infection or by immunization of animals with the modified virus. The precise nature and mode of action of the antibodies are unknown, but the available evidence indicates a close resemblance to the bacterial antibodies. It has been suggested that they act on the tissue-cells rather than on the actual virus, but there appears to be little justification for this view.

The demonstration of antibodies can be carried out by means of the neutralization test. In this test a potent virus is mixed and left in contact with (a) the serum and (b) normal saline; the various mixtures are then injected into suitable animals or tissue-cultures. Those receiving the mixture (b) should exhibit the characteristic lesions, whereas those receiving mixture (a) remain unaffected if antibodies are present in the serum.

Complement-fixation, haemagglutination and flocculation tests are also carried out with many viruses, e.g. those of vaccinia, herpes, influenza, meningitis and psittacosis. Two samples of serum are required, one collected in the acute stage (about third day) and the other in convalescence; a definite increase in titre is essential to establish or confirm a diagnosis.

Viruses, in certain cases at least, appear to possess a complex antigenic structure. With suspensions of the vaccinia, psittacosis and influenza viruses, it has been found that the antigen involved in the complement-fixation test is a soluble substance and is distinct from that concerned in the neutralization test.

INTERFERENCE

It has been found that animals inoculated with one virus may become resistant to inoculation with another, and perhaps more virulent, type of virus. The term "interference" has been given to this phenomenon. In the case of yellow fever, intramuscular inoculation of a susceptible monkey with the relatively mild neurotropic virus was found to afford protection against a later inoculation with a highly virulent pantropic strain. The second inoculation was made a short time after the first before any humoral immunity could have developed. Similar results were obtained with the viruses of Rift Valley fever and influenza; in some cases the viruses had actually been inactivated. The mechanism of this reaction is not fully understood but it is considered that the first virus inactivates or destroys the receptors of the susceptible tissue-cells, and in consequence the second virus is unable to establish firm contact with the cells, which is probably a necessary step in their penetration by the virus. The "interference phenomenon" has also been demonstrated in tissue-cultures of certain viruses.

An interesting development has been the demonstration that, on mixing influenza virus and certain other viruses with chick chorio-allantoic membrane or monkey kidney cells, a substance is formed which can induce the interference phenomenon. This substance has been called "Interferon". It is active against a number of viruses, including the influenza and vaccinia viruses, and appears to exercise its activity inside the cell and does not simply prevent the virus from entering the cell.

PATHOGENICITY

Filterable viruses are responsible for a number of important human and animal diseases. They appear to be essentially parasites; true saprophytic members have never been described.

The incrimination of a virus in any disease-process may be a matter of considerable difficulty. The reproduction of the condition in an experimental animal and subsequent passage in series are essential. Failure to obtain passage of the infection indicates the action of a toxin rather than of a virus. The adaptation of a virus to a new host may be difficult, and during adaptation the virus may become greatly modified, e.g. the street and fixed viruses of rabies. It is also important, in adapting viruses to new hosts, to exclude the possibility of irregularities arising from the presence of spontaneous disease in the experimental animal.

Certain viruses produce disease only in man, experimental animals are quite refractory; the participation of a virus in these conditions is concluded mainly by the elimination of other disease-agents, the cultivation of the virus in the chick-embryo and, or, the reproduction of the disease in human volunteers.

TRANSMISSION

The mode of spread of viruses is similar to that of the bacteria. In many cases, e.g. influenza, measles, mumps, variola and chicken-pox, spread is mainly by droplet or dust infection. In other cases, e.g. rabies and psittacosis, there may be direct contact with infected animals or birds; while in lymphogranuloma venereum and warts there is direct human contact; arthropods are concerned in the transmission of a number of virus diseases, e.g. the mosquito in yellow-fever, dengue and St. Louis encephalitis, ticks in louping-ill and mites in equine encephalomyelitis. The arthropod-borne viruses are sometimes referred to as the *Arbor* group of viruses. In most conditions the viruses actually infect the arthropods, which are not just passive vectors.

Ingestion is considered to be responsible for the spread of the Enteroviruses such as poliomyelitis, ECHO, and Coxsackie viruses.

CLASSIFICATION

Our knowledge of viruses and the diseases produced by them is far from complete. Divergent views have been expressed as to the nature of viruses; some workers considered that viruses approximate to the enzymes, and are dead, and others that they were essentially living bodies. In the case of at least the larger viruses the available data, particularly their particulate nature, morphology and power to multiply, indicate that in many respects there is a close relationship between the viruses and the bacteria. Most workers accept this relationship and consider that viruses are living, particulate bodies comparable with the bacteria, i.e. viruses are small forms of bacteria. As the viruses decrease in size, there is progressive loss of metabolic activity and a greater dependence on the enzyme systems of the tissue-cells.

It has now been established that viruses contain protein and nucleic acid, either ribonucleic acid (RNA) or deoxyribonucleic acid (DNA), which appears to be the primary genetic material and is responsible for genetic specificity. In the case of the tobacco mosaic virus, the virus activity appears to be closely related to, if not identical with, the RNA and is retained in spite of considerable chemical and physical treatment. Moreover, the RNA is pure form and without protein is infective.

Any final attempt at the classification of either the viruses or the virus diseases is quite impossible at the present time and an officially accepted scheme is not available. Some viruses exhibit an affinity to certain tissues, and the terms "neurotropic", "dermotropic", "viscerotropic" and "pantropic" viruses are frequently used in a loose descriptive sense to indicate various classes of these agents. It has been suggested that, in view of their large size, the viruses of lymphogranuloma venereum, psittacosis, and certain animal pneumonias form a distinct and separate group, which has been designated the "L.G.V.-psittacosis" Group. The viruses which may be isolated from the intestinal contents, i.e. Poliovirus, Coxsackie and ECHO viruses have been grouped as the *Enteroviruses*, whilst those causing the "pox" diseases—smallpox, vaccinia, cowpox and others have been termed the *Poxviruses*. It has already been mentioned that the arthropod-borne viruses have been tentatively grouped as the *Arbor* viruses.

POLIOMYELITIS

Acute anterior poliomyelitis or infantile paralysis is an acute infectious disease involving mainly the nerve-cells of the spinal cord and medulla. It was described in detail by Heine (1840) and Medin (1890), and is sometimes referred to as the Heine-Medin disease. The condition was first reproduced in the monkey by Landsteiner and Popper (1908).

The Virus. The aetiological agent of poliomyelitis is one of the smallest viruses; by ultra-filtration methods its dimensions have been estimated at about 8–12mμ, but electron microscopy suggests a rather larger size of 27mμ; it readily passes through the bacterial filters. It grows readily in tissue culture using kidney tissue or human amnion.

The only available experimental animal for general use is the monkey, in which the condition can be produced in order of reliability by intracerebral, intraneural, intrathecal and intranasal injection of virus-containing material. The experimental disease is comparable to the severe, fatal form in humans; when the virus has become adapted to the monkey the mortality rate in these animals is approximately 95–100 per cent.

Three main antigenic types of poliomyelitis virus have so far been isolated—Type 1 (Brunhilde); Type 2 (Lansing) and Type 3 (Leon). Type 1 is the usual type in this country, but all have been isolated; the Lansing type can produce infection in cotton rats and mice on intracerebral injection.

The following properties of the virus have been established; it is extremely resistant to freezing, drying, pure glycerine and 0·5 per cent phenol; it is, however, readily destroyed by heat over 45°C and oxidizing agents. In nervous tissue the virus can be conserved for many years at 0°C in 50 per cent glycerine.

DISTRIBUTION AND PATHOGENESIS

Poliomyelitis is an acute infectious disease primarily of children, which occurs sporadically and epidemically. Infection is spread by human association. Epidemics are commonest in the late summer and autumn in temperate climates while in the tropics cases occur throughout the year. In the past, poliomyelitis has been a disease of infancy and childhood. Recently in some epidemics a number of adults have been affected. During the past ten years it has been noticed that only a small proportion of those infected develop paralytic symptoms. Some of the factors which appear to influence the occurrence of paralysis are the biological character of the strain of the virus, and the genetic constitution of the host, pregnancy and other endocrine factors, infections of the throat and intestinal tract, over-exertion and surgical trauma such as tonsillectomy.

There is epidemiological and experimental evidence to indicate that the disease is transmitted by droplet infection from cases or carriers. Some workers consider that the main route of entry is the intestinal tract as the virus can be isolated from the faeces of cases and carriers. It is possible that both routes are involved in the transmission of the disease.

The route taken by the virus after implantation in the alimentary tract is not definitely known. In the experimental disease the initial lesions are found in the nerve-cells and not in the interstitial tissue, i.e. the virus is essentially neurotropic. The lesions consist of degeneration, vacuolation and neuronophagia of the nerve-cells with a well-marked secondary perivascular and meningeal mononuclear infiltration. Evidence has been produced to suggest that in the human disease there may be a more generalized infection of the tissues with a viraemia; the virus then localizes and produces damage in the central nervous system.

The involvement of nerve-cells is indicated by the appearance of paresis and paralysis of the associated muscles. The mortality rate is variable, but tends to be about 10 per cent, and is due to damage to the respiratory and cardiac centres. When recovery occurs there is usually some associated permanent muscular impairment.

IMMUNITY

Recovery from the disease is followed by a high degree of immunity; second attacks are exceedingly rare. This resistance is often associated with the presence in the serum of antiviral bodies, which can be demonstrated without difficulty by means of the neutralization test. In regions where the disease is endemic adults are seldom affected and antibodies can be demonstrated in the serum of many adults who have displayed no previous manifestation of the disease. In contrast to this, it has been observed that in rural or isolated communities, where poliomyelitis is seldom seen, the introduction of the virus frequently gives rise to outbreaks involving individuals of all ages. Moreover, antibodies are rarely present in the serum of the adult population. These facts indicate that in districts where the condition is endemic, sub-clinical infection is a common feature, and as a result the adult population becomes immunized. There is a well-recognized "abortive" form of the disease, in which muscular weakness is not found.

The theory of specific immunization by sub-clinical infection is not universally accepted. A few workers consider that the resistance of adults is due to a non-specific physiological maturation of the tissues. The evidence put forward to support this theory is, however, far from conclusive.

DIAGNOSIS

Laboratory methods are of little value in establishing a clinical diagnosis and are reserved for epidemiological purposes. The most important test is the isolation of the virus from nervous tissue collected at autopsy or from faeces or throat washings by either the tissue-culture technique, using human or monkey tissue, or the intracerebral inoculation of monkeys. The virus is not found in the cerebrospinal fluid of cases, but this shows definite changes, viz., lymphocytic pleocytosis with increased protein content but no alteration in the sugar or chloride values.

PROPHYLAXIS

In recent years two methods of specific prophylaxis—active and passive—have been attempted.

Passive Immunization. Although human gammaglobulin has been shown to give passive protection against paralysis in monkeys, injection of this material is of no practical value in the protection of close human contacts as they have already been infected for several days when the case is recognized.

Two methods of active immunization have been developed in recent years: living attenuated vaccine made from Sabin strains of polio-virus; and killed vaccine. Living vaccine is given orally and killed vaccine by injection.

Living vaccine mimics the natural infection. It produces local resistance to reinfection of the gut, and circulating antibodies which inhibit the spread of the virus to the central nervous system. Living vaccine therefore gives immunity to the individual and also reduces the spread of infection by reducing the amount of virus excreted.

Killed vaccines also produce a high level of circulating antibody, but do not confer the same degree of gut immunity.

At the present time attenuated poliomyelitis vaccine is in general use both for routine prophylaxis and epidemic control. Its main disadvantages are that its immunizing action may be inhibited by interfering viruses, especially in developing countries, and that on very rare occasions it may produce paralysis in vaccinated subjects. However, the disadvantages are outweighed by the cheapness of the vaccine and the ease of administration.

In the presence of an outbreak of poliomyelitis, it is recommended that over-exertion and fatigue, as well as the prophylactic inoculations of whooping-cough and diphtheria vaccines should be avoided.

Specific treatment with gamma globulin is of no value. Although a number of substances have been found to have an inhibitory effect on the multiplication of poliomyelitis virus in tissue culture, none has been found to have any marked effect when symptoms are already present.

Important non-specific therapeutic measures are rest in bed and the treatment of respiratory complications.

COXSACKIE VIRUS

During investigations of poliomyelitis in U.S.A. in 1948, a virus was isolated from the faeces of two children with typical symptoms of paralytic poliomyelitis in a town called Coxsackie in the State of New York. Type 1 poliovirus was also isolated from the same specimens. The new virus was found to produce weakness, ataxia, paralysis and death on intramuscular or intracerebral inoculation of suckling mice. This agent was later isolated from the faeces, blood and oral washings of non-paralytic cases of poliomyelitis by appropriate tissue-culture techniques. Further work indicates that the virus is widely distributed and is quite distinct from that of poliomyelitis; both have been isolated from the same patients and also from the faeces of normal individuals.

Many serological types of the Coxsackie virus have now been identified and these fall into two main groups—A and B. Group A viruses have been mainly associated with a febrile throat condition, herpangina, which is characterized by vesicular lesions on the mucosa of the throat; it has also been isolated from cases of aseptic meningitis and more recently in U.S.A. from an outbreak of vesicular stomatitis and vesicular exanthem. Group B viruses have been isolated with some frequency from cases of aseptic meningitis, epidemic myalgia (Bornholm disease), acute myocarditis and from some suspected cases of poliomyelitis.

ECHO VIRUSES

During recent investigations of faeces from cases of poliomyelitis, some viruses were isolated which produced cytopathic changes in tissue-culture preparations but which were not neutralized by poliomyelitis antisera and were not pathogenic to suckling mice. As these viruses were also isolated from the faeces of normal individuals and did not appear to be responsible for human disease, they were designated *Enteric-Cytopathogenic-Human-Orphan* or ECHO viruses. Many distinct serological types have since been isolated and it has now been established that they may be responsible for cases of aseptic meningitis in man. The virus has been isolated from the C.S.F. of such cases as well as from faeces and throat washings. It has also been established that *Echo* viruses may be responsible for outbreaks of acute diarrhoea in children and also for short-term acute fevers.

LYMPHOCYTIC CHORIO-MENINGITIS (L.C.M.)

A virus has been isolated in America and this country from certain cases of so-called aseptic meningitis. This condition is normally non-fatal and may run a long course, perhaps several months; complications are unusual. The disease can be transmitted to mice, guinea-pigs and monkeys by intracerebral inoculation of the cerebrospinal fluid collected from cases of the condition.

A leptomeningitis develops after an incubation period of 7–14 days and the cerebrospinal fluid shows a definite pleocytosis with large numbers of lymphocytes.

The virus can be grown by cultivation on egg-membranes. Antibodies, demonstrated by neutralization tests, are present in the serum of humans or animals recovering or recovered from the infection. Irregular results may, however, be obtained from these tests as several antigenically distinct strains of the virus have been isolated. Complement-fixation tests are also useful.

The virus has also been isolated from spontaneous disease in mice and consequently experiments in which mice are concerned must be strictly controlled.

RABIES

Rabies, or hydrophobia, is an inflammatory disease of the central nervous system transmitted to man by the bites from infected animals, usually dogs.

THE VIRUS

Rabies virus is 100–150mμ in diameter as measured in electron micrographs. It is inactivated by sunlight, by exposure to ultraviolet light, by heat at 60°C for 30 minutes or by dilute formalin. It is unaffected by antibiotics. The virus will survive for many weeks at 4°C and many months at lower temperatures. In neutral glycerol it remains active for several weeks at room temperature and many months at 4°C.

Man and all warm-blooded animals including baby chicks and chick embryos are susceptible. The virus is found constantly in the nervous system and in saliva. Freshly isolated strains are known as "street virus" and are characterized by a long and variable incubation period in experimentally infected animals, 10–90 days in rabbits inoculated intracerebrally, and by frequent production of cytoplasmic inclusion bodies termed Negri bodies in the cells of the nervous system, particularly in those of Ammon's horn of the hippocampus. After a number of rabbit passages the incubation period is reduced to 4–6 days while the diagnostic Negri bodies become more difficult to find. After repeated intracerebral inoculations the rabies virus becomes "fixed" for the central nervous system of the rabbit and no longer multiplies in other tissues.

Animals can be infected by intradermal, intramuscular or intracerebral inoculation. Recently, animals exposed in cages in a cave infested with bats became infected presumably by the intranasal route. Two workmen who had frequented this cave also died of rabies although there was no history of exposure to bites. Growth of the virus has been obtained in tissue culture of embryo mouse brain cells and in developing chick embryos following inoculation on to the chorio-allantoic membrane.

DISTRIBUTION AND INFECTIVITY

Rabies is primarily a disesae of canine animals, from the bites of which man becomes infected. The virus is present in the central nervous system and the saliva of rabid animals. The incidence of the human disease is therefore dependent on the presence of infected animals. In Great Britain the disease has been eradicated by strict quarantine laws, but in other countries, such as India, China and Russia, the disease is still prevalent.

The virus is introduced with the saliva into the subcutaneous tissues and is probably conveyed to the central nervous system by the regional nerve-axones. Rabies has been found to result in only some 35 per cent of non-immunized individuals bitten by proved rabid animals; once developed, however, the disease is invariably fatal.

The incubation period is long and subject to much variation (27–64 days); it depends largely on the proximity of the bite to the central nervous system. The main symptoms are irritation around the bite, fever, difficulty in swallowing and respiration, marked salivation and later delirium and paralysis.

The main lesions are found in the central nervous system. The nerve-cells show signs of degeneration and neuronophagia, while some, particularly those of the hippocampus major, exhibit the characteristic cytoplasmic inclusions first described by Negri in 1903. Perivascular infiltration is also present.

DIAGNOSIS

The diagnosis is definitely established at autopsy by the demonstration of Negri bodies in the brains of infected individuals or animals. Negri bodies are demonstrable in 90 per cent of animals and 70 per cent of patients who have died of rabies. Mice should be inoculated intracerebrally with 10 per cent brain suspension in cases where no Negri bodies can be found. A mouse test may be positive six days after inoculation.

The complement-fixation test with known rabies antiserum and antigen prepared from tissue suspected to contain rabies virus, and the fluorescent antibody test in which the suspected tissue is placed in contact with rabies antiserum conjugated with fluorescent dye, may both be used to confirm the diagnosis.

PROPHYLAXIS

Although the disease, once established, is always fatal, the development of symptoms can be prevented if artificial immunization is carried out during the early stages of the incubation period. This was one of the many important discoveries made by Pasteur. In view of the long incubation period of rabies after infection with the street virus, inoculation with graduated doses of the attenuated fixed virus is able to produce an immunity which aborts the original infection. Many modifications of the original Pasteur method, mainly in the mode of attenuating the fixed virus, have since been introduced. The inoculations are usually given subcutaneously on 14 or more successive days.

(1) Pasteur's Method. The spinal cord of rabbits inoculated with fixed virus is dried over KOH for varying periods of time, and when required saline suspensions are made of the cord. The most attenuated virus, i.e. one dried for the longest time (14 days), is used for the first injections, this is followed by less attenuated virus until finally virus dried for only five days is used.

(2) Hogye's Method. Active virus diluted to 1/10,000 is used for the initial injections and the amount is gradually increased for the subsequent doses.

(3) Babes' Method, in which heated fixed virus is used; at 80°C for the first injection, then at 60°C, 45°C, and finally unheated.

(4) Fermi and Semple's Method. Virus inactivated by 1 per cent phenol is used.

(5) Remlinger's Method, in which the virus is attenuated by treatment with ether.

The introduction of active immunization has decreased the incidence of rabies in a striking manner. There is, however, no general agreement as to the most satisfactory type of vaccine. More recently fixed virus grown in duck embryos and then killed with β-propriolactone has been used. This has given satisfactory results with a lower incidence of complications than the phenol inactivated virus. In the use of any preparations of rabies vaccine many repeated doses are necessary to ensure adequate immunity.

VARIOLA OR SMALL-POX

Small-pox is a widely distributed eruptive disease which is frequently present in epidemic form.

THE VIRUS

The virus of variola can be demonstrated in the vesicular fluid of the specific lesions by suitable staining methods. Its infectivity for experimental animals is low; the most suitable animal being the monkey, in which, when injected subcutaneously, a well-marked local lesion and constitutional signs are produced. The inoculation of the rabbit merely results in a slight local lesion with little, if any, general reaction. After several passages through the monkey the infectivity of the virus is increased considerably and well-marked lesions can be produced in the lower animals. As a result of adaptation to these animals the virus is definitely modified and is identical with the "vaccinia" virus; i.e. the vaccinia virus is probably the virus of variola modified by passage through the lower animals.

The vaccinia virus readily produces infection in the ordinary laboratory animals on cutaneous or intradermal inoculation, and it has consequently been the subject of extensive investigations. It is relatively large, 0·125–0·175μ, and therefore does not readily traverse bacterial filters; it is resistant to glycerine, phenol, ether and low temperatures, but is readily destroyed by heat over 55°C and potassium permanganate; it grows readily *in vitro* in the presence of living or surviving cells. In the vesicular fluid of the lesions small, coccal, elementary or Paschen bodies can be demonstrated, and these are considered to represent the virus (Plate VB).

Inclusions, termed *Guarnieri bodies*, are present in the cytoplasm of the epithelial cells of the lesions. On the intracerebral inoculation of a potent virus a meningo-encephalitis can be produced in rabbits and monkeys. The neurotropic strains are sometimes referred to as "neurovaccine". Following the inoculation of a potent virus by any route, it tends to generalize and lesions may be observed in various tissues.

DISTRIBUTION AND INFECTIVITY

Small-pox was at one time one of the dread scourges. Large outbreaks with a high mortality rate were frequently reported; the lesions were severe, and on recovery pitting and scarring were the usual features. Following the introduction of prophylactic vaccination by Jenner there was a progressive decrease in the incidence and severity of the condition and at the present time the severe form is rarely encountered in this country.

The exact mode of transmission is not known, but most workers consider that the infection occurs through the agency of air-borne particles which settle in the upper respiratory tract. The virus is then carried by the blood stream to the skin tissues, where it settles and multiplies in the capillary circulation.

The incubation period is about 9–15 days; the onset is characterized by fever, headache, vomiting and pains in the back. The rash appears on the third day in the form of discrete shotty papules mainly situated on the face, hands and legs; in three days the papules become vesicular, then pustular, finally drying up and after decrustation scars are left. Severe forms may occur and the mortality rate varies in different outbreaks (10–40 per cent); it is high in patients with a confluent or haemorrhagic rash.

ALASTRIM

A mild form of variola has been observed in many countries. This disease is seldom fatal and has been termed *alastrim* or *variola minor*. Some doubt was at first felt concerning the identity of this condition with the true variola as the disease breeds true and is invariably mild. Experimental work has, however, shown that the viruses of these conditions are immunologically identical. They do however differ in their growth on the chorio-allantoic membrane of the chick embryo; smallpox virus will multiply at 38°C after 2–3 days, alastrim virus will not. Alastrim is now accepted as an attenuated form of smallpox. Both conditions are closely related to cow-pox.

IMMUNITY

An attack of smallpox or alastrim is followed by a high degree of immunity; second attacks are extremely rare. A high degree of immunity can also be produced by artificial immunization. The immunity is associated with the presence of antibodies in the serum; these can be demonstrated by neutralization, complement-fixation, agglutination and flocculation tests. The practice of active immunization by means of the vaccinia virus was introduced as a general measure mainly through the efforts of Jenner, who investigated the farm-workers' contention that an attack of cow-pox rendered them immune to small-pox. Vaccination, as the process is now known, has been carried out largely under State control; it was first officially adopted in England in 1853. The living virus is employed, but the type employed in different countries varies; in this country calf-lymph is used, on the Continent the potent neurovaccine is frequently employed, and in America virus cultured *in vitro* has recently been used; the last preparation has the great advantage in that it can be readily obtained in a sterile condition.

In England the preparation of calf-lymph is controlled by various Acts and Regulations. Attention is directed to the condition of the calves and also to the purity and potency of the final product. The vaccinia virus is seeded on the scarified abdomen of young healthy calves, 4–6 months old; on the fifth day this area is scraped off and triturated with 50 per cent glycerine and distilled water. The pulp is sealed and kept at —11°C; cultures, aerobic and anaerobic, are made at intervals. The lymph is ready for issue only if the potency tests are satisfactory, if streptococci and anaerobes are absent, and if the aerobes are less than 5,000 per ml.

Vaccination in man should be carried out in infancy by cutaneous scarification or the multiple pressure technique; the lymph is placed on the cleansed skin and light scarification made through it or gentle pressure applied. A local vesicular lesion is produced in five or six days; this passes through a pustular phase and finally dries up, leaving a scar. The immunity produced by this form of vaccination is sufficient to protect against the slight risk of infection in this country; re-vaccination at intervals is desirable and should invariably be carried out in the event of an outbreak in the region. Service personnel, who are likely to visit countries where small-pox is endemic, are re-vaccinated at frequent intervals (3–5 years); in the presence of an outbreak re-vaccination with three insertions of lymph is recommended.

In immunized individuals, successful re-vaccination is followed by three types of reaction depending on the immunity state of the individual: (1) *primary reaction* which is the classical type and indicates lack of immunity, (2) *accelerated* or *vaccinoid reaction*, in which there is a mild form of reaction, complete in 8–10 days; this indicates a moderate degree of immunity, and (3) *immediate reaction* in which a raised area of erythema appears, reaches a maximum within 72 hours and disappears rapidly; this is indicative of a high degree of immunity. It has recently been suggested that "Reaction of Immunity" is an unfortunate term as, in many instances, the reaction may be due to increased sensitivity to other products in the lymph and not to any specific immunity.

Complications are unusual, but occasionally after primary vaccination of adults an encephalo-myelitis develops some 11–16 days after the vaccination. While there is little doubt that the nervous condition is definitely associated with the process of vaccination, it appears equally certain that the vaccinia virus is not the direct cause of the lesions in the nervous system. In order to reduce the risk of such complications, the Ministry of Health (1928) recommended *inter alia* that one insertion with the minimum of trauma should be given; primary vaccination should be performed in infancy.

Owing to the possible occurrence of these serious complications and of the mild form of small-pox, some authorities recommend that vaccination should no longer be enforced; they consider that the natural infection is less hazardous than the immunization process. In consequence only 20–40 per cent of children in this country are now vaccinated in infancy. It is, however, possible for an unimmunized community to develop under such conditions, and the introduction of a virulent virus into such a population may have serious consequences. The latter event is rendered more probable by the increase and development of air transport, as it is now easier for individuals incubating the infection to gain admittance into the country. Several small outbreaks have recently occurred here as a result of the arrival from overseas of men either during the incubation period of the disease or exhibiting mild atypical clinical manifestations. The disease developed, as a rule, in persons not vaccinated for some time and the fatal cases in those not previously vaccinated. It is therefore important that the vaccination of infants should be universally practised and that persons likely to come in contact with the disease should be re-vaccinated at intervals.

DIAGNOSIS

The early diagnosis of small-pox is important and as the clinical signs may be atypical and perhaps resemble those of severe chicken-pox, laboratory tests have proved invaluable. These consist of (1) the examination of pocks for the presence of elementary bodies and for antigenic activity in complement-fixation tests with known antisera; (2) the cultivation of virus in the chick embryo (Plate VB) and (3) the detection of antibodies in the patient's serum.

TREATMENT

Sulphonamides and antibiotics have proved of little value in the treatment of the variola infection but they are useful in preventing secondary infection of the pustules.

HERPES SIMPLEX

Herpes simplex is an acute eruptive disease of the skin and mucous membranes, occurring frequently on the face and genitalia.

The disease is very mild; a crop of vesicles appears after local irritation, the serous fluid exudes and forms a crust, which falls off, leaving no scar. Herpes frequently recurs in certain individuals who appear to harbour the virus in the naso-pharynx; it is activated on stimulation from varied sources.

THE VIRUS

The virus of herpes simplex is probably more widely distributed and more constantly present in man than any other pathogenic virus; more than 60 per cent of the population are infected with the virus and remain as carriers throughout life. In all infected persons antibody is constantly present in the blood but this does not lead to the elimination of the virus, which, for the most part remains latent probably in the epithelial cells of the buccal or nasal mucosa.

Herpes can be reproduced experimentally in most laboratory animals by corneal, cutaneous or intradermal injection of the virus. The following properties of the virus have been determined: it is resistant to 50 per cent glycerine and low temperatures, but is readily destroyed by heat over 50°C. It is relatively large, 100–150mμ, and does not readily pass through bacterial filters; it can be cultivated *in vitro* in the presence of living cells.

With some strains a meningo-encephalitis has been produced in rabbits; this suggested to some workers that two distinct strains of the virus existed, one being dermotropic and the other neurotropic. The general opinion is, however, that the affinity of the virus for different tissues is merely a question of virulence and does not indicate the presence of two distinct types of the virus. The so-called "neurotropic" strains have been shown to be more virulent than the "dermotropic" strains.

IMMUNITY

The development of immunity in man is uncertain. A second crop of vesicles may occur within a short interval of the first crop. Antibodies have been demonstrated in the serum of many individuals, and the presence of these bodies appears to exercise little influence on the development of the vesicles. The action of the virus is possibly too superficial to be affected by them. A relatively high degree of immunity can, however, be produced in rabbits by active immunization.

No specific effect was obtained in the treatment of recurrent herpes by the repeated injection of a formalized herpes virus vaccine.

5-iodo-2-deoxyuridine has proved of some value in the treatment of herpes simplex. It has been reported to give good results on repeated local application by eye drops in cases of herpetic keratitis and the more rapid healing of skin lesions following application using a spray gun or some agent which will facilitate penetration such as dimethyl sulphoxide.

YELLOW FEVER

Yellow fever is an acute febrile disease occurring usually in tropical regions in epidemic and endemic forms.

THE VIRUS

The American Yellow Fever Commission (1900) found that the infecting agent was filterable and was conveyed by a mosquito, the *Aedes aegypti*. Stokes, Bauer and Hudson (1928) reproduced the disease in Asiatic monkeys and more detailed examination of the condition became possible. The size of the virus has been estimated to lie between 17–28mμ; it passes through the Berkefeld "V" and "N" filters, but not the "W" type. The virus is resistant to the action of 50 per cent glycerine, drying and low temperatures, but is readily destroyed by heat over 55°C and formalin. Theiler (1930) established infection in mice by intracerebral inoculation and from this discovery the mouse protection test was developed for the detection of neutralizing antibody in human serum.

DISTRIBUTION AND INFECTIVITY

Yellow fever is widespread in tropical and sub-tropical countries, being particularly common in South America and West Africa. The disease is usually transmitted from man to man by the mosquito, *Aedes aegypti* or allied species. The eradication of *Aedes aegypti* in urban communities and the widespread use of vaccine has brought the disease more or less under control in most regions. However, the virus still maintains itself in forest and other areas by a mammal-mosquito-mammal cycle in which man may be infected.

Laboratory infection has been reported following the handling of infected material; the virus either penetrates the unbroken skin or enters through minute abrasions.

White people are particularly susceptible, and the disease usually runs a typical course. After an incubation period of 3–6 days there is a sudden onset with fever, headache and muscular pains, which last about three days; then the temperature falls to about 98°F, while jaundice, albuminuria and haemorrhages into the intestinal tract are observed. The mortality rate varies from 20–90 per cent; death may occur in the early stages, but it is more common on the fifth or sixth day.

On autopsy, jaundice, haemorrhages into the skin and mucous membranes with fatty and necrotic changes in the liver and kidney are found; nuclear inclusions are present in the liver-cells.

The virus is present in the blood during the late incubation period and the first three or four days of fever. Mosquitoes only become infected when they suck the blood during these stages, but they do not usually become infective for man for at least 12 days; the time varies with the atmospheric temperature. Once infected the mosquito remains so indefinitely.

IMMUNITY

An attack of yellow fever is followed by a high degree of immunity, second attacks being rare; this state is associated with the presence of antibodies in the serum which can be demonstrated by the neutralization or protection test in mice. Neutralization tests have been used extensively in epidemiological surveys of yellow fever and results indicate that the condition is widespread in Africa and South America. It is well recognized that the inhabitants of regions where the condition is endemic are resistant to infection. Recent studies have shown that the sera of these natives contain the antiviral bodies; this suggests that they undergo mild sub-clinical infections during childhood. Antibodies have also been demonstrated in the serum of monkeys in regions where the condition is endemic, and this suggests that monkeys may act as alternative hosts for the virus.

PROPHYLAXIS

Important preventive measures are the suppression of mosquitoes and their exclusion from districts which are free from the disease. The increasing use of air transport is a possible danger in tropical regions as a means of introducing not only infected mosquitoes, but also infected humans incubating the disease into countries at present free from yellow fever.

Active immunization is now essential for individuals visiting certain tropical regions such as W. Africa; various vaccines have been tried; one widely used, is an attenuated pantropic strain, 17D grown in a chick-embryo culture medium, and the results have been highly satisfactory. The reactions are not severe and the degree of immunity induced appears to be effective for a number of years.

PSITTACOSIS

Psittacosis is primarily an infectious disease of birds, particularly parrots and budgerigars, which is transmissible to man, giving rise to severe pulmonary symptoms. A closely related disease, ornithosis, occurs in a wide range of domestic birds including pigeons, canaries, fowls and ducks.

THE VIRUS

Bedson, Western and Simpson (1930) found that the responsible agent was a filterable virus. These workers reproduced the disease in budgerigars by injecting material obtained from infected humans and parrots. These observations were confirmed and experimental infection was later produced in a number of animals, including mice, guinea-pigs and monkeys. The virus is relatively large (200–300mμ) and only traverses the coarse bacterial filters with difficulty; it is resistant to glycerine and low temperatures. Elementary bodies are found in the specific lesions and also in tissue-cultures; various developmental stages have been reported.

DISTRIBUTION AND INFECTIVITY

The association of human psittacosis with parrots has long been recognized. The majority of the human cases give a history of close contact with sick birds, infection probably occurring by inhalation. The virus apparently possesses a high degree of infectivity for man as, although transmission of infection from man to man is not common, many instances of laboratory infection have been recorded. In this country the disease practically disappeared as a result of the ban on the importation of parrots and budgerigars; when the ban was recently lifted, several cases occurred and it has now been reimposed.

After an incubation period of 7–14 days, there is an abrupt onset with fever, malaise and headache, a pneumonia of the influenzal type frequently develops. The mortality rate in different outbreaks has averaged about 20 per cent.

DIAGNOSIS

The diagnosis is usually confirmed by two methods. In one infection can be produced experimentally in mice, parrots, budgerigars and monkeys by the intraperitoneal inoculation of blood or sputum collected during the acute phase of the disease. In the other the complement-fixation test is used, preferably using two distinct samples of serum; one collected in the early stages of the illness and the other 10–14 days later. In a positive case there is a sharp rise in titre of antibody.

IMMUNITY

An attack of psittacosis is usually followed by a high degree of immunity; second attacks in man have not been reported, while recovered parrots are resistant to re-infection. Antibodies have been demonstrated in the serum of recovered cases by means of the complement-fixation test. Active immunization of mice has been carried out by the use of the formalized virus.

TREATMENT

The wide spectrum antibiotics such as the tetracyclines have proved valuable.

LYMPHOGRANULOMA VENEREUM: CLIMATIC BUBO

This is a venereal disease, which is characterized by adenitis of the inguinal glands and the presence of a small primary lesion on the external genitalia. Although rare in this country it is common on the Continent and in tropical regions. Material obtained from infected glands and found to be bacteriologically sterile gives rise to a meningo-encephalitis on intracerebral inoculation into monkeys and mice. Findlay carried out detailed investigations and considered that climatic bubo and lymphogranuloma venereum are identical conditions and are both caused by a filterable virus. It is now firmly established that the aetiological agent is a virus of relatively large size (300mμ) which can be readily stained by the Giemsa or Macchiavello technique.

DIAGNOSIS

A useful diagnostic measure is the Frei Test. This test is carried out by the intradermal injection of a saline extract of infected lymph nodes, the brains of infected mice, or culture virus heated at 60°C. The most satisfactory antigen is prepared from virus grown in yolk-sac culture. In a positive reaction there is local erythema after 24 hours which persists for 5–10 days. Like the Mantoux test for tuberculosis a positive Frei test does not necessarily indicate active infection.

Other useful diagnostic measures are (1), the isolation of the virus by the intracerebral inoculation of mice or by the injection of the yolk-sac of fertile eggs or (2), serological methods using the complement-fixation test with infected yolk-sacs as antigen.

TREATMENT

The tetracyclines and sulphonamides have both given satisfactory results in the treatment of lymphogranuloma venereum.

RIFT VALLEY FEVER

A disease causing a heavy mortality in newly born lambs appeared in Rift Valley, Kenya, during 1930. On investigation the aetiological agent was found to be a filterable virus which was demonstrated in the blood, liver and spleen of infected animals by the intravenous inoculation of the tissue suspensions into healthy sheep. Man appears to be highly susceptible, as during the outbreak the majority of natives engaged in herding the sheep experienced a febrile illness with joint pains. Many laboratory workers have experienced similar attacks following the examination of infected material. A high degree of immunity appears to follow an attack. The virus has been isolated and grown in egg and tissue cultures. It is transmitted from animal to man by the bite of the mosquito.

EPIDEMIC INFLUENZA

Influenza is an acute febrile disease of the upper respiratory tract, which may appear in pandemic, epidemic and sporadic form. It is characterized by its sudden onset, marked prostration of the patient and the tendency to complications, such as broncho-pneumonia. It has a world-wide distribution. The pandemic of 1918–1919 is considered to have been responsible for at least 10,000,000 deaths. In spite of much effort to discover the cause of influenza it was not until 1933 that a way was found to recover and study the influenza virus in the laboratory. Smith, Andrewes and Laidlaw were able to transmit in series a febrile disease to ferrets by nasal instillation of throat washings from influenzal patients. It was later found that the intranasal instillation of the ferret virus into mice under ether anaesthesia produced a bronchopneumonia which usually ended fatally in 3–5 days. Direct passage of the virus from human to mouse tissue was not possible.

It was decided to designate this virus as the influenza A virus. Antigenic variation was early recognized among the influenza viruses recovered from different epidemics but the differences were relatively minor before 1946 when a new antigenic variant of influenza A appeared in Australia and spread slowly throughout the world. For the next eleven years epidemics were associated with viruses of this new A1 family but none of these epidemics reached pandemic proportions. In 1957 the Asian influenza virus was recognized in China and was found to be antigenically distinct from any of the A or A1 strains and was therefore grouped as the A2 family. The A2 viruses contained, however, the same soluble complement-fixing antigen as that found in all viruses of the earlier A and A1 families.

In 1940 Francis and his colleagues isolated a virus from outbreaks of influenza in New York which was antigenically distinct from the A strains and was termed influenza B. Further antigenic variations of the B strain have been found since 1940 but none has been associated with pandemics. Another virus C has been isolated from cases of minor respiratory disease but has not been associated with an epidemic.

Attempts to infect ferrets with material from sporadic cases have been unsuccessful. This is not altogether surprising, as in many cases of so-called "sporadic influenza" the diagnosis is made without much justification.

The influenza virus is relatively large, being about 100mμ in its largest diameter. It grows readily in tissue culture or on the chorio-allantoic membrane of chick embryos. Antibodies can be readily demonstrated by either the neutralization test or the complement-fixation test in the serum not only of convalescents but also of many normal adults. In the neutralization test the elementary bodies, i.e. the virus bodies, are involved; while in the complement-fixation test the antigen is a soluble substance present in virus suspension and apparently produced by the virus during multiplication. A further method is the erythrocyte agglutination test, which was described by Hirst (1941) and is now extensively used; the influenza virus is able to agglutinate the red blood corpuscles of the chick through adsorption of the virus on to the cells. This action is neutralized by serum containing the homologous antibodies. This reaction is now being used in studies of the interference phenomenon.

DIAGNOSIS

Tissue cultures of monkey kidney cells are increasingly being used to recover the virus from human garglings and throat swabs. Satisfactory results, particularly for epidemiological purposes, can be obtained by the various serological techniques. Samples of serum are collected from the patient during the initial stages of the illness and later during convalescence. Tests are then carried out against antigens prepared with both A and B strains of the virus; a definite rise in titre indicates the nature and type of infecting agent.

PROPHYLAXIS

Work on specific prophylaxis and therapy is still in the preliminary stages and a final opinion is not possible. From results so far obtained it appears that convalescent serum or immune horse serum is not of great value either as a prophylactic or therapeutic agent. Mice can be immunized with the virus killed by heat at 57°C or by formalin The results of active immunization in man are difficult to interpret. The injection of the vaccine undoubtedly stimulates antibody production and recent trials indicate that they also afford some protection against epidemic influenza. Where an epidemic is due to a known strain of virus, as in the "Asian" influenza epidemic, a vaccine prepared from this virus is undoubtedly effective. Polyvalent vaccines are still undergoing trial.

ADENOVIRUSES

There is a group of ill-defined diseases of the upper respiratory tract which tend to resemble influenza and for which the term "febrile catarrh" has been used. By means of human volunteers and more recently tissue-culture techniques, several viruses have been isolated from throat washings of these cases; "Hela" cell cultures have proved particularly useful in diagnosis. These viruses appear to be closely related and have been referred to at different times as A.P.C. (adenoidal-pharyngeal-conjunctival), A.R.D. (acute respiratory disease), and R.I. (respiratory illness) viruses. These have now been grouped together as the Adeno-Viruses.

There appear to be at least 23 serological types; types 1, 2, 3 and 5 tend to be responsible for infections of the tonsils and adenoids in children; types 3, 7 and 14 for pharyngo-conjunctival fever, particularly in children, and types 4 and 7 for acute upper respiratory disease in Servicemen; types 4 and 7 for virus pneumonia.

The viruses are not pathogenic for laboratory animals but produce definite cytopathic changes in tissue-culture. They possess a common complement-fixing antigen but can be differentiated by the neutralization test. They are quite distinct from the viruses of influenza and the common cold.

Prophylactic vaccines have been used in U.S.A. with some success.

MUMPS

Mumps is an acute febrile condition characteristically involving the parotid glands and occurring most frequently in childhood; symptoms of encephalitis may occasionally be found. Mumps is transmitted mainly by droplet infection. The incubation period is long (14–21 days) and the initial symptoms are pyrexia and headache with swelling and tenderness of one or both parotid glands. The condition usually subsides in a few days, but such complications as orchitis, pneumonia, endocarditis and local suppuration may supervene. Meningitis is a recognized complication.

There is now overwhelming evidence to incriminate a virus as the aetiological agent of mumps. Earlier work was suggestive, and Wollstein (1916), by the intratesticular and intraparotid inoculation of cats with the filtered saliva from cases of mumps, produced local swelling and tenderness with pyrexia; this condition was passed serially through cats. Johnson and Goodpasture (1934) injected filtered saliva from cases of mumps into the parotid glands of monkeys and reproduced a condition comparable to mumps in man. They later found that the virus responsible for the monkey condition was identical with that causing human mumps.

It has recently been established by several workers that the virus will propagate in the chick-embryo and that complement-fixation tests can be successfully carried out by using virus particles or extracts of infected chick-embryos as the antigen. Two separate antigen preparations S and V may be used to detect two distinct antibodies. The S antibody occurs early and lasts for only a limited time after infection whilst the V antibody persists, often for years. It is thus possible to differentiate between recent and old cases of mumps.

An attack of mumps usually confers a high degree of immunity, but second attacks have occasionally been reported. A single intradermal injection of a live attenuated virus vaccine prepared from infected cultures of chick-embryo fibroblasts has been found to give good protection both in the U.S.S.R. and in the U.S.A.

MEASLES

Measles is an acute febrile disease of extreme infectivity occurring generally in children in epidemic form. The infection is transmitted mainly by droplet infection. The condition tends to take the following course: after an incubation period of about 14 days, the onset is characterized by fever, sore-throat and coryza, with the appearance of white spots, termed "Koplik spots", on the buccal mucous membrane. A papular rash appears on the fourth day, first around the neck and spreading over the body; the tongue is thickly coated. Three or four days later the symptoms subside and desquamation commences. The mortality rate depends largely on the incidence of complications, the most frequent of which are bronchopneumonia and otitis. Signs of encephalitis may appear after the seventh day of the illness in about 1 in 1,000 cases. The mortality from this complication may be as high as 10 per cent and there may be nervous sequelae in those who recover.

AETIOLOGY

The identification of the aetiological agent of measles has been rendered very difficult by the failure to obtain an experimental animal in which the typical disease can be reproduced. Monkeys have been used by many workers with varying degrees of success; other animals appear to be unsuitable. It is therefore not surprising that various agents have been incriminated in the aetiology of the disease. Blake and Trask (1921) reproduced a measles-like condition in monkeys by inoculating into the upper respiratory tract of the animals human naso-pharyngeal washings and filtrates collected in the early stages of the disease. The condition also developed in susceptible humans and monkeys following the intravenous injection of sterile human blood collected from early cases of measles. These observations suggest the participation of a filterable virus.

The causal relationship of a virus was settled by the isolation of a virus from the blood and throat washings of cases by the inoculation of tissue cultures prepared with monkey kidney cells by Enders and Peebles in 1954. The measles virus has also been recovered from the lungs of fatal cases of giant-cell pneumonia in debilitated children who showed no eruption during their illness.

IMMUNITY

Measles is followed by a high degree of immunity; second attacks are extremely rare. In urban communities most adults are immune, and as a number usually give no history of a previous attack of measles, this suggests that sub-clinical infection is a common feature in such communities. The introduction of measles into a virgin population generally results in a widespread outbreak involving individuals of all ages. In this country measles exhibits a definite biennial epidemicity. Antibodies have been demonstrated in the serum of persons with a history of a definite attack of measles by neutralization and by complement-fixation techniques using infected tissue-cultures as antigen.

PROPHYLAXIS

Antibodies are present in convalescent or adult serum and the use of *gamma* globulin has proved satisfactory as a prophylactic measure.

Gamma globulins are more potent than adult sera and have given good results. They should be administered to contacts within 72 hours of the appearance of the rash in the primary case; the dose varies with the age of the patient and also with the desire for modification or prevention of the attack. Prevention should be reserved for children whose life might be endangered by an attack of measles.

MEASLES VACCINE

Since 1958 trials have been made on the prophylactic value of various kinds of measles vaccine. Killed vaccines usually produced a good initial rise in antibody but this titre rapidly diminished after a few months. Some of the earlier live vaccines produced severe reactions with high temperatures and rash. Recently, in this country a live vaccine made from the attenuated Schwartz strain was given after a preliminary injection of killed Edminston vaccine and the results of this combination showed good protection and an absence of severe reactions.

THERAPY

Convincing results of the value of serum and gamma globulin after the onset of symptoms have not been forthcoming. Antibiotics should be given to control the secondary bacterial complications such as otitis and bronchopneumonia.

GERMAN MEASLES (RUBELLA)

German measles is one of the mildest of the common infectious diseases but it has been the subject of intensive investigation since Gregg's discovery in 1941 of the relationship between maternal rubella and congenital defects. The mechanism by which the infection is transmitted by the mother to the foetus has been clarified as a result of the cultivation of the virus by two groups of American workers in 1962.

German measles is a mild illness associated with a slight rise in temperature, a macular rash and a characteristic enlargement and tenderness of the lymph nodes of the occipital and post-auricular group. The rash usually starts on the face, scalp and neck and spreads downwards to affect the skin of the whole body. The mortality from German measles is negligible. The incubation period is 16–18 days.

The malformations associated with congenital rubella include cataract, congenital heart abnormalities, deafness, jaundice, hepatitis, pneumonitis, defective ossification of the long bones and mental retardation. They appear to result from early infection of the embryo with the rubella virus. The virus has been isolated from the majority of infants showing the rubella syndrome at birth and from the placenta or foetus in a proportion of cases of women in whom pregnancy had been terminated because of rubella in the first three months of pregnancy. When a pregnant woman is infected with rubella virus in the first three months of her pregnancy the virus passes through the placenta and produces a general infection of the foetus.

The virus was successfully grown on tissue culture in 1962 and it was found that it could be neutralized by convalescent serum but not by serum from acute cases of rubella. An effective inactivated vaccine is not yet available. Gamma globulin prepared from pooled serum containing neutralizing antibody gives some measure of protection to contacts during outbreaks of German measles.

Therapeutic abortion may be carried out in women suffering from rubella during the first three months of pregnancy. In women exposed to rubella during this period of their pregnancy the risk of infection of the foetus depends upon the presence or absence of rubella antibodies in the maternal serum at the time of exposure. If antibodies are present there is no danger to the foetus. If no antibodies are present the foetus may be in danger, and large doses of gamma globulin should be given. A second specimen of serum should be obtained four weeks later, and if at this time no antibody is present the danger to the foetus is passed; if antibody can be demonstrated at this time the foetus is in danger of infection and the advisability of termination of pregnancy should be considered.

DENGUE FEVER

Dengue fever is a mosquito-borne disease occurring in the tropics. Alarming epidemics have been reported, but in spite of the severity of the symptoms the mortality rate is very low. The condition is transmitted from man to man by the agency of mosquitoes, principally *Aedes aegypti.*

Attempts to reproduce the disease in experimental animals were unsuccessful until recently when the virus was established in the mouse by intracerebral injection. The disease has been transmitted to man by means of infected mosquitoes and also by the inoculation of blood collected during the febrile stages of the condition. It has also been shown that the aetiological agent passes through Berkefeld filters.

The degree of immunity following an attack of dengue fever is uncertain. It is usually considered to be of a low order and of short duration. The dengue virus is a poor antigen and attempts to make a satisfactory vaccine have been unsuccessful.

SAND-FLY OR PAPPATACI FEVER

Sand-fly fever is a tropical disease transmitted by sand-flies and characterized by its sudden onset and short duration. Recovery is as a rule rapid and complete, the disease is never fatal.

It can be propagated by intracerebral inoculation in suckling mice. Experiments on human volunteers suggest that the aetiological agent is a virus which is present in the blood during the febrile period.

An attack is followed by a high degree of immunity which persists for at least two years.

ENCEPHALITIS

Many forms of encephalitis and encephalomyelitis of virus origin have been described in different parts of the world. A virus has been isolated in many instances and, while some of these appear to be related, the majority have proved to be distinct types. In other cases a virus has not been obtained but the clinical and pathological pictures indicate that the causative agent is a virus. Several attempts have been made to classify the various forms of encephalitis but a satisfactory scheme has not yet been produced.

EPIDEMIC ENCEPHALITIS (ENCEPHALITIS LETHARGICA)

Epidemic encephalitis is an infectious disease of the central nervous system involving mainly the grey matter of the brain. It was first recognized as a clinical entity in 1917 by v. Economo, who described a small outbreak in Vienna; lethargy was a prominent feature, and he termed the disease *encephalitis lethargica*. Subsequently outbreaks of varying severity were reported in most parts of the world; in recent years the incidence has decreased considerably and the condition is now rarely seen.

The onset may be sudden or gradual. Prominent symptoms are pyrexia, headache, lethargy during the day, nocturnal insomnia, visual disturbances such as diplopia and blurred vision; there may also be delirium, and choreiform movements. The mortality rate is about 30–40 per cent.

On autopsy the chief lesions are generally found in the grey matter of the mid-brain, pons and medulla, and consist of perivascular and diffuse infiltration with degeneration of the nerve-cells and occasional neuronophagia.

AETIOLOGY

In spite of much investigation by workers in different countries, the nature of the aetiological agent remains unknown.

MURRAY VALLEY ENCEPHALITIS

In 1917 an outbreak of encephalitis occurred in the Murray Valley in Victoria, New South Wales and Queensland. A virus was isolated from the brains of fatal cases by inoculation of monkeys and later passed to sheep. In 1925 ten more cases occurred at Broken Hill in New South Wales, six of which were fatal and again the virus was isolated by intracerebral inoculation of sheep. No further cases occurred and all strains of the virus were lost.

In 1951 and again in 1956 cases of encephalitis occurred in the Murray Valley and again a virus was isolated. The question of the same virus being responsible for the 1917, 1951 and 1956 outbreaks must remain unresolved.

ST. LOUIS ENCEPHALITIS

The first known epidemic occurred in St. Louis County, Missouri, in 1933 when there were over 1,000 cases and 200 deaths. The virus was isolated from man during this epidemic. The disease is transmitted by mosquitoes and has since reappeared in the St. Louis district on several occasions. The virus produces an encephalitis in mice and monkeys on intracerebral inoculation. It is considered to be closely related immunologically to Japanese B encephalitis and West Nile Viruses.

JAPANESE B ENCEPHALITIS

The first large epidemic of summer encephalitis occurred in Japan in 1924 when there were over 6,000 cases in Tokyo and over 3,500 deaths. The condition was named B encephalitis to distinguish it from von Economo's Encephalitis Lethargica or A Encephalitis. The disease is epidemic mainly in Japan and the seaboard of China. The infection builds up in the spring and early summer by intense transmission in heron colonies between nesting herons and mosquitoes (*Culex tritaeniorhynchus*). The infected mosquitoes may bite man and so transmit the infection, but they also bite pigs in which a viraemia occurs and in due course the mosquito population around the piggeries becomes infected. These mosquitoes are more likely to bite man and it has been found that a high incidence of infection in pigs precedes an outbreak in man by two to three weeks. At the present time widespread vaccination of pigs with inactivated and live attenuated vaccines has been carried out in an attempt to protect the intermediate host from infection and so prevent the seasonal epidemics.

EQUINE ENCEPHALOMYELITIS

Small outbreaks of encephalitis have been reported in U.S.A.; in one of these 46 cases were observed with 33 deaths; approximately half the cases occurred in children under 5 years of age. A virus was isolated from the brains of fatal cases by the inoculation of mice. This virus was later found to be identical with the Eastern type of equine encephalomyelitis virus. From other outbreaks the Western type has been isolated. These diseases are transmitted by mosquitoes or mites and the viruses are quite distinct; horses and mules are the primary hosts. The onset of the disease is sudden with pyrexia, irritability, neck rigidity, convulsions; most patients become semi-comatose while fatal cases developed deep coma. Pathological changes are found in the brain and cord, consisting of nerve cell destruction with perivascular and meningeal infiltration with polymorphonuclear and large mononuclear cells.

A Venezuelan type of virus has also been isolated.

LOUPING-ILL

Louping-ill is a widespread disease of sheep. The aetiological agent is a small virus which can be readily transmitted to mice, monkeys and hamsters but not to rabbits and guinea-pigs. The disease is transmitted by the tick, *Ixodes ricinus*. Several cases have been reported in man, usually in laboratory workers; the disease takes the form of a mild encephalitis with recovery the rule. The mode of transmission in these cases is uncertain. The complement-fixation test has proved a useful diagnostic measure.

OTHER FORMS OF ENCEPHALITIS

Mild forms of encephalitis have been encountered in natives of East Africa and the West Indies and described as West Nile, Mengo, Bwamba and Ilheus fevers or encephalitis. In each case a distinct virus has been isolated. Transmission is by mosquitoes or other arthropods. These diseases have not occurred in this country.

HERPES ZOSTER OR SHINGLES

Herpes zoster is an acute infection characterized by a vesicular dermal eruption with pain around the infected region. The infectivity is low; small outbreaks have been reported but they are rare. The vesicular eruption is frequently unilateral and is often preceded by local hyperaesthesia, itching or pain, which is referable to one or more nerve-roots corresponding to the level and site of the eruption. Perivascular and diffuse mononuclear infiltrations are present in the posterior root ganglia and the posterior horns of the spinal cord.

The causative agent of zoster is a virus, identical with that of varicella but unrelated to that of herpes simplex. It has been grown on cultures of human tissues and produces cytopathic changes. Laboratory animals are refractory but the lesions have been reproduced in children by the inoculation of the vesicular fluid, in which elementary bodies have been demonstrated. By means of the complement-fixation test it has been found that the serum of individuals recovered from an attack of varicella reacted equally well with antigens prepared from both zoster and varicellar lesions. The inoculation of susceptible children with zoster material resulted in the development of chicken-pox, whereas in children immune to chicken-pox no lesions developed. Although there is a close relationship between these two viruses, there is a marked difference in their individual infectivity. Varicella is extremely infectious and occurs in children, while zoster has a low infectivity and is usually met in adults where it produces only a localized skin rash.

VARICELLA OR CHICKEN-POX

Chicken-pox is a highly contagious disease frequently found in children and characterized by a cutaneous eruption. The condition is usually mild, complete recovery being the rule. The identity of the aetiological agent with that causing herpes zoster has been discussed above. Nuclear inclusion bodies have been demonstrated in the epidermal and endothelial cells of the skin. The condition has been reproduced in children by the injection of the bacteriologically sterile vesicular fluid obtained from the specific lesions.

A high degree of immunity follows an attack of chicken-pox; convalescent serum has consequently been suggested as a possible therapeutic measure.

ACUTE CORYZA OR THE COMMON COLD

Acute coryza is an acute infectious disease affecting the mucous membrane of the upper respiratory tract. The symptoms are relatively trivial but they give rise to much local discomfort. It is only within recent years as a result of intensive studies carried out on volunteers that evidence about viruses responsible for the common cold has become available.

These viruses, now termed Rhinoviruses, were first recovered in tissue cultures, either of human embryo kidney or of human embryo fibroblasts. The Rhinovirus group is now known to contain at least 30 serologically distinct viruses. They are ether stable, inactivated at pH 3 in 1–3 hours, and are of the same order of density as the polio-virus. Infection of volunteres with nasal secretions containing Rhinoviruses with resulting colds has been found to occur almost entirely in those with low levels of pre-existing antibody. It has also been found that Rhinoviruses of different serological types induce during infection antibody formation specific for each type; but the duration of homologous immunity to each of these infections is not known.

There is no specific form of prophylaxis or treatment for the common cold. Work with specific Rhinovirus vaccines has been in progress for some time but no protection was found against an antigenically different Rhinovirus, and in view of the antigenic diversity of this virus group vaccination does not offer much hope of cold control. General measures directed against the spread of air-borne infection such as ultra-violet light or chemical aerosols have failed to produce a reduction in the incidence of colds.

In spite of the early belief in the prophylactic value of mixed bacterial vaccines prepared from the organisms found in the nasopharynx, all carefully controlled trials of such vaccines have failed to establish their value in the prevention of the common cold.

CROUP

Obstructive laryngotracheobronchitis or croup is a well-known disease of young children. The disease is of virus origin and several types of virus have been isolated. These are now known as the para-influenza viruses 1, 2, 3 and 4. Serological evidence suggests that infection with these viruses is widespread and is acquired early in life.

INFECTIVE HEPATITIS

Infective hepatitis tends to be particularly prominent under war-time conditions. Sporadic cases and small outbreaks are usual in peacetime and the disease is often termed "catarrhal jaundice". Extensive outbreaks are encountered not infrequently among Service personnel; during the Middle East campaign of World War II the disease was widespread.

Infective hepatitis is characterized by general malaise, headache, nausea, vomiting and pyrexia with the appearance of jaundice some seven days after the onset. The mortality rate is low. Histological studies have shown definite inflammatory changes in the liver.

The incubation period is long, 4–6 weeks in many cases, and the mode of transmission is uncertain. Some workers consider that the virus is spread by droplet infection, but attempts to isolate the virus have been mainly unsuccessful as laboratory animals have given negative results. A few tests on human volunteers indicate that the virus is present in the naso-pharynx during the early stage of the disease. The virus is also present in the faeces and serum of cases.

A closely related, if not identical, form of infective hepatitis has followed the administration of human serum or blood in the form of transfusions, convalescent serum or yellow fever vaccine; in these cases the blood was originally collected from individuals harbouring the virus. This disease is referred to as homologous serum hepatitis. The disease has also resulted from routine arsenical treatment in V.D. clinics; there is much evidence to indicate that infection in these cases was due to the contamination of the syringe with infected blood and incomplete sterilization before subsequent use. There is a long incubation period, usually 60–120 days.

It has been recommended that the two forms of the hepatitis virus should be designated—(1) hepatitis virus A (epidemic hepatitis) and (2) hepatitis virus B (homologous serum jaundice). It is however uncertain whether these are to be considered as two distinct viruses or simply variants of the same virus. Evidence at present available would suggest that they are distinct.

PRIMARY ATYPICAL PNEUMONIA

Primary atypical pneumonia is a disease characterized by sudden onset, prostration, pyrexia with evidence of lung involvement which is mainly of an interstitial nature. There is good evidence to believe that several agents may produce this syndrome. The pneumonia of Q fever due to *R. burnetti* is of this type as is the pneumonia caused by psittacosis virus. Adenovirus (especially type 4) has been isolated from cases of primary atypical pneumonia as have the influenza viruses. A large group of cases still remain which are not due to these agents, and are due to infection with a PPLO *Mycoplasma pneumoniae*.

Two non-specific tests are of value in the diagnosis of primary atypical pneumonia. These are the detection of agglutinins for a non-haemolytic streptococcus designated M.G. and tests for the presence of non-specific cold agglutinins. The latter may be demonstrated by incubating patients' serum with a suspension of Group O cells at 4°C; the agglutination is reversible at 37°C. Streptococcus M.G. is not responsible for primary atypical pneumonia but may be agglutinated by serum from some cases. This is probably due to the presence of common antigens between the streptococcus and the aetiological agent of the pneumonia.

VERRUCAE OR WARTS

Common warts occur most commonly on the skin of the hands but may be found on the soles of the feet, the face and the genitalia. On histological examination the affected skin is seen to contain intranuclear basophilic inclusions in the cells of the stratum granulosum of the epidermis. Electron microscopy has shown that virus particles are directly associated with these inclusions and as with Molluscum contagiosum the virus grows best in the degenerating rather than in the actively proliferating layers of the epidermis.

Human warts are not easily transmitted to experimental animals and to date no reliable method of cultivation of the virus in tissue culture has been developed. Spontaneous cure is the rule and it is only in cases of painful warts, for example, on the sole of the foot that resort to freezing with CO_2 snow or curettage may be necessary.

MOLLUSCUM CONTAGIOSUM

Molluscum contagiosum is a world-wide virus infection of the epidermis of the skin, in which painful nodules develop particularly over the face, the genitalia and the skin of the hands and feet. The skin of the palms and soles is not affected. On histological examination the cells of the germinal layer of the epidermis are seen to be enlarged and to contain within their cytoplasm acidophilic granular inclusions with a diameter of 20–35μ filled with virus particles. Examination of the superficial layers of the epidermis reveal aggregations of virus embedded in keratin and it would appear that virus proliferation is associated particularly with degenerating cells. No susceptible host other than man is known and to date no tissue culture has been found which will support the growth of this virus.

Spontaneous healing usually occurs but it may take as long as several months. Enucleation of the centre of the skin lesion and application of silver nitrate or cautery is claimed to hasten recovery; care must be taken to avoid unnecessary scarring and possible secondary infection.

TRACHOMA

Trachoma is a specific disease characterized by inflammation and hypertrophy of the conjunctiva. It is usually found in communities where hygienic conditions are poor.

The presence of inclusion bodies in affected conjunctival epithelial cells and the demonstration of elementary bodies in the conjunctival epithelium suggests a virus infection.

The virus has been successfully grown in the yolk-sac of the chick-embryo and has been transferred to the conjunctiva of a human volunteer. An experimental disease has been produced in baboons. Trachoma virus is large (130–200mμ) and appears to be related to the Psittacosis-L.G.V. group of viruses.

The sulphonamides and the tetracyclines have proved to be of great value in the treatment of trachoma.

Inclusion Blennorrhoea is considered by some workers to be closely related to trachoma and to have a similar aetiology.

GLANDULAR FEVER

Glandular fever or infectious mononucleosis is an acute infectious disease characterized by fever, sore throat, enlargement of the lymph nodes and a lymphocytic blood picture. Heterophil antibodies are present in the blood. Liver function tests suggest that some degree of hepatic damage is present in all cases.

The disease is commonest between the ages of 15–25, the incubation period is said to be 5–10 days and the illness lasts from 1–6 weeks. No organism has been consistently demonstrated in the blood or tissues of glandular fever patients and it has become widely accepted that the disease is due to a generalized virus infection although positive evidence of the existence of such a virus is still lacking. No virus has been isolated in tissue culture; laboratory animals have not been successfully infected and even human volunteers have failed to develop the signs of glandular fever following injection of nasal washings or material from abnormal lymph nodes obtained from active cases. This difficulty in experimental transmission to human volunteers is both disappointing and surprising in view of the apparent infective nature of the condition as shown by occasional definite outbreaks of the disease in households, schools and hospitals.

LABORATORY DIAGNOSIS

The agglutination of sheep red cells by the serum of glandular fever patients was observed by Paul and Bunnell in 1932 and forms the basis of a valuable diagnostic test associated with their names. The laboratory diagnosis also depends on the presence of characteristic changes in the blood picture. These consist of an increase in lymphocytes and the presence of abnormal mononuclear cells which are believed to be derived from lymphocytes. Histological examination of excised lymph nodes show a disappearance of the normal architecture of the node and a replacement with areas of marked pleomorphic mononuclear hyperplasia.

TREATMENT

There is at present no specific therapy for this condition. Extremely sore throats and very tender lymph nodes have been relieved by rapidly decreasing doses of corticosteroids but great care should be exercised in using this form of symptomatic therapy because of the dangers of intercurrent bacterial infection.

VIRUSES AND TUMOURS

A number of viruses have been isolated from naturally occurring tumours of birds and animals. These include the avian lymphomatosis virus of fowl leukaemia, the Rous sarcoma virus of malignant fowl sarcoma, the papilloma virus from the wild cottontail rabbit, the renal carcinoma virus from the leopard frog, and the subcutaneous fibroma virus of the wild deer. All these viruses have been called comprehensive or "do-all" tumour viruses by Rous to differentiate them from other tumour-producing viruses which do not produce tumours in nature and can only stimulate cell proliferation in experimental animals. In general the "do-all" tumour viruses are species specific and so produce a constant histological lesion in a particular tissue of a particular species. In this way they differ from the polyoma virus which was first isolated from mice with leukaemia and was subsequently cultivated in tissue cultures. The polyoma virus when injected into new-born mice gave rise to a wide range of histologically different tumours, from which the virus could subsequently be recovered. It also caused tumour formation in new-born hamsters, rats, rabbits and ferrets. When polyoma virus was added to a tissue culture of normal mouse fibroblasts, changes comparable to a malignant transformation took place in these infected cells. These cells lost the property of contact inhibition, grew in heaped-up masses and produced sarcomata when injected into three-week-old hamsters.

Another potential tumour-producing agent was isolated from Rhesus monkey kidney tissue cultures, and was known as the vacuolating virus or SV40 and was not at first associated with disease in spite of its DNA content and its similarity in both size and structure to the polyoma virus. However, when injected into new-born hamsters SV40 produced sarcomata and these sarcomata are capable of serial transplantation. The transplanted SV40 induced tumours will continue to grow vigorously even when the free virus can no longer be recovered from the tissue. In fact, both polyoma and SV40 viruses act more like chemical carcinogens than actively infecting neoplastic agents.

The search for viruses in human tumours continues and many agents which act like viruses in experimental animals and tissue culture have been isolated.

The agent cultivated from leukaemic bone marrow in tissue culture in 1964 and thought at first to be a virus was later shown to be a mycoplasma. The Burkitt lymphosarcoma is found in children in those parts of Africa infested with the *Aedes* mosquito. The tumours occur on the face and may, it was suggested, be infective in origin. A number of viruses including herpes and a reovirus have been isolated from biopsies of tumour material, and lymphoblasts derived from Burkitt tumours have been grown in tissue culture and have been shown to contain both in their cytoplasm and in their nuclei virus-like particles; these same lymphoblasts have shown themselves resistant to infection with the vesicular stomatitis virus, suggesting that they were already infected with another but unidentified virus.

It is improbable that man, unlike many other mammals, should be entirely free from virus induced tumours; equally it is improbable in view of the small number of comprehensive "do-all" viruses that have been isolated from naturally occurring animal tumours that any but a very small number of human neoplasms would be due to naturally occurring viruses.

CAT SCRATCH DISEASE—BENIGN LYMPHORETICULOSIS

This is an infectious disease associated with fatigue, anorexia, a low grade fever and marked enlargement and tenderness of the lymph nodes particularly the epitrochlear and axillary groups. The condition almost always develops one to three weeks after the patient has been scratched or bitten by a cat.

Attempts to demonstrate the infective agent in the affected glands by microscopy, cultural methods or animal inoculation have proved inconclusive. However, a tuberculin-type sensitivity is developed in cat scratch disease to the agent present in the inflamed lymph nodes. The history of a bite or scratch by a cat, the benign clinical course, the histology of the affected glands and a positive skin test to cat scratch antigen may serve to establish the diagnosis.

MYCOPLASMA

Mycoplasma or pleuro-pneumonia-like organisms (PPLO) are the smallest microorganisms with a free living existence, and are responsible for an atypical pneumonia in man associated with the presence of cold agglutinins. Mycoplasma are a common source of contamination in tissue cultures and can often be found in cases of malignant disease in man.

PROPERTIES OF THE ORGANISM

Mycoplasma can grow on agar with sterol and protein supplementation. In size they resemble the myxovirus, ranging from 125–150mμ in diameter. Mycoplasma are absolutely resistant to penicillin but are sensitive to tetracycline.

Chapter XXXII

The Rickettsiae

Small rod-shaped bodies, frequently present in the alimentary tract of arthropods, have been found in association with certain diseases of man and animals. These bodies were designated *Rickettsia* by Da Rocha-Lima (1916) in honour of Ricketts, who first described them in 1909 in cases of Rocky Mountain spotted fever.

The *Rickettsiae* are generally accepted as minute forms of bacteria representing a stage of biological evolution intermediary between the viruses and the bacteria.

GENERAL PROPERTIES

The *Rickettsiae* are minute bodies of an average size, $0{\cdot}5 \times 0.3\mu$. The majority of the species are non-motile. These bodies are difficult to stain and when stained their outline is somewhat indefinite; they are Gram-negative. Giemsa is one of the most suitable methods of staining; the bodies are purplish and bipolar staining is a common feature.

The *Rickettsiae* do not grow on the ordinary bacteriological media; living cells are necessary and the yolk-sac of the chick-embryo has given good results. Cultures should be harvested immediately prior to the death of the embryo.

Another satisfactory method of isolating the *Rickettsiae* is the inoculation of susceptible laboratory animals, of which the guinea-pig is the most valuable. Blood from suspected cases is generally injected by the intraperitoneal route; a fever with a temperature of about 104°F is produced in 5–12 days; the animal should be killed during this phase and the *Rickettsiae* are found in large numbers in the spleen and brain.

The *Rickettsiae* do not pass through the usual bacterial filters, and their resistance to bactericidal agents is similar to that exhibited by the vegetative bacteria.

The serological behaviour of these bodies has recently been studied. Two distinct types of test have been used: (1) the *Weil-Felix Test* which is non-specific, the serum of the patient being tested against "O" suspensions of *Proteus* organisms, and (2) specific agglutination or complement-fixation tests in which the *Rickettsiae* are used as the antigen.

The *Rickettsiae* have been found in close association with the following diseases of man of which they are considered by most bacteriologists to be the aetiological agents (Table 1.XXXII).

Table 1.XXXII
Human Rickettsial Diseases

Disease	*Rickettsia*	*Insect vector*	*Weil–Felix reaction*	
			OX 19	*OX K*
Epidemic or exanthematic typhus	*R. prowazeki*	Louse	+++	(+)
Endemic or murine typhus	*R. mooseri*	Rat flea	+++	(+)
Scrub typhus (Tsutsugamushi fever)	*R. tsutsugamushi*	Mite larva	0	+++
Rocky Mountain spotted fever	*R. rickettsi*	Tick	++	++
Fièvre boutonneuse	*R. rickettsi* (*conori*)	Dog tick	++	++
Q fever	*R. burneti*	Tick	0	0

TYPHUS FEVER

Typhus fever, known also as gaol fever, camp fever, endemic or urban typhus, is a disease associated with insanitary conditions, overcrowding, poverty or famine and at one time was widespread, epidemics being frequently encountered. At present it is uncommon in most civilized communities where marked improvement in sanitation has taken place; in this country it is mainly of historic interest although cases occurred among British troops in Africa and the Far East during the past war. Typhus fever has a world-wide distribution and is most prevalent in the colder months of the year, when people tend to crowd together.

The classical or epidemic disease is transmitted by the agency of the louse, generally *Pediculus humanus corporis*, which becomes infected by sucking the blood of patients. The period of infectivity in lice varies from 2–10 days. Man is infected either by the bite of an infected louse or through wounds produced by scratching skin contaminated by the excreta of the insect; there is an incubation period of from 5–20 days, with an average of 14 days. The onset is sudden and is characterized by fever, headache, vomiting and on the fourth or fifth day by the appearance of a generalized macular or haemorrhagic rash. After 14 days the temperature usually falls by crisis and convalescence, which should be prolonged, ensues. The mortality rate varies and depends on the age and condition of the patient; it may be as low as 2 per cent or as high as 30 per cent. A high degree of immunity follows an attack.

BACTERIOLOGY

R. prowazeki is generally accepted as the aetiological agent of epidemic typhus fever. This relationship was first described by Da Rocha-Lima, who observed large numbers of the organisms in the intestinal tract and the lining epithelial cells in the majority of lice collected from typhus patients. They were not found in lice obtained from healthy individuals. It was later shown that lice containing *R. prowazeki* were able to produce infection in guinea-pigs and monkeys; *Rickettsiae*-free lice were not infective. The demonstration of the organisms in human patients is difficult, but they have been found in various vascular lesions.

Monkeys and guinea-pigs can be infected by the subcutaneous inoculation of blood collected from cases during the febrile period. The condition is similar to the spontaneous human disease. In some cases the guinea-pig, while becoming infective, does not give any reaction; to this condition Nicolle and his colleagues have given the term "masked typhus".

DIAGNOSIS

The demonstration of *R. prowazeki*, either in lice caught on typhus cases or in human lesions, is not widely employed in practice as a diagnostic measure owing to technical difficulties.

A serological test discovered by Weil and Felix in 1916, and referred to as the Weil-Felix reaction, is now widely used and gives constant and reliable results. The reaction is an agglutination by the serum of typhus patients of a strain of *Proteus* organism isolated from the urine of a case of typhus fever. Several strains of this organism were isolated, and the one giving the most specific results was termed *Proteus* OX 19. The agglutinins appear in the patient's serum on the fourth or fifth day, increase until the second or third week, and then decrease until by the end of the fifth month they may have disappeared. Titres as high as 1/50,000 are obtained during the febrile stages, and the agglutination is invariably of the O type.

The constancy of this reaction raised the question of the relationship of the *Proteus* OX 19 strain to typhus fever. Attempts to incriminate them as the aetiological agent have been, however, unsuccessful. The organisms could only be isolated from a few cases; they did not produce either the experimental infection or immunity in guinea-pigs to infection with *Rickettsiae*. The agglutination with *Proteus* organisms is due to these organisms having antigens in common with the *Rickettsiae*.

Agglutination, or complement-fixation tests, with specific rickettsial suspensions is now used to establish the identity of the infecting organism.

PROPHYLAXIS

The prevention of typhus fever depends mainly on the control and eradication of the louse, with attention to general hygienic conditions, personal hygiene and cleanliness. In the 1939–45 War dusting with insecticides, such as D.D.T., was widely used with great success.

Active immunization, produced by inoculation of living, attenuated or killed *Rickettsiae* has been widely employed. The main difficulty originally was to obtain an adequate supply of the *Rickettsiae* but this difficulty has been overcome by the use of yolk-sac cultures or the lungs of infected rabbits or mice. War-time experience suggests that such vaccines were at least effective in reducing the severity of a subsequent attack of typhus.

THERAPY

Chloramphenicol and the tetracyclines have given good results in the treatment of typhus fever; 2 g per diem, in divided doses, has been the usual schedule.

MURINE TYPHUS (ENDEMIC TYPHUS)

Murine typhus (endemic typhus), the aetiological agent of which is *R. mooseri*, is a similar but milder disease, which can be differentiated from epidemic or exanthematic typhus by serological tests with rickettsial suspensions. *R. mooseri* is antigenically related to *R. prowazeki* and is usually transmitted to man from rodents, particularly the rat, by the flea, *Xenopsylla cheopis*. It has a world-wide distribution.

The control of murine typhus depends upon adequate methods of rat destruction and rat control by ratproofing of buildings. Dusts containing D.D.T. applied to rat runs and rat-infested premises diminish the infestation of rats by fleas and thereby help to prevent the transmission of the disease to man. The methods of laboratory diagnosis and treatment of murine typhus are essentially similar to those in use for epidemic typhus.

ROCKY MOUNTAIN SPOTTED FEVER

Rocky Mountain spotted fever is a condition resembling typhus fever and found in certain districts of America. It is the disease in which the *Rickettsiae* were first detected by Ricketts in 1909. Ricketts had previously reproduced the disease in healthy individuals and guinea-pigs by the injection of the defibrinated blood of cases. The organisms were later found in ticks that had been feeding on cases. The ticks serve not only as vectors, but also as reservoirs for *R. rickettsiae*, and the dog is the important host for the adult tick. The Weil-Felix Test is a useful diagnostic procedure.

The tetracycline antibiotics and chloramphenicol are effective remedies for the treatment of human infections and are best given during the early stages of the disease.

FIÈVRE BOUTONNEUSE

This is a mild disease found mainly in Tunisia and characterized by the appearance of a local nodule at the site of the insect bite. The aetiological agent is a *Rickettsia* which appears to be identical with *R. rickettsiae*, the causative agent of Rocky Mountain spotted fever. It is transmitted from dogs by the dog tick.

SCRUB TYPHUS

This disease is identical with the Japanese River or Tsutsugamushi fever. The clinical course is similar to that of epidemic typhus but there is usually definite evidence of a mite bite. It is transmitted by a larval mite, *Trombicula deliensis*, which is found in the scrub district of S.E. Asia; the main reservoir of the organism is the rat. A positive Weil-Felix reaction with OX K strains of *Proteus* is obtained in both conditions. The mortality rate varies considerably in different localities (10–60 per cent). It was an important disease among troops in the Far East, where anti-mite measures such as clearing scrub and the use of protective clothing proved effective. Tetracycline and chloramphenicol have proved useful therapeutic agents.

Q FEVER

Q fever was originally described in Australia among workers in dairies and abattoirs; it is an acute illness of 7–24 days' duration with headache, fever, myalgia and later a dry cough. The disease is widely distributed and cases have recently been encountered in this country. It is generally acquired from cattle by the consumption of infected milk or by the inhalation of contaminated dust. The mortality rate is low.

The causative agent closely resembles the *Rickettsiae* with which it is often classified as *R. burneti*. However, it is much more resistant to desiccation than the *Rickettsiae* and, unlike them, will survive in milk or in water for long periods. The diagnosis is established either by the isolation of this organism or by the demonstration of specific antibodies by the complement-fixation test. Agglutination is not given with *Proteus* suspensions (Table 1.XXXII). Treatment with chloramphenicol or tetracycline antibiotics has proved of value in the majority of cases.

Chapter XXXIII

Bacteriophage

Twort, in 1915, observed that some colonies of a white micrococcus became clear and glassy. No micrococci were found in the clear colonies. When fresh colonies of the micrococcus were inoculated with the glassy material, they too became clear and glassy. This lytic action could be transmitted indefinitely. In 1917 D'Herelle observed a comparable phenomenon. When faeces from dysentery cases were emulsified in broth and filtered, the filtrate lysed young broth cultures of *Sh. shigae.* This phenomenon could be carried on indefinitely by adding the filtrate of a lysed culture to fresh young cultures of the same organism. This "Twort-D'Herelle Phenomenon" is now known to be due to a bacteriophage, or simply phage.

Phage has been the subject of a great deal of investigation in recent years, and it is now known to be a bacterial virus. Many different phages have been described, and their properties investigated. The size of different strains of phage varies from 8mμ to 75mμ. As seen in electron micrographs, most phages have a well-defined head and a relatively long thin tail. The head contains nucleic acid. Phage can be propagated only by growing in a susceptible bacterial cell. Some phages have a relatively broad range of activity on bacteria of the same general type, and others will attack only a specific strain. The tip of the tail of the phage particle attaches to a receptor on the bacterial surface. The DNA of the head is then injected into the bacterial cell, and after a few minutes replication of phage takes place. Several hundred new phage particles may be produced within the bacterium which then bursts and releases the phage. The enormous increase in the number of phage particles can be demonstrated by diluting the culture filtrate and spreading it on a plate seeded with the susceptible organism. After incubation colonies of phage, known as plaques, or *taches vièrges* appear as clear spots in the otherwise confluent growth of the bacterium. Only young, living cultures are attacked by phage. Dead organisms may absorb the phage, but no lysis occurs.

Phage is resistant to glycerine, alcohol and low temperatures but is destroyed by ultra-violet rays, and heat over 70°C. Phage contained in sealed tubes and kept in the dark at room temperature retains its activity for years.

Bacteriophage is widely distributed in nature. It can be isolated from faeces, sewage, water and soil. Phages have been found for bacteria of almost every variety.

When a bacterial culture is attacked by phage, there are usually some surviving organisms, which may then multiply. These survivors and their progeny are resistant to the phage.

Phage is antigenic, and antisera can be prepared by the usual methods in rabbits. The antiserum inhibits the action of the phage, probably by attachment to the tail which is then prevented from attachment to the receptors on the bacterial cell surface.

PHAGE TYPING

The susceptibility of various cultures of *S. typhi* to different phages was first applied by Craigie as an epidemiological tool to trace the source of infection. This work has been greatly expanded by other workers. Phage typing is now applied to several species of salmonellae, and to pathogenic staphylococci. It is possible by the appropriate selection of phages to differentiate strains of an organism which have identical biochemical and serological reactions.

Chapter XXXIV

The Bacteriology of Water, Milk and Shellfish

WATER

Water is the main constituent of the body and is consequently a necessary article of diet. Water for drinking purposes is obtained from various sources and is divided according to its origin into two main types—(a) surface and (b) deep.

(a) Surface waters comprise those collected at or near the surface of the ground and include rain, lakes, rivers and reservoirs. This type is exposed to contamination from many sources and invariably contains bacteria.

(b) Deep waters are collected from an underground source, such as wells and springs. The water undergoes a process of filtration during percolation through the soil and is usually bacteria-free at the source.

Surface waters are the main source of drinking-water on account of their accessibility. Such water always contains bacteria, the common saprophytes being invariably present. Lakes and rivers are often heavily contaminated as they receive the drainage from the neighbouring land and are sometimes used for the disposal of sewage. When sewage is disposed of by other means the sewage and water systems may run close together and fractures and leakages in the pipes have resulted in contamination of the water supply. It therefore follows that the presence of pathogenic organisms is not uncommon in this type of water. If it is consumed in an untreated state, it may be responsible for serious outbreaks of infections, the most important of which are those involving the intestinal tract, especially enteric fever, dysentery and cholera. Many such epidemics have been reported in which the drinking-water has been the vehicle of infection.

The water supply of large towns is either brought from remote regions many miles away or is collected from local rivers, and is subjected to strict control. In rural districts, however, drinking-water is frequently obtained from shallow wells situated close to dwelling-houses and exposed to faecal contamination; these supplies do not usually conform with the accepted bacteriological standards for drinking-water.

There is thus a definite necessity for the rigid control of drinking-water. This is carried out by attention not only to the bacterial content, but also to the physical appearance and the chemical analysis.

The bacteriological examination constitutes an important part of the control of water, but a final judgment of the potability of a water should never be given from the bacteriological examination alone; in all cases detailed information of the topography of the drainage area should be available. It is also important to note that a single bacteriological investigation, unless quite unsatisfactory, is of very limited value. The detection of the important pathogenic organisms is difficult and often impossible, and this consequently does not form part of the routine test, which is devised to determine (1) the total number of living bacteria and (2) the presence of coliform organisms.

The latter provides the most important information as contamination with sewage and excreta constitutes the greatest danger associated with drinking-water. It has been shown that the total number of bacteria in a sample has little practical significance and this investigation is often omitted from routine tests.

A representative sample of the water is collected, with due precaution to avoid contamination, in a sterile, stoppered, 200-ml bottle, and then placed in a special ice-box and submitted to the laboratory without delay. Two or three successive tenfold dilutions of the sample are then made in sterile water, using a fresh sterile pipette for each dilution and mixing well at each stage. Using a fresh pipette, 1 ml of each dilution is added to a Petri dish with melted agar and the contents well mixed; the number of dishes required depends largely on the nature of the original sample. On cooling, the agar plates are incubated at 37°C for two days and a colony count is then made. The results are averaged and expressed as so many bacteria per cubic centimetre or millilitre of the original sample.

The examination for coliforms is usually carried out by adding various amounts of the sample, e.g. 50 ml, 10 ml and 1 ml, to tubes of fluid MacConkey's medium, which contains lactose, bile salts and neutral red. These are incubated at 37°C for two days; subcultures are made on solid MacConkey's medium from the highest dilutions showing acid and gas formation. Two or three lactose-fermenting colonies are then selected from each plate and identified. The presence of *Escherichia coli* is more suggestive of faecal contamination than is that of the other *Enterobacteriacae*.

A simple test to differentiate faecal from non-faecal coliforms is based on the production of acid and gas in MacConkey broth at 44°C by faecal coliforms. Subcultures into MacConkey broth are made from the original MacConkey broth tubes showing acid and gas; the subcultures are incubated in a water-bath at 44°C—the temperature must be carefully controlled as variation may lead to irregular results. After 24 hours' incubation tubes showing acid and gas are accepted as containing faecal coliforms. This test has given satisfactory results.

Statistical tables giving the probable number of coliforms per 100 ml have been prepared according to the number of positive results given by the various quantities of water used in the test and the final report is a figure giving the probable number of coliform organisms per 100 ml of the water under examination.

The presence of members of the *E. coli* group, or faecal coliforms as they are generally termed, is of the greatest significance as this is indicative of recent faecal pollution. Most of the other *Enterobacteriacae* isolated are of less significance as they are considered to be introduced mainly from the local soil and vegetation; they have, however, been isolated on certain occasions from the faeces of man and animals and this division is therefore not strictly correct. As these organisms tend to be more resistant than the accepted faecal coliforms, their presence in water, when faecal coliforms are absent, may be due to more remote faecal contamination.

The results of a single examination are generally of doubtful significance, and a tentative report only can be given in such cases unless there is evidence of gross faecal pollution. When, however, repeated examinations of one water supply are made, rough standards can be fixed and any marked deviations from these should be viewed with suspicion.

The presence of *E. coli* is of greatest practical importance, as this is the most direct available evidence of faecal pollution. The following working rule is adopted: the greater the faecal pollution the greater the chance of the presence of pathogenic intestinal organisms. Therefore water with a relatively high *E. coli* count, ten or more per 100 ml, is considered to be unsuitable for drinking purposes, while a count of three to ten per 100 ml is regarded as suspicious. Further tests to indicate faecal pollution are the examination for *Cl. welchii* and *Strept. faecalis*, but these are not usually performed in routine work.

An examination for the presence of *S. typhi* or *V. cholerae* is sometimes required and the special selective media are used. Many methods have been tried, the most suitable being the concentration of the bacteria by centrifuging and seeding on special media, such as Wilson and Blair's, or adding large quantities of the water to equal quantities of a special broth, e.g. Selenite broth.

Water from contaminated sources can be rendered potable by various processes which destroy or remove bacteria. Those commonly employed are storage, sand-filtration, chemical treatment, e.g. chlorination, and treatment with ultra-violet light. The success of these measures has been reported in numerous public health records, and is indicated by the rarity of water-borne epidemics in civilized communities at the present time. In the household the purification of water is carried out either by boiling or by the use of a bacterial filter.

SWIMMING-BATH WATER

The great increase in the popularity of bathing during recent years has directed attention to the bacteriology of swimming-bath water. There is little doubt that the untreated water of bathing pools may become a suitable medium for bacterial multiplication, and during hot weather when the baths are crowded there are numerous opportunities for contamination with human excreta. Diseases stated with some justification to have been contracted in swimming pools include otitis media, infections of the upper respiratory tract and enteric fever. It is therefore natural that the condition of the water should be subjected to some control; the measures taken in this direction include continuous filtration of the water, treatment of the water with chemicals, such as ozone or chlorine (0·3 parts per million), and repeated bacteriological examinations. There are unfortunately no accepted standards for the bacterial content of such waters. It is usual to carry out an examination similar to that used for drinking-water, i.e. a determination of the bacterial content and a coliform test. The interpretation of the results presents many difficulties and the bacteriologist can only suggest that the condition of the water is either satisfactory or not.

It has been suggested that overcrowding in the dressing-rooms and the communal use of towels are probably more important in the spread of infection than the swimming-bath water.

MILK

Milk is an important food, particularly for young children. It is also a good bacteriological medium and it is therefore essential that the standard of milk for human consumption should be rigorously controlled. Four methods of examination are employed for this purpose: (1) physical—taste, odour and appearance; (2)

chemical—content of fat and total solids; (3) veterinary examination of the cows; (4) bacteriological. Only the bacteriological examination will be considered here.

Milk, even though collected with the best aseptic precautions from healthy cows is rarely sterile. Non-pathogenic staphylococci from the teat canal are almost invariably present. *Streptococcus lactis* is also present in small numbers. This is an important organism for the dairy industry; it is the organism responsible for the normal souring of milk, and for the initial acidity of milk during the process of cheese manufacture. Lactobacilli may also be present, and continue the process after the initial acidification by the streptococci.

All other organisms in milk from healthy cows must be considered as contaminants. These may be pathogens or saprophytes. Infected cows may contribute a pathogenic flora. The most important of these organisms are *Myco. bovis* and *Br. abortus*. Cows infected with Salmonella organisms—chiefly *S. dublin*, *S. enteritidis*, or *S. typhimurium*—may excrete these organisms in the milk. Cows suffering from mastitis may excrete pathogenic haemolytic streptococci or staphylococci in the milk. Other pathogenic organisms may be contributed to the milk by the milker or dairy staff by contamination from hands, or droplet contamination from the respiratory tract by coughing or sneezing. Many epidemics of streptococcal infection, diphtheria and enteric diseases have been traced to contamination of milk by infected dairy staff.

The keeping quality of the milk, which is usually of the greatest interest to the average consumer, depends on the contamination with saprophytic organisms during the course of milk production. There may be some contamination from cow dung on the udder unless it is carefully washed before milking. The greatest source of contamination, however, is unsterile utensils—milking machine, milk churns and other utensils. Even a small number of saprophytic organisms in an unsterile milk churn may multiply to astronomical numbers during warm weather before the milk reaches the retailer and the consumer.

PASTEURIZATION

Pathogenic organisms likely to be present in milk are non-sporing varieties which are relatively easily destroyed by heat. The main object of the process of heat treatment by pasteurization is to make a clean milk safe for consumption. It should not be applied to preserve the keeping quality of dirty milk. Various methods of heat treatment have been authorized, and tests to determine the efficiency of such methods have been prescribed. Untreated milk may also be sold, subject to certain regulations.

The Milk (Special Designation) Regulations 1963 and the Milk (Special Designation) (Amendment) Regulations 1965, have classified milk as (1) Pasteurized, (2) Sterilized, (3) Ultra Heat Treated, and (4) Untreated, and have stipulated the various methods of treatment and the tests to which each grade of milk should be submitted.

(1) Pasteurized Milk. The milk shall be retained at a temperature of not less than 145°F and not more than 150°F for at least 30 minutes and immediately cooled to a temperature of not more than 50°F *or*, the milk shall be retained at a temperature of not less than 161°F for at least 15 seconds and immediately cooled to a temperature of not more than 50°F.

Pasteurized milk must satisfy the phosphatase test. The methylene blue test is also applied to pasteurized milk.

(2) Sterilized Milk. The milk shall be sterilized, that is to say, filtered or clarified, homogenized and thereafter heated to and maintained at such a temperature, not less than 212°F for such a period as to ensure that it will comply with the prescribed turbidity test.

(3) Ultra Heat Treated Milk. The milk shall be treated by the ultra high temperature method, that is to say, retained at a temperature of not less than 270°F for not less than one second. A colony count test is prescribed for U.H.T. milk.

(4) Untreated Milk. Any cow which shows evidence of any disease which is likely to affect the milk injuriously shall be segregated from the rest of the herd, and the milk from such cows is not to be sold as untreated milk. Untreated milk is subjected to the methylene blue test.

The Methylene Blue Test for Untreated and Pasteurized Milk. Samples taken during the period from 1st May to 31st October shall be kept at atmospheric shade temperature until 9.30 a.m. on the following day. Samples taken during the period from 1st November to 30th April shall be kept at atmospheric shade temperature until 5.00 p.m. on the day of sampling, and thereafter at a constant temperature of 65°F until 9.30 a.m. on the following day. If the atmospheric shade temperature at any time exceeds 70°F the test is void.

The sample is thoroughly mixed and transferred to a test tube up to the 10 ml mark. One ml of a methylene blue solution is added to give a final concentration of 1/300,000. The tube is closed with a sterile rubber bung and placed in a waterbath at 37°C. The milk is regarded as decolourized when the whole column is decolourized up to within 5 mm of the surface. The test is satisfied by milk which fails to decolourize the methylene blue in 30 minutes. The methylene blue test gives an indication of the number of viable bacteria in the milk.

The Phosphatase Test. This test is used to check the efficiency of the pasteurizing process, which should inactivate the enzyme phosphatase which is normally present in milk. The reagent is p-nitrophenyl phosphate in a buffered solution. This is incubated with the milk. The enzyme, if present, reacts on the substrate and releases p-nitrophenol which is detected colorimetrically.

The Turbidity Test. Ammonium sulphate is added to the milk and the precipitate is filtered off. The clear filtrate, after boiling for five minutes should show no turbidity.

The Colony Count Test. 0·01 ml of milk, measured by a standard platinum loop of 4 mm diameter is inoculated into 5 ml of melted yeastrel agar at 45°C to 50°C. After mixing, the tube is sloped until solidified and then incubated for 48 hours. The colonies developing are counted. The test is satisfactory if less than 10 colonies are found.

Antibiotics in Milk. Since the introduction of antibiotics in veterinary medicine cows suffering from mastitis and other infections are frequently treated with penicillin or other antibiotics. These drugs are excreted in the milk, and even in very small amounts constitute a danger to consumers who may be allergic to them. It is now illegal to sell milk containing antibiotics and tests for the presence of these substances in milk are prescribed.

Pathogenicity Tests. Milk is frequently examined for the presence of human pathogens, particularly the tubercle bacillus. Several methods may be used to detect tubercle bacilli in milk—microscopy, culture and pathogenicity tests.

The usual procedure is to take two samples of the milk and centrifuge in special centrifuge tubes, by means of which the deposit can be readily separated from the supernatant fluid. The deposit, 2 ml, is injected into the groin of a guinea-pig, i.e. two animals (A and B) are used for each sample of milk. The A guinea-pig is killed after three weeks and the B guinea-pig at the end of five weeks. An autopsy is made and any tuberculous lesions are noted and carefully examined for the presence of the tubercle bacillus.

This method is far from satisfactory, as it is costly and requires several weeks before the result is known. Many attempts have been made to replace it by a simpler test. In recent years it has been claimed that both microscopical and cultural methods can give equally satisfactory results. For the cultural method a modification of Löwenstein's method is employed. The milk deposit is treated with either NaOH or H_2SO_4, neutralized and seeded on to a selective medium. After incubation for about three weeks it is stated that not only can the presence of the tubercle bacillus be detected, but also its type can be determined.

In the microscopical method the milk deposit is stained by the Ziehl-Neelsen method and is examined by the low-power magnification for the presence of cell groups, which are collections of endothelial cells. When these are found the field is examined by the oil-immersion lens for tubercle bacilli. It is stated that the tubercle bacilli are usually present in or around these cells.

While these tests appear to yield useful results, they do not appear to be sufficiently reliable entirely to replace the more costly pathogenicity tests.

The examination of milk for the presence of *Br. abortus* is becoming increasingly important. Investigations may be carried out by culture, serological methods or pathogenicity tests. In the pathogenicity test about 2–4 ml of the sample are injected subcutaneously or intraperitoneally into a guinea-pig. The animal is killed at the end of 4–5 weeks; blood and material from the spleen and regional lymph glands are seeded on liver-infusion agar and placed in an atmosphere containing 10 per cent CO_2 at 37°C. A similar medium containing 1 in 100,000 gentian-violet and adjusted to pH 6·6 is used in the cultural method. Serological tests depend on the presence of agglutinins in the milk and are less reliable than the isolation of the organism.

The examination for other pathogens is more complicated, special technique is required and positive results are frequently difficult to obtain.

ICE-CREAM

Ice-cream has been responsible for many serious outbreaks of enteric fever and food poisoning. In consequence special Regulations were introduced in 1947 to control its preparation and subsequent treatment; these require that after initial preparation, by heat treatment and rapid freezing, ice-cream must be protected from contamination and kept at a temperature not more than 28°F. Bacteriological standards have not been officially adopted, as at present, they are not considered sufficiently reliable. A modified methylene blue test is, however, recommended as a simple, practical test of bacterial cleanliness and, as a tentative measure, four grades have been defined according to the time taken to reduce the methylene blue; any sample failing to reduce completely in four hours is classified as Grade 1. The coliform test was found to be unreliable.

SHELLFISH

Shellfish, such as oysters, mussels, cockles, periwinkles, whelks, and escallops are frequently grown in polluted streams and estuaries, and may be contaminated, particularly with enteric pathogens. They may be cleaned by placing them in clean sea water for two or three days, but they are by no means sterile. Many cases of enteric fever have been due to the consumption of oysters.

Bacteriological control of oysters and mussels is difficult, and many methods of examination have been devised. The method commonly used is to cut up the tissue of a number of oysters or mussels, mix well with water or saline, and test for faecal coli by a modification of the test used for coliform organisms in water. The results are expressed quantitatively as the number of faecal coli per ml of tissue. The standards proposed are:

Grade I. Not more than 5 faecal coli per ml of tissue.
Grade II. 6–15 faecal coli per ml of tissue.
Grade III. More than 15 faecal coli per ml of tissue.

Part III

GENERAL TECHNIQUE

Chapter XXXV

The Microscope: Staining Methods

The medical student does not require more than a limited knowledge of bacteriological technique, and for this reason only a brief account of the more important methods is given. Several excellent books have been devoted solely to technique and the reader is advised to consult these if more detailed information is required. The following methods, while perhaps not those described precisely by the originators, have been found to give satisfactory results.

THE MICROSCOPE

The microscope has been used on many occasions during the earlier studies of the student, but in dealing with bacteria certain points require special attention. Owing to the minute size of bacteria it is essential to use a microscope possessing two eye-pieces, 5 times and 10 times, three objectives, $\frac{2}{3}$, $\frac{1}{6}$ and $\frac{1}{12}$ in., a sub-stage condenser, flat and concave mirrors, and some form of mechanical stage.

The source of light is also important. Artificial light, such as an electric bulb, is more suitable than daylight, as it is constant and does not change from day to day. The $\frac{1}{12}$-in. lens is used with immersion oil, which has the same refractive index as glass; this lens is frequently referred to as the oil-immersion lens. The oil is placed on the slide or cover slip, and the lens is carefully racked down with the coarse adjustment so that the oil is flattened out without the lens coming in contact with the slide. The eye is now applied to the eye-piece and the objective is gradually raised by the coarse adjustment until the smear or sample is brought into view; the final focusing is then carried out with the fine adjustment. When the examination is completed the oil, if moist, should be wiped off the lens with a silk cloth, if dry, xylol is used; alcohol must never be employed for this purpose, as it dissolves the cement in which the lenses are fixed. The oil-immersion lens is used with the plane mirror, an open diaphragm and a raised condenser.

The $\frac{1}{6}$ and $\frac{2}{3}$-in. lenses are mainly employed in the examination of sections and unstained material, as in the hanging-drop technique for the detection of motility.

With the usual working conditions the range of microscopic resolution is generally about 0·25–0·275μ, but it can be increased by the use of ultra-violet light, quartz lenses and photography.

Dark-ground illumination is required for the examination of objects with a low refractive index, such as the spirochaetes. By this method direct rays of light are cut out and the specimen is illuminated by oblique rays. This is made possible by the use of a special condenser of the paraboloid or of the concentric reflecting type. An illuminant of high intensity, such as the "pointolite" lamp, is necessary to obtain satisfactory results. If an oil-immersion lens is used a funnel stop is required, as the numerical aperture of the objective must not be greater than 1·0.

In the preparation of the specimen, special slides 1·0–1·1 mm thick and thoroughly clean are used; a thin film is made between slide and cover-slip. A drop of oil is placed on the condenser, the under-surface of the slide and the cover-slip; the condenser is racked up until the drops on the condenser and the under-surface of the slide meet; the objective is then lowered into the oil on the cover-slip and the specimen is focused. Particulate bodies appear as bright objects against a dark background. When much dark-ground work is carried out it is advisable to have a special microscope and the source of illumination arranged in fixed positions.

STAINING METHODS

Slides must be well cleaned, free from grease and stored in a stoppered jar containing 50 per cent methylated spirit. If the slides are greasy it is impossible to make a thin uniform film. The dried smears or films are usually fixed by passing slowly through a bunsen flame. Sections are prepared and treated in a similar manner to those required for histological examination.

SIMPLE STAINS

LOEFFLER'S METHYLENE-BLUE

1 per cent aqueous potassium hydrate .	1 ml
Water	99 ml
Saturated alcoholic methylene-blue .	30 ml

Films are stained for five minutes, washed well with water and dried. Sections are stained for at least ten minutes.

CARBOL-FUCHSIN

Basic fuchsin	1 g
Absolute alcohol	10 ml
5 per cent aqueous phenol . . .	100 ml

Carbol-fuchsin when diluted with 19 volumes of water is a useful simple stain, which should be applied for 15–20 seconds only as it tends to overstain.

DIFFERENTIAL STAINS

JENSEN'S MODIFICATION OF GRAM'S METHOD

(1) 0·5 per cent aqueous methyl-violet (6B), one minute.

(2) Pour off excess of stain and wash away remains with Lugol's iodine solution, which is allowed to act for one minute.

Iodine	1 g
Potassium iodide	2 gm
Distilled water	100 ml

N.B.—Lugol's iodine is three times stronger than Gram's iodine.

(3) Wash off iodine with spirit and continue until no more violet is removed from the film. The time required for this process varies with the nature of the film; pus takes much longer to decolourize than a bacterial culture.

(4) Wash well with water.

(5) Counter-stain with dilute carbol-fuchsin (1/20) for 20 seconds. Some workers prefer to substitute neutral red (1/1,000) as the counter-stain.

(6) Wash in water and dry between blotting-paper. Gram-positive bacteria retain the violet stain and Gram-negative organisms and tissue-cells take the counter-stain.

LILLIE'S GRAM STAIN FOR ORGANISMS IN TISSUES

(1) Tissues are embedded in paraffin in the usual way, sections are cut and brought to water by treatment with xylol and alcohol.

(2) Treat the section with ammonium oxalate crystal-violet solution for 30 seconds.

Crystal-violet	2 g
95 per cent alcohol	20 ml
1 per cent aqueous ammonium oxalate .	80 ml

Dissolve and filter.

(3) Wash in water.

(4) Treat with Lugol's iodine for 30 seconds.

(5) Wash in water.

(6) Decolourize with acetone (10–15 seconds).

(7) Wash in water.

(8) Counter-stain with 0·5 per cent aqueous safranin for 30 seconds.

(9) Wash in water.

(10) Dehydrate and differentiate with acetone (or alcohol); this removes the excess of safranin and leaves the cell-nuclei deep red and the cytoplasm pinkish, while the organisms are either Gram-positive or -negative; clear with xylol and mount in Canada balsam.

ZIEHL-NEELSEN METHOD FOR TUBERCLE BACILLI

(1) Cover the slide with carbol-fuchsin, heat until steam rises; the preparation is stained for five minutes, heat being applied at intervals. The stain must not be allowed to dry on the slide.

(2) Wash well in water.

(3) Decolourize with 20 per cent sulphuric acid; the red colour changes to yellowish-brown, but returns on washing in water; decolourization is continued until after washing the smear is a faint pink.

(4) Wash well in water.

(5) Treat with alcohol for two minutes.

(6) Wash in water.

(7) Counter-stain with methylene-blue or malachite green for 30 seconds.

Tubercle bacilli are stained red, while most other bacteria and the tissue-cells take the counter-stain.

Steps (5) and (6) are frequently omitted in the examination of sputum. Treatment with alcohol tends to decolourize some of the non-pathogenic members of the acid-fast group and is particularly useful in the examination of urine and faeces; 3 per cent hydrochloric acid in alcohol (acid-alcohol) is frequently used as a decolourizing agent.

NEISSER'S METHOD (MODIFIED) FOR DIPHTHERIA BACILLI

Solution A

Methylene-blue	1 g
Absolute alcohol	50 ml
Glacial acetic acid	50 ml
Distilled water	1,000 ml

Solution B

Crystal-violet	1 g
Absolute alcohol	10 ml
Distilled water	300 ml

(1) Mix two parts of solution A and one part of solution B and stain film for 30–60 seconds.
(2) Wash in water.
(3) Stain with Bismarck-brown (0·2 per cent) for 30–60 seconds.
(4) Wash in water and dry.

The granules of the bacilli are bluish-black and the protoplasm is light brown in colour.

MOELLER'S METHOD FOR SPORES

(1) Carbol-fuchsin, steam for five minutes.
(2) Wash in water.
(3) Decolourize by treatment with spirit for a few seconds.
(4) Wash in water.
(5) Counter-stain with methylene-blue for two minutes.
(6) Wash in water and dry.

The spores are stained bright red and the bacterial protoplasm blue.

STAINING OF CAPSULES

HISS'S METHOD

(1) Add gentian-violet solution, heat until steam rises and leave for 20–30 seconds.

Saturated alcoholic solution of gentian-violet	1 part
Distilled water	19 parts

(2) Wash off the stain with 20 per cent copper sulphate solution and dry, without washing, between blotting-paper.

The capsules are stained pale bluish-purple and the bacteria deep violet.

MUIR'S METHOD

(1) Stain film with carbol-fuchsin for one minute, heating gently.
(2) Wash slightly with spirit and then in water.
(3) Treat with the mordant for 30 seconds.

Saturated solution of corrosive sublimate	2 parts
20 per cent solution of tannic acid .	2 parts
Saturated solution of potassium alum .	5 parts

(4) Wash well in water.
(5) Treat with spirit for about 20–30 seconds.
(6) Wash well in water.
(7) Counter-stain with methylene-blue for 30–60 seconds.
(8) Wash in water and dry.

The capsules are stained blue and the bacteria red.

CRAIGIE'S METHOD FOR STAINING FLAGELLA

Pipette off the condensation water of a very young agar culture into 2 per cent formol-saline and allow to fix overnight. Dilute with distilled water and make a thin film on a clean slide; dry and fix by heat at 100°C for five minutes.

Mordant with the following solution for 5–10 minutes at 100°C.

Tannic acid (light and pure) . . .	10 g
Tartar emetic (aqueous 5 per cent solution)	30 ml
Distilled water	200 ml

Add a crystal of thymol.

Wash thoroughly in tap water and then in distilled water. Treat with the silver solution, which is prepared in the following manner:

Silver sulphate	1 g
Distilled water	250 ml

To 20 ml of this solution add 33 per cent mono-ethylamine until the precipitate produced just clears. The slide is flooded with this and heated gently until the preparation becomes brown or black.

Wash in distilled water.

Tone with gold chloride solution for 30 minutes in daylight.

Gold chloride (1 per cent) . . .	20 drops
Distilled water	20 ml

Wash and dry.

METHODS FOR DEMONSTRATING SPIROCHAETES

FONTANA'S METHOD

(1) Treat the film three times, 30 seconds each time, with the fixative:

Acetic acid	1 ml
Formalin	20 ml
Distilled water	100 ml

(2) Wash off with absolute alcohol, allowing it to act for three minutes; drain off excess and burn off the remainder until the film is dry.

(3) Add the mordant, heat gently until steam rises and allow to act for 30 seconds.

Mordant

Carbolic acid	1 g
Tannic acid	5 g
Distilled water	100 ml

(4) Wash well in water and again dry the slide.

(5) Treat with the silver solution, heating until steam rises, for 30 seconds, when the film becomes brown.

Silver solution. Add 10 per cent ammonia to 0·25 per cent solution of silver nitrate drop by drop until the precipitate which forms does not dissolve.

(6) Wash well in water and dry.

The spirochaetes are stained brownish-black on a yellowish background.

LEVADITI'S METHOD OF STAINING SPIROCHAETES IN TISSUES

(1) Fix small pieces of tissue in 10 per cent formalin for 24 hours.

(2) Wash the tissue for one hour in water and then place in 96–98 per cent alcohol for 24 hours.

(3) Place in 1·5 per cent silver nitrate solution in a dark bottle and keep in the incubator for three days.

(4) Wash in water for 30 minutes.

(5) Place the tissue in the following reducing solution:

Pyrogallic acid	4 g
Formalin	5 ml
Water	100 ml

and keep in a dark bottle at room-temperature for 48 hours.

(6) Wash well in water, dehydrate the tissue with increasing strengths of alcohol and embed in paraffin. Sections are cut when required and mounted in the usual way.

The spirochaetes are stained black and the tissues brown.

ROMANOWSKY STAINS

Several modifications of the original Romanowsky stains particularly Leishman's and Giemsa's, are sometimes used for staining bacteria. These stains can be obtained ready for use They contain a mixture of special compounds of methylene-blue and eosin; these form a precipitate which is washed, dried and dissolved in methyl-alcohol and glycerol.

Giemsa's stain is used mainly for staining spirochaetes, malaria parasites and the rickettsiae. For blood films, the smear is fixed by treatment with methyl-alcohol for three minutes; diluted stain (1 drop stain to 1 ml buffer solution) is allowed to act for ½–4 hours; the film is then washed well with distilled water and dried.

In the examination of exudates containing bacteria, good results may be obtained by covering a thin film with the concentrated stain for 1 minute, and then adding about 5 volumes of buffer solution, mixing well and, after 5 minutes, washing well with distilled water.

A slow method, in which the stain is allowed to act overnight, is more suitable for the staining of certain spirochaetes, e.g. *Trep. pallidum.*

Leishman's stain is usually obtained ready for use and is added to the unfixed smear; the methyl-alcohol acts as a fixative, and is allowed to act for 1–2 minutes, two volumes of distilled water or buffer solution are then added, mixed thoroughly and allowed to act for 5–7 minutes; finally it is washed well in distilled water and dried.

Chapter XXXVI

Preparation of Culture Media

It has already been mentioned that not only do the growth requirements of different organisms vary, but also that individual species vary according to their age and habitat. Many strictly parasitic organisms can be adapted to multiply on the simple media by serial subculture on media of decreasing complexity. The selection of a suitable medium, particularly for the isolation of pathogenic organisms from the tissues, thus constitutes a matter of great importance. It is therefore not surprising to find that a wide range of culture media is available; in fact, it has been estimated that no less than 7,000 different types have been described. It is obviously impossible to be acquainted with even a small proportion of these, but as the majority are essentially modifications and elaborations of certain basic media, it appears most satisfactory to describe the latter in some detail.

In recent years much attention has been directed to the study of bacterial nutrition, and relatively simple *synthetic* media, containing only known chemical compounds, have been introduced. These have proved useful in the study of some pathogenic bacteria, particularly with regard to the production of toxin or enzymes.

For general application culture media are usually prepared from an extract of some animal protein, such as lean muscle. When required different substances, such as peptone, salts, serum, blood, etc. are added to the meat extract. In this way more elaborate media are prepared for special purposes.

After preparation it is necessary to standardize the reaction of the medium to the required pH, which is generally about 7·2–7·6. This is carried out by testing colorimetrically 5 or 10 ml of the medium against a fixed standard, using a suitable indicator, such as phenol-red. If the medium is too acid or alkaline, decinormal NaOH or HCl respectively is added drop by drop from a burette until the colour agrees with that of the standard. The amount required to correct the reaction of 10 ml is obtained, and from this figure the volume of normal NaOH or HCl necessary to adjust the bulk of the medium can be estimated.

The cultivation is carried out in glassware of various types, usually in the form of test-tubes, screw-capped bottles, flasks and Petri dishes. These must be thoroughly cleansed before use. New glassware is treated with hydrochloric or nitric acid to remove the free alkali, which is often present. It is then treated with weak caustic soda and well washed in hot water. Old glassware is boiled in a solution of a detergent, preferably one which rinses easily, and then thoroughly washed in water. When dry the tubes and flasks are stoppered with cotton wool plugs and all glassware is sterilized in the hot-air oven at 160°C for one hour. Inhibitory substances present in some types of cotton wool may interfere with the growth of delicate organisms. After the medium has been added it is usually sterilized either by steaming on three successive days or by treatment in the autoclave at 120°C for 20–30 minutes.

Meat Infusion Broth. Mince 1 lb of lean beef, freed from fat, mix with 1,000 ml of water and leave overnight in a cold place. Then skim the fat off the surface, strain through muslin, squeezing the residue, and after boiling for 15 minutes filter through filter paper; add water to bring the volume back to 1,000 ml.

This constitutes "meat infusion".

Add 10 g of peptone and 5 g of sodium chloride.

Adjust the reaction to required pH.

Place in tubes or flasks and sterilize in the autoclave.

Infusion Agar. Add 10–20 g of powdered agar to 1,000 ml of infusion broth; steam for two hours; adjust the reaction. If a clear medium is desired, add egg-albumen beaten up in water, steam for 30 minutes, when the egg-albumen is coagulated, bring the volume to 1,000 ml with water, and filter through paper in the steamer or jacketed funnel. Tube and sterilize in the autoclave.

Gelatin. Add 10–15 per cent of fine sheet-gelatin to infusion broth, steam for 30 minutes, adjust the reaction, clear with egg-albumen as in the preparation of agar, and filter through paper. Tube and sterilize by the intermittent method or in the autoclave at a relatively low pressure (e.g. 5 lb). If the temperature is too high, the setting property of gelatin may be interfered with.

Peptone-water

Peptone	10 g
NaCl	5 g
Water	1,000 ml

Peptone and salt are dissolved by boiling; adjust the reaction; distribute and sterilize in the autoclave.

Peptone-water is used chiefly in testing for the production of indole and as a basis for the fermentation tests.

Hartley's Digest Broth

Lean beef (freed from fat and minced) .	1,500 g
Water	2,500 ml
Mix and heat to a temperature of 80°C.	
Add 0·8 per cent sodium carbonate solution	2,500 ml
Cool to 45°C.	
Add pancreatic extract . . .	50 ml
Chloroform	50 ml

Incubate at 37°C for 6 hours, stirring frequently.

When digestion is completed, add 40 ml of strong hydrochloric acid, steam for 30 minutes and filter.

Adjust the reaction so as to be neutral to phenolphthalein and steam for 20 minutes.

When cool add 0·25 per cent of chloroform and store in the cold-room.

When required, steam for 20 minutes to remove the chloroform, filter and adjust to the required reaction.

Hartley's broth is a very useful medium, which is widely used for the production of diphtheria toxin.

Loeffler's Serum

Ox, sheep or horse serum . . .	3 parts
I per cent glucose broth . . .	1 part

Mix, add to sterile test-tubes and inspissate in a sloped position at 75°C for six hours, when the serum coagulates and becomes a yellowish solid.

Sterilize by steaming for 20 minutes on three successive days; this is unnecessary if the serum was originally sterile.

Loeffler's serum is mainly used for the cultivation of the diphtheria bacillus.

Serum Agar. Serum, horse, rabbit, etc., should be obtained with sterile precautions. Agar (2 per cent) is melted and allowed to cool to 45°–50°C, when the sterile serum (5–10 per cent) is added with a sterile pipette; after mixing the medium is allowed to set either in a Petri dish or as a slope in a test-tube.

Blood Agar. Sterile defibrinated blood is added to 2 per cent melted agar cooled to 45°–50°C to give a concentration of 5–10 per cent; it is allowed to cool in either a Petri dish or a tube in a sloped position.

Chocolate Agar (Boiled blood agar). Blood-agar medium is heated by immersing in boiling water for long enough to convert the medium to a chocolate-brown colour; it is then allowed to solidify either in a Petri dish or in a test-tube as a slope.

Chocolate agar is an excellent medium for the cultivation of *H. influenzae*; heating converts the haemoglobin into haematin and, by breaking down the corpuscles, disperses their contents.

Serum- and Blood-broth. Sterile serum or blood is added to broth by means of a sterile pipette to give a final concentraton of 5–10 per cent.

Dorset's Egg Medium. Four fresh eggs are beaten and mixed with 25 ml of distilled water; the mixture is strained through muslin into test-tubes and then inspissated in a sloped position at 75°C. It is sterilized by steaming on three successive days, but should be incubated at 37°C before use to ensure sterility.

Dorset's medium is used for the cultivation of the tubercle bacillus; 5 per cent of glycerine may be added.

Potato Medium. Large potatoes are peeled and thoroughly cleaned; a cylinder is removed by means of a borer and is washed in running water to remove excess of starch. The cylinder is cut obliquely into two portions, each of which is placed in a sterile test-tube so that the thick end rests on the bottom. A special type of tube is generally used; this has a constriction a short distance from the bottom, the potato rests on this and some water or broth is placed in the bottom portion to retard drying. The tubes are filled with sterile water and steamed for 30 minutes, the water is subsequently poured off and the tubes are autoclaved for 20 minutes.

Potato, to which 5 per cent glycerine has been added, is sometimes used for the cultivation of the tubercle bacillus.

Robertson's Meat Medium. Mince 500 gm of fresh bullock's heart and place in 500 ml of boiling N/20 NaOH; allow to simmer for 20 minutes. The lactic acid is thus neutralized and the pH becomes about 7·5. Drain off the fluid and partially dry the meat on a cloth or filter paper. Then place the meat in sterile test-tubes to fill about 5 cm of the tube and add broth (pH 7·5) until the meat is covered by about 1 cm. Sterilize by autoclaving.

Robertson's medium is extensively used for the cultivation of the anaerobes.

MacConkey's Medium

Peptone	2 g
Sodium taurocholate (commercial) .	0·5 g
Water (tap)	100 ml

Steam for two hours, filter when hot, allow to stand for 24 hours and filter again.

Dissolve 2 per cent agar in the solution.

Add a 1 per cent solution of neutral red to give a reddish-brown colour, and 1 per cent lactose.

Sterilize by steaming on three successive days.

MacConkey's medium is of great value in the isolation of members of the coli-typhoid group, particularly in the examination of faeces. The bile-salt inhibits many other species, while the lactose-fermenters, such as *E. coli*, produce pinkish colonies and the non-lactose-fermenters, e.g. *S. typhi*, are greyish.

Dieudonné's Medium. Equal parts of defibrinated ox blood and normal caustic soda are heated in the steamer.

Three parts of this mixture are added to seven parts of 3 per cent agar.

Dieudonné's medium is used for the isolation of the cholera vibrio.

Hiss's Serum-water. One part of serum is mixed with three parts of distilled water and a suitable indicator, such as phenol red, and the sugar solution (1 per cent) are added.

The mixture is sterilized by steaming for 20 minutes on three successive days.

Hiss's serum-water is used in carrying out fermentation tests with certain pathogenic organisms, such as the pneumococcus, which do not grow well in the peptone-water solutions. Fermentation is indicated by the change in the indicator by the production of acid, which may also coagulate the serum.

Litmus-milk. Fresh milk is steamed for 20 minutes and then allowed to stand for 24 hours in order to separate off the cream. The milk is siphoned off and coloured with litmus. It is placed in tubes and sterilized by steaming on successive days.

Litmus-milk is used for biochemical tests. Acid production, due to the fermentation of lactose, turns the medium pink and may also coagulate it; proteolytic activity is indicated by digestion and alkali formation.

McLeod's Chocolate Tellurite Medium. Add to broth, sterilized by filtration, an equal volume of 5 per cent agar in water, potassium tellurite to make a final concentration of 0·04 per cent, and 7–10 per cent of freshly drawn defibrinated rabbit's blood; the completed medium is heated at 75°C for 10–15 minutes and poured into Petri dishes.

This medium is used for the typing of diphtheria bacilli and their isolation from contaminated material, e.g. throat swabs. The potassium tellurite inhibits many organisms, while the diphtheria bacilli produce greyish-black colonies, which can be readily differentiated by an experienced worker from those formed by other organisms.

Tellurite Media. Many other tellurite media are available for the isolation and typing of diphtheria bacilli. In all the inhibitory substance is potassium tellurite and it has been found that the highest concentration not interfering with the growth of the diphtheria bacillus is 0·04 per cent. The enriching substances in general use are blood and serum and it is important to note that the origin of these is a matter of great importance in type-differentiation. Blood and serum from the rabbit and guinea-pig are most useful; following these in order of merit are sheep, goat, horse and the ox. Horse blood and ox blood, apart from human blood, are least satisfactory but these are considerably improved if the red cells are lysed by freezing and thawing or other means. The usual concentration of the enriching substances is from 5–10 per cent. For routine work on a large scale the use of rabbit or guinea-pig blood is impracticable and consequently sheep blood or serum or lysed horse blood is generally employed.

Semi-solid Medium for the Cultivation of Certain Spirochaetes. Mix at 55°C 8 ml of sterile saline or Ringer's solution and 1 ml of melted agar; add 20 drops of blood direct from the ear vein of a rabbit; allow to become semi-solid without shaking. Test for sterility by incubating at 37°C for 24 hours.

This medium is suitable for the cultivation of *Lept. icterohaemorrhagiae*

Jensen's Medium for the Cultivation of the Tubercle Bacillus. (1) *Mineral Salt Solution.* Potassium sulphate, 0·4 per cent; magnesium sulphate, 0·4 per cent; magnesium citrate, 0·1 per cent; asparagin, 0·6 per cent; glycerine, 2 per cent; distilled water.

Heat the mixture to dissolve the ingredients, then heat at 100°C for two hours.

(2) *Starch Solution.* To every 600 ml of solution (1) add 30 g of potato flour; mix carefully over a water-bath with constant stirring for 15–20 minutes; a paste is produced and is placed in a water-bath at 56°C for one hour.

(3) *Egg-fluid.* Fresh eggs are washed in 5 per cent soft-soap solution and running water; 20 eggs are required for every 600 ml of the medium. The contents of the eggs are mixed by hand for 10 minutes and filtered through sterile gauze. The salt-starch solution is added to the egg-fluid in the proportion of 600 ml to 1 litre of egg-fluid.

(4) *Malachite Green.* To each 1,600 ml of the egg-salt-starch solution add 20 ml of a 2 per cent solution of malachite green. The medium is now distributed before it cools, and the tubes are placed in an inspissator and heated at 85°C for 30 minutes; after

standing overnight at room-temperature they are heated at 75°C for 30 minutes.

The medium has a rich green colour and is particularly useful for the isolation of tubercle bacilli from contaminated material, such as sputum and faeces. These are subjected to a preliminary treatment with 4 per cent NaOH (or 6 per cent H_2SO_4) for about 20 minutes at 37°C; after which they are centrifuged and the deposit, which may be rendered neutral, is seeded on the medium.

Kirchner's Medium (modified) for Cultivation of Tubercle Bacilli

Na_2HPO_4, 12 H_2O	19·0 g
KH_2PO_4	2·0 g
Magnesium sulphate	0·6 g
Sodium citrate	2·5 g
Asparagin	5·0 g
Glycerol	10 ml
Distilled water	1,000 ml
Phenol red 0·4 per cent	3 ml

Dissolve all the salts before adding the glycerol.
Dispense in 9 ml and 45 ml amounts in screw-capped bottles.
Autoclave at 10 lb pressure for 10 minutes.

Prepare sterile horse serum containing 1,000 units penicillin per ml.

Add 1 ml of the serum to each 9 ml of the medium.

Dubos Medium for Tubercle Bacilli

M.R.C. modification

This medium gives a rapid and diffuse growth of tubercle bacilli and is used for estimating streptomycin sensitivity.

Potassium dihydrogen phosphate (KH_2PO_4)	1·0 g
Disodium hydrogen phosphate ($Na_2HPO_4 . 12H_2O$)	6·25 g
Sodium citrate	1·5 g
Magnesium sulphate (Mg. $SO_4 . 7H_2O$)	0·6 g

Dissolve separately in glass-distilled water. Add:

Tween 80 (10 per cent soln.)	5 ml
Casein hydrolysate (20 per cent soln.)	10 ml
Glass-distilled water to	1,000 ml

Adjust pH to 7·2; add 2·5 ml to small screw-cap bottles (bijou type); autoclave at 10 lb pressure for 10 minutes; before use add 0·1 ml of a 9 per cent solution of bovine albumen (fraction

V), Seitz-filtered, to each bottle. In testing for streptomycin sensitivity, 0·2 ml amounts of the streptomycin dilution and the culture of tubercle bacilli are added to make the final volume 3 ml.

Wilson and Blair's Medium (for the Salmonella group).

(a) Dissolve 6 g bismuth ammonio-citrate in 50 ml boiling distilled water, neutralize with NaOH and cool.

(b) Dissolve 20 g sodium sulphite crystals in 100 ml boiling distilled water.

(c) Mix (a) and (b), add 10 g anhydrous Na_2HPO_4, boil for two minutes and cool.

Solution 1. Add to (c) 10 g glucose dissolved in 50 ml water.

Solution 2. Eight per cent aqueous ferrous sulphate.

Solution 3. One per cent aqueous solution of brilliant green.

To 100 ml agar, melted and cooled to 80°C, add 20 c.c. No. 1 solution, 1 ml No. 2 solution, and 0·5 c.c. No. 3 solution and pour plates. Coliforms tend to be inhibited, while the colonies of *Salm. typhi* are black.

Deoxycholate-Citrate Medium (for the dysentery and salmonella groups). Hynes' modification of Leifson's medium.

Agar Base. Dissolve 20 g lab. lemco in 200 ml water; add 50 per cent NaOH until just alkaline to phenolphthalein, boil and filter; adjust to pH 7·3, restore volume to 200 ml and add 20 g Difco proteose peptone (Evans' peptone is also satisfactory).

Add 90 g agar to 3,700 ml distilled water, steam for one hour; filter and add the lab. lemco-proteose solution.

Add 5 ml 2 per cent neutral red and 40 gm lactose.

Bottle in 100 ml quantities and sterilize in steam for one hour.

Solution A

Sodium citrate A.R. $Na_3C_6H_5O_7 . 2H_2O$.	17 g
Sodium thiosulphate A.R. $Na_2S_2O_3 . 5H_2O$	17 g
Ferric citrate (scales)	2 g
Distilled water	100 ml

Solution is effected by heat or standing at room-temperature for two days.

Solution B

Sodium deoxycholate	10 g
Distilled water	100 ml

These solutions need not be sterilized.

For use, add with separate pipettes 5 ml of solution A and then 5 ml of solution B to 100 ml of the melted agar base; mix well, pour plates and dry.

Selenite Medium (for the salmonella group)

Sodium acid selenite	4 g
Peptone	5 g
Lactose	4 g
Disodium hydrogen phosphate	9·5 g
Sodium dihydrogen phosphate	0·65 g
Water	to 1,000 ml

Adjust to pH 7·1 and sterilize by steaming.

Tetrathionate Broth (for the salmonella group).

Solution A

Sodium thiosulphate	60 g
Water (sterile)	100 ml

Iodine Solution

Iodine	25 g
Pot. iodide	20 g
Water	100 ml

Preparation

Solution A	1·0 ml
Iodine solution	0·2 ml
Chalk (sterilized and dry)	0·25 g
Broth	9·0 ml

This medium should be prepared immediately before use and distributed in 5–10 ml amounts in tubes.

Thioglycollate Medium for Anaerobes (modified after Brewer).

Hartley broth	1,000 ml
Peptone	20 g
Yeast extract	2 g
Glucose	5 g
1. cystine	0·75 g
Hepamino	1 g
Thiogycollic acid	1·0 ml

Adjust the pH to 7·8 with NaOH, boil and filter. Add agar to 0·1 per cent, distribute into tubes or screw-capped bottles and sterilize by steaming.

Christensen's Urea Medium

The basic medium is prepared in the following manner:

Peptone	1 g
Sodium chloride	5 g
Monopotassium phosphate	2 g
Phenol red	0·012 g
Agar	20 g
Distilled water	1 litre

Filter, add glucose (1 g), and prepare in 5 ml bottles and autoclave.

For use melt 5 ml and cool to 50°C and add 0·5 ml of 20 per cent urea solution, sterilized by filtration. Solidify as a slope.

For use, the organism is seeded on the surface of the medium, which is then incubated. If urea is split, ammonia is formed and the medium turns purple-pink.

Proteus organisms give a positive result in a few hours; most coliforms, salmonellae and paracolons do not.

Koser's Citrate Medium

Sodium ammonium hydrogen phosphate	1·5 g
Potassium dihydrogen phosphate	1·0 g
Magnesium sulphate	0·2 g
Sodium citrate	2·0 g
Water	to 1,000 ml

Growth in citrate medium indicates that the organism is able to utilize citrate as a sole source of carbon.

Berger's Medium for the Isolation of Neisseria

A.	Bacto GC Medium (Difco)	7·2 g
	Aqua dest.	100 ml

Dissolve in warm water and autoclave at 120°C.

B.	Bacto Haemoglobin (Difco)	2·0 g
	Aqua dest.	100 ml

Dissolve in cold water and autoclave at 120°C.

Bacto Supplement B (Difco)	2·0 ml
Lincomycin-HCl solution	1,000 γ/ml
Polymyxin B solution	1,000 γ/ml

To the autoclaved solution *A* cooled to 50°–60°C add 2·0 ml Bacto Supplement B. Then add the Haemoglobin solution *B*, and

2·0 ml of each of the antibiotic solutions. Shake the flask well after each addition.

Pour plates, using 25 ml for each 9 cm plate.

Vancomycin may be used instead of Lincomycin.

For the isolation of gonococci, colimycin sulphate at a concentration of 25 γ/ml may be added.

Inoculate plates, and incubate at 37°C in an atmosphere of 5 per cent CO_2. Examine at 24 and 48 hours.

Chapter XXXVII

Serological Technique

Serological reactions can be applied in two ways:

(1) In the diagnosis of disease by testing the serum of the patient against known antigens.

(2) To identify an organism (or other antigen) by testing it against known antisera.

The serological reactions most widely used are agglutination, precipitation and complement-fixation, the latter mainly in the form of the Wassermann reaction. The main features and the general application of these reactions have already been considered and we are now only concerned with the technical details. Other tests, such as the opsonic index and the bactericidal test, have also been used, but they have a more limited application.

AGGLUTINATION

Agglutination is probably the most extensively applied of the serological reactions. The test may be carried out by either the macroscopic or microscopic methods; the latter are, however, less reliable than the former.

Macroscopic Method. The test is carried out in small (3 in. × $\frac{3}{8}$ in.) rimless tubes with round bottoms. These are arranged in a metal rack and are numbered from the left. Table 1.XXXVII.

Serial dilutions of the serum may be prepared as follows: Add to the first tube 0·9 ml of saline and to all the others 0·5 ml; to the first tube add 0·1 ml of the serum and mix, then transfer 0·5 ml of the mixture to tube 2, mix and transfer 0·5 ml of this mixture to tube 3, and repeat until tube 10 is reached, then after mixing discard 0·5 ml of the mixture. When several suspensions are being tested against the same serum, it is more satisfactory to prepare larger quantities of the dilutions separately and to add 0·5 ml of each dilution to the corresponding tube of the test. Next, using a fresh pipette, add to all tubes 0·5 ml of the bacterial suspension, mix by shaking and place in a water-bath at 50°C so that only half the column of fluid is immersed. Examine after varying intervals of time, usually 1, 4 and 24 hours, preferably by holding in front of a shaded lamp set against a dark back-

Table 1.XXXVII

Scheme of Agglutination Technique

TUBE

Contents	*1*	*2*	*3*	*4*	*5*	*6*	*7*	*8*	*9*	*10*	*11 (control)*
Saline (ml)	0·9 } →	0·5 } →	0·5 } →	0·5 } →	0·5 } →	0·5 } →	0·5 } →	0·5 } →	0·5 } →	0·5 } →	0·5
Serum (ml)	0·1	0·5	0·5	0·5	0·5	0·5	0·5	0·5	0·5	0·5 discard 0·5 ml	—
Bacterial suspension (ml)	0·5	0·5	0·5	0·5	0·5	0·5	0·5	0·5	0·5	0·5	0·5
Total volume (ml)	1	1	1	1	1	1	1	1	1	1	1
Serum dilution (final)	1/20	1/40	1/80	1/160	1/320	1/640	1/1,280	1/2,560	1/5,120	1/10,240	—

ground; the differences in the reactions given by O and H antigens have already been noted.

Reading. Tube 11 is the control tube, in which the bacterial suspension is tested for spontaneous agglutination. This is examined first; clumping, i.e. spontaneous agglutination, should not occur, as in such an event the test would be unsatisfactory. The other tubes are then examined and the following points noted: The presence and type of agglutination and its time of appearance—H agglutination is rapid and floccular; the O type is slow and granular. All tubes should be examined, as the prozone phenomenon may be present. In such cases agglutination may not be present in tubes 1 and 2, i.e. those containing the greatest concentration of serum, but may be well marked in other tubes. The titre is taken to be the highest dilution at which definite agglutination takes place.

The Microscopic or Slide Method. A drop of a heavy bacterial suspension is mixed with a drop of the serum (undiluted or diluted 1/10); after standing a few minutes the mixture is examined with the hand lens or the low power of the microscope. In a positive reaction clumping is well marked.

PRECIPITATION

In this test the antigen is diluted and the antiserum kept constant. The mixtures are prepared as for the agglutination reaction, but the tubes are incubated at 37°C. The test is frequently used for the identification of blood stains. It is also applied in bacteriology, e.g. in typing pneumococci in cases of pneumonia; the urine of these cases frequently contains the specific carbohydrate, and using this as the antigen it may be tested against the type sera.

A modified precipitation or flocculation test is used extensively in the diagnosis of syphilis; many forms have been described, e.g. Sachs-Georgi, Meinicke, Sigma, Kahn, etc.; of these the Kahn test is probably the most satisfactory and reliable.

Kahn Test. If the complete details are required, reference should be made to the original monograph. The following reagents are required: (1) antigen, (2) patient's serum, and (3) saline.

(1) The "antigen" is an alcoholic extract of the ether-insoluble lipoids of beef heart. This is prepared by treating repeatedly the dried muscle with ether; the dried muscle is extracted with alcohol, the extract is then cholesterolized (0·6 per cent) and filtered. Before use it must be standardized. A standard antigen can be readily obtained. When required it is mixed with saline according

to the titre, the usual proportion being 1 ml of antigen to 1·2 ml of saline, in special tubes by pouring the saline into the antigen; the pouring operation is repeated twelve times. The diluted antigen is allowed to stand for ten minutes, but must be used within 30 minutes of preparation.

(2) The patient's serum is heated at 56°C for 30 minutes.

The test is carried out in special small tubes, 7·5 × 1·0 cm, according to the scheme given in Table 2.XXXVII. Owing to the minute quantities involved special graduated 1·0 ml and 0·1 ml pipettes are required. Known positive and negative sera should be included in each series for control purposes.

Table 2.XXXVII

The Khan Test

	Tube 1	*Tube 2*	*Tube 3*	*Tube 4*
Antigen (ml.)	0·05	0·025	0·0125	0·05
Serum (ml)	0·15	0·15	0·15	—
	Shake vigorously, at about 280 oscillations per minute, for three minutes and then add the saline.			
Saline (ml)	1·0	0·5	0·5	1·0

After the addition of the saline, which is added to simplify the detection of flocculation, the tubes are lightly shaken and read. The degree of flocculation is indicated by a scale varying from ++++, indicating the formation of visible particles in a completely clear fluid, to — denoting uniform opalescence, which should be given by control tube 4. The standard test is not strictly quantitative, each of the three tubes is equally significant, and for the final interpretation of the result an average of the three tubes is taken. A scale indicating the value of the various readings has been adopted by Kahn; a positive result is one in which the average reading is ++ or over. A quantitative method is also available.

The Kahn test is sensitive and relatively simple; however, while results of ++++, +++ and — values are easy to read, those of ++ and + values may require considerable experience to detect; a hand lens and a special viewing box should be used. The technique required for the examination of cerebrospinal fluid is much more complicated than that required for serum.

Kahn Verification Test. A positive reaction is sometimes given with the Kahn test when the diagnosis of syphilis is most improbable. The explanation of the "false" reactions is unknown; they may be associated with either a pathological condition or some constitutional factor, e.g. pregnancy.

In order to distinguish these false or biological reactions from the true positive results, Kahn introduced a series of Verification tests in which the usual technique is modified by various means. In one method the diluting fluid is varied; three separate tests are done in the same sample of serum, each with a different diluting fluid—distilled water, normal saline or 2·5 per cent saline; a false reaction is most marked in the distilled water series. Another method is to vary the temperature at which the test is carried out — 37°, 16° and 1°C; false reactions are most marked at the lowest temperature, while syphilitic sera react strongest at 37°C.

COMPLEMENT-FIXATION

Complement-fixation is used mainly in the diagnosis of syphilis, as the Wassermann test. It is also applied in the diagnosis of chronic gonorrhoea, and virus diseases; in these cases the general technique is similar to that of the Wassermann test, the chief difference being in the nature of the antigen.

The Wassermann Reaction. The principles of the reaction have already been discussed. Sensitized sheep cells are used to indicate the fixation of complement by the antigen and the patient's serum. If complement is fixed, i.e. the test is positive, the cells are not lysed; if the test is negative the complement is free and the blood-cells are lysed.

The reagents required for this test are (1) antigen, (2) patient's serum, (3) complement, (4) sheep corpuscles, (5) antisheep haemolysin, and (6) saline.

(1) The *antigen* is a cholesterolized alcoholic extract of heart muscle. To prepare, macerate 10 g of lean heart muscle in 100 ml of absolute alcohol; filter through paper and store in the dark. Prepare a 1 per cent solution of cholesterol in alcohol. These are kept in separate bottles and are mixed on the day of the test in the following manner: three parts of heart extract and two parts of cholesterol solution are mixed in a large test-tube; a 1/15 dilution of the mixture is made by the rapid addition of 14 parts of saline.

(2) The *patient's serum* is carefully labelled and heated at 58°C for 20 minutes on the morning of the test to destroy the complement; it should be free from haemoglobin.

(3) *Complement* is obtained by bleeding guinea-pigs when required; the blood is allowed to clot and the serum collected. Complement is unstable and can only be used on the day of the test unless kept in a frozen or dried state or preserved by the addition of various salt solutions.

(4) *Sheep corpuscles.* Sheep are bled from the jugular vein into a vessel containing beads; the blood is defibrinated by shaking vigorously for 5–10 minutes, strained through muslin, washed in saline and centrifuged three to five times, finally a 5 or 10 per cent suspension is made from the deposit. The concentrated cells keep about three days at 0°C.

(5) *Antisheep haemolysin* is prepared by injecting rabbits with sheep corpuscles: the injections should be continued until a trial test indicates that a satisfactory serum, i.e. one with an M.H.D. of 1/5,000 or more, has been obtained. The animal is then bled, the serum collected and stored at 0°C. The haemolysin is very stable and keeps for years with only a slight reduction in titre.

Standardization of Reagents. Before the actual test can be performed certain preliminary titrations must be carried out. The antigen and haemolysin exhibit little variation from day to day and, in routine work, it is only necessary to make a preliminary titration of each batch of complement.

Titration of Complement. This is done to determine the M.H.D. (minimal haemolytic dose) of the complement, i.e. the smallest amount which completely lyses the unit volume of sensitized cells in 1 hour at 37°C. Serial dilutions of complement are tested against constant amounts of sensitized cells. In all preliminary titrations and the actual test a unit volume of each reagent (usually 0·5 ml) is used.

A 1/5 dilution of the fresh guinea-pig serum is prepared by adding 0·5 ml of the serum to 2 ml of saline. Set up a row of tubes and add, in succession from the first tube, the following amounts of saline in ml: 0·75, 1, 1·25, 1·50, 1·75, 2·25, 2·75, 3·25. Using a fresh pipette, add to each tube 0·25 ml of the 1/5 dilution of guinea-pig serum. This gives the following series of dilutions: 1/20, 1/25, 1/30, 1/35, 1/40, 1/50, 1/60, 1/70.

Set up a rack containing eight special Wassermann tubes and number from the left. With a fresh pipette add 0·5 ml of the 1/70 dilution of complement to tube 8; then mix the 1/60 dilution and add 0·5 ml to tube 7; continue to the 1/20 dilution.

Using a fresh pipette for each reagent, add to each tube 1 ml of saline, 0·5 ml of haemolysin (3 M.H.D.) and 0·5 ml of a 5 per cent suspension of red blood corpuscles. Incubate at 37°C and read after one hour; the dilution of complement producing complete haemolysis is the M.H.D. In the final test 3 M.H.D. of complement are required; if 1 M.H.D. is a dilution of 1/45, 3 M.H.D. represents a dilution of 1/15.

Titration of Antigen. Each fresh batch of antigen is tested to ensure that it is not too anticomplementary. In carrying out this test two complement titrations are necessary, the standard method and another in which 0·5 ml of saline is replaced by 0·5 ml of the diluted antigen.

A satisfactory antigen is one with slight anticomplementary activity but with a high degree of sensitivity to syphilitic sera.

Titration of Haemolysin. It is necessary to test the haemolysin at regular intervals. This is carried out in a similar manner to the titration of complement. The main difference is that the dose of complement is kept constant at 3 M.H.D. and various dilutions of haemolysin are prepared. The dilution giving complete haemolysis after one hour at 37°C is the M.H.D.

The Wassermann Test. Set up four tubes in a rack and label carefully. The first is the serum control (dilution 1/2) and does not contain antigen, saline being the substitute; the others are for the serum dilutions 1/2, 1/8 and 1/32 and require antigen. Complement is added to all, and the tubes incubated at 37°C for one hour. Unit volumes of haemolysin and corpuscles are then added and the tubes again incubated at 37°C for one hour (see Table 3.XXXVII).

Results. Readings are easier to make if the tubes have been left overnight in the refrigerator. The absence of lysis indicates a positive, complete lysis a negative, and incomplete lysis a doubtful result. A quantitative reading is obtained by readings of the different serum dilutions. The control tube should show complete lysis, otherwise the serum is anticomplementary and therefore unsuitable for examination. When only one serum is being tested an antigen control must also be made. This is usually omitted in routine work as the antigen is known to be satisfactory, and, when many tests are done, some are negative and these serve as controls. Only a slight modification is required for the examination of cerebrospinal fluid, this is in the preparation of the dilutions of the fluid. Many modifications of the same basic tests are used in different laboratories.

Clinicians usually prefer a quantitative system of recording the result, such as +1/32, +1/4, or doubtful 1/2. The League of Nations Committee, however, recommended that the results should be recorded as + (positive), ± (doubtful), and — (negative).

The test shows a high degree of disease specificity but false

Table 3.XXXVII

The Wassermann Reaction

	Tube 1	*Tube 2*	*Tube 3*	*Tube 4*
Serum (inactivated) (ml)	0·5 (1/2)	0·5 (1/2)	0·5 (1/8)	0·5 (1/32)
Complement (3 M.H.D.) (ml)	0·5	0·5	0·5	0·5
Antigen (1/15) (ml)	—	0·5	0·5	0·5
Saline (ml)	0·5	—	—	—
	Incubate at 37°C for 1 hour			
Haemolysin (3 M.H.D.) (ml)	0·5	0·5	0·5	0·5
Corpuscles (5 per cent) (ml)	0·5	0·5	0·5	0·5
	Incubate at 37°C for 1 hour and then leave overnight at 0°–4°C.			

positives may be found in leprosy, malignancy, malaria and other febrile conditions, after vaccination and in some apparently normal individuals (biological false positive).

Reagents for the complement-fixation test and various precipitation tests are obtainable from commercial sources.

Standardized agglutinable cultures of enteric bacteria, and standardized agglutinating serum are also obtainable from commercial sources.

Chapter XXXVIII

Assay Methods in Chemotherapy

Laboratory methods play an important *rôle* in the control of chemotherapy. They are mainly used for two purposes: (1) to determine the amount of antibiotic or other agent in the blood and other body fluids, and (2) to determine the sensitivity of an organism to the various chemotherapeutic agents, which fall into two main classes—the sulphonamides and other chemical agents and the antibiotics (cf. Chapter VII).

SULPHONAMIDE ASSAYS

The amount of sulphonamide in the various body fluids is determined by chemical methods and the technical detail need not be considered here.

Various methods have been used to determine the sensitivity of bacteria to the different sulphonamides. The same general principles apply in all techniques but it is important to note that the results may be influenced by many factors. The media must not contain any substances, e.g. certain peptones, which may be antagonistic to the sulphonamides (the addition of lysed horse blood tends to neutralize such antagonistic substances); the *p*H of the medium should be 7·2–7·6; the duration of incubation must be carefully controlled (usual time 18–24 hours); a low concentration of organisms is necessary—this varies with the different bacteria (usually 1/500–1/20,000 dilution of an overnight broth culture).

The organisms are usually tested by culturing on solid media containing the sulphonamide either incorporated in the whole medium or as a ditch (about 1–10 mg per 100 ml) or added to discs of filter paper (0·1–1 mg). Inhibition of growth indicates sensitivity to the drugs.

THE ANTIBIOTICS

Chemical methods are not used to assay the amount of antibiotic in the tissues except in the case of chloramphenicol, when the test is of limited value as it does not differentiate between the inert and active products. It is therefore necessary to carry out biological tests, which may be used to determine either the amount of antibiotic in the tissue-fluids or the sensitivity of an organism to the various agents.

Several different methods are available for these purposes including (1) serial dilution in fluid media, and (2) diffusion tests with solid media. The general principles are the same in both types of test: two main factors are involved—the organism and the fluid containing the antibiotic; either of these may be the unknown factor. Originally special organisms, e.g. Oxford *Staphylococcus*, *K. pneumoniae* (*K*.41) and *B. cereus*, were the standard test organisms but any satisfactory sensitive organism may be used.

(1) SERIAL DILUTION TECHNIQUE

One drop containing 0·02 ml of an 18–24-hour broth culture of the test organism is added to 1 ml of broth or serum-broth containing twofold dilutions of the serum or other fluid in sterile ($3 \times \frac{3}{8}$ in.) tubes; a control series of known dilutions of the particular antibiotic is also prepared; after incubation for 18–24 hours, the highest dilution in each series inhibiting growth is obtained and the antibiotic level of the serum determined. The Oxford *Staphylococcus* is usually inhibited by 1/32–64 units/ml penicillin.

In testing the sensitivity of an unknown organism by this technique two known series of antibiotic dilutions are prepared; one is seeded with the standard test organism and the other with the organism under review; after incubation the sensitivity can be determined as above.

(2) DIFFUSION TESTS

The Cylinder Technique has been used for penicillin assays. Glass or porcelain cylinders are placed on flood plates of the test organism. Dilutions of penicillin, or of body fluids are filled in the cylinders. After incubation, the zones of inhibition of growth are measured, compared with the zones produced by standard penicillin solutions, and the concentrations of the unknown solutions are calculated.

The Disc Technique is used to determine the sensitivity of organisms to the different antibiotics. They may be prepared by cutting discs (6 mm in diameter) from blotting paper or filter paper, sterilizing them, and impregnating them with a measured amount of a standard antibiotic solution. Commercially prepared "multidiscs" are more convenient. Circles of filter paper are cut with six or eight discs projecting from the edge, and each impregnated with a different antibiotic. Plates are seeded heavily with the organism to be tested by flooding with a broth culture. The "multidisc", or individually prepared discs are placed on the plate, and after incubation the inhibition zones indicate the sensitivity of the organism.

SENSITIVITY TESTS FOR TUBERCLE BACILLI

The serial dilution technique is the usual method using the Dubos medium (cf. p. 413) in screw-cap bottles. A standard amount of a culture of the tubercle bacilli in Dubos medium is added to a series of twofold dilutions of streptomycin; a control series seeded with the standard test strain *H* 37 *Rv* is also prepared. Readings are made after 14 days' incubation at 37°C; the *H* 37 *Rv* strain is usually inhibited by concentration of $\frac{1}{2}$–$\frac{1}{8}$ μg per ml of streptomycin.

A simple method for routine practice is to add a standard of streptomycin (2–5 μg/ml) to a solid medium (e.g. Herrold's) and inoculate with a culture of the tubercle bacillus or, when large numbers of organisms are found, with the original specimen (e.g. sputum).

INDEX

Printed in Dunstable, United Kingdom

91336869R00258